CAMBRIDGE LIBRARY COLLECTION

Books of enduring scholarly value

Earth Sciences

In the nineteenth century, geology emerged as a distinct academic discipline. It pointed the way towards the theory of evolution, as scientists including Gideon Mantell, Adam Sedgwick, Charles Lyell and Roderick Murchison began to use the evidence of minerals, rock formations and fossils to demonstrate that the earth was older by millions of years than the conventional, Bible-based wisdom had supposed. They argued convincingly that the climate, flora and fauna of the distant past could be deduced from geological evidence. Volcanic activity, the formation of mountains, and the action of glaciers and rivers, tides and ocean currents also became better understood. This series includes landmark publications by pioneers of the modern earth sciences, who advanced the scientific understanding of our planet and the processes by which it is constantly re-shaped.

Letters and Extracts from the Addresses and Occasional Writings of J. Beete Jukes

The geologist Joseph Beete Jukes (1811–1869) studied at Cambridge under Adam Sedgwick (1785–1873). Between 1839 and 1846, he participated in expeditions to the coasts of Newfoundland, Australia and New Guinea. This posthumous book, edited by Jukes' sister and published in 1871, contains selected highlights from Jukes' professional and personal writings. After a short biographical sketch, the material is arranged chronologically, beginning with Newfoundland and Australia and then focusing on Britain and Ireland, where Jukes was involved with national geological surveys. Linking passages by the editor explain the contexts of the pieces, and the book includes a list of Jukes' publications. Jukes lived during an exciting period that saw far-reaching discoveries and advances in his field, and his energy and enthusiasm permeates even his technical communications. Modern readers, like the original audience, will form a vivid impression of Jukes' lively personality, his 'penetrating glance and his sturdy step'.

Cambridge University Press has long been a pioneer in the reissuing of out-of-print titles from its own backlist, producing digital reprints of books that are still sought after by scholars and students but could not be reprinted economically using traditional technology. The Cambridge Library Collection extends this activity to a wider range of books which are still of importance to researchers and professionals, either for the source material they contain, or as landmarks in the history of their academic discipline.

Drawing from the world-renowned collections in the Cambridge University Library, and guided by the advice of experts in each subject area, Cambridge University Press is using state-of-the-art scanning machines in its own Printing House to capture the content of each book selected for inclusion. The files are processed to give a consistently clear, crisp image, and the books finished to the high quality standard for which the Press is recognised around the world. The latest print-on-demand technology ensures that the books will remain available indefinitely, and that orders for single or multiple copies can quickly be supplied.

The Cambridge Library Collection will bring back to life books of enduring scholarly value (including out-of-copyright works originally issued by other publishers) across a wide range of disciplines in the humanities and social sciences and in science and technology.

Letters and Extracts from the Addresses and Occasional Writings of J. Beete Jukes

EDITED BY C.A. BROWNE

CAMBRIDGE
UNIVERSITY PRESS

CAMBRIDGE UNIVERSITY PRESS

Cambridge, New York, Melbourne, Madrid, Cape Town,
Singapore, São Paolo, Delhi, Tokyo, Mexico City

Published in the United States of America by Cambridge University Press, New York

www.cambridge.org
Information on this title: www.cambridge.org/9781108030991

This edition first published 1871
This digitally printed version 2011

ISBN 978-1-108-03099-1 Paperback

LETTERS AND EXTRACTS

FROM

THE WRITINGS OF J. BEETE JUKES.

J. Beete Jukes

LETTERS

AND

EXTRACTS FROM THE ADDRESSES AND OCCASIONAL WRITINGS

OF

J. BEETE JUKES, M.A. F.R.S. F.G.S.

LATE LOCAL DIRECTOR OF THE GEOLOGICAL SURVEY OF IRELAND,
ETC. ETC. ETC.

Edited, with connecting Memorial Notes,

BY HIS SISTER.

'To live in hearts we leave behind
Is not to die.'   CAMPBELL.

LONDON:

CHAPMAN AND HALL, 193 PICCADILLY.

1871.

TO THE

REV. ADAM SEDGWICK, M.A. F.R.S.

WOODWARDIAN PROFESSOR IN THE UNIVERSITY OF CAMBRIDGE, ETC. ETC. ETC.

In dedicating to you this memorial of a beloved Brother, I would recall his own words on a like occasion, and 'acknowledge your great kindness shown to him as an Undergraduate at Cambridge, and subsequently renewed at many periods of his life. Upon the course of that life your influence was as great as it was beneficial; and you were always most gratefully and affectionately remembered by your attached pupil.'

Permit me to join in these sentiments of regard for the Master under whom my Brother not only learnt the Science he loved so well, but whose example he followed in devoting his life to its exposition and advancement.

I beg to subscribe myself, with much respect and esteem,

Very faithfully yours,

C. A. BROWNE.

PREFACE.

THE first part of this work consists of Letters written from Newfoundland and Australia to members of Mr. Jukes' own family, relating principally to the incidents of his voyages, and to his work as surveyor and naturalist in those countries.

The Letters addressed to Professor Ramsay, forming so large a part of the collection, exhibit the working of the Geological Survey, and depict the kind of life led by the field geologists. They convey also Mr. Jukes' views on those questions to which he had paid special attention, and in particular the classification of the Devonian rocks. He considered that his work in the south-west of Ireland had given him the key to the true structure of Devonshire, and led him to form a different conclusion from some of his contemporaries.

Although the chief topic upon which these Letters dwell is the science that absorbed so much of his thoughts and energies there will also be found some Letters relating to subjects of more general interest, especially those to his cousin Dr. Ingleby.

The Editor desires here to record her great obligation to Professor Ramsay for the trust he reposed in her by placing her Brother's ample correspondence with him in her hands, and for his valuable assistance in selecting and, when necessary, annotating the Letters most suitable for publication.

Her thanks are also due to Mr. Whitaker of the Geological Survey, for so kindly furnishing the complete chronological list of Mr. Jukes' works, which has been added by way of appendix.

September 1871.

CONTENTS.

CONTENTS. xv

PAGE
Marriage of Mr. Jukes, and the death of his Mother . . . 437
Letters: To Prof. Ramsay. On improvements in the department of 'Woods and Forests' 438

1850.

To the same, Merchlyn. On the Caradoc and 'pale slates' . . 439
To the same. On submarine volcanic rocks 441
To the same, Bala. Section work and the six-inch-map question 442
To the same, Ffestiniog. Running sections over mountains and valleys 444
To the same, Bala. On section tables and a proposed new book by Sedgwick 445
To the same, Llangollen. On finishing Staffordshire work, and appointment to the Irish survey 447
To the same, Rugeley. On the Rhetic beds in Staffordshire . 449
To the same, Rugeley. On finishing the coalfield—Report on water-works at the Clent Hills—Shooting at Beaudesert—Faults in the coalfield, and appointment to Ireland 451
Note by Prof. Ramsay 454

IRISH SURVEY. 1851.

Editor's Notes 457
Letters: To Prof. Ramsay, Dublin. On the Old Red Sandstone at Llangollen, and the six-inch maps 460
To the same, Dungarven. Depression of health and spirits. Age of valleys 462
To the same, Monkstown. Memorial of S. Staffordshire—Electro-biology 464
Visit of E. Forbes to Mr. and Mrs. Jukes at Monkstown . . 465
Letter to Prof. Ramsay, Dublin. On the practical uses of Geological Survey, and the loss incurred by bad mining . . 467
Note on the same subject 470

1852.

Letters: To Prof. Ramsay, Museum, London. On the Clent Hills trap and pebble beds—Lectures on the gold regions . . 473
To the same, Cork. On Ludlow and Wenlock rocks . . . 476
To the same, Macroom. On the necessity of maps being accompanied by explanatory memoirs 478
To the same, Macroom. On his own feelings of non-resentment 479

1853.

To the same, Dublin. On Silurian and Cambrian . . . 481
Extract from a paper on the same subject by J. B. Jukes and Andrew Wiley 482

xvi

CONTENTS.

INTRODUCTORY NOTE.

It is not until we have experienced the pain of being parted by death from one whom we have greatly loved and reverenced, that we can fully realise and understand that feeling of our common nature which finds its expression in hero-worship and saint-adoration. When the voice can no more utter, the pen no longer convey the living thoughts, then we turn to what has been said or written in years gone by, and we find those words which the memory retains or which greet us in letters assume a fuller power and significance than they originally possessed. The consoling conviction dawns upon us, that it is the good and the true which remain in our memories; the small defects of character, the trifling disagreements, which at the time may have unduly vexed, now dwindle into utter insignificance, or entirely die out of mind. But we recall the kindly actions, the lively sympathies, the charitable judgments, and we feel our hearts burn within us to emulate these virtues. Thus a true 'spiritualism' influences us, and the loved one, 'being dead, yet speaketh' in language more forcible than that of earth. It is not surprising that near relatives and partial friends frequently fail as biographers; for to them 'thoughts breathe and words burn' with vivid meaning, while to the outer

world they may seem comparatively lifeless and cold. This
consideration might have prevented the present attempt, had
the biography of a beloved Brother been contemplated ; but
a humbler office is undertaken, viz. to form a slight framework in which may be set sketches drawn by his own hand
with a clear and correct outline, often brightly and freshly
coloured.

His Letters are most characteristic of the man. His was
a true and earnest spirit, impressing itself on all he did and
all he wrote, of a nature peculiarly English, which in youth
finds its chief pleasure in visiting distant scenes and traversing foreign countries, and at the same time clings closely in
affection to home and friends left behind, which shrinks from
no amount of toil and danger, and yet encounters both in the
quietest and most unpretending manner.

' A clear true-hearted man, unselfish in his labours, and
looking steadily to the greatest and noblest ends of science.
The great cause of science suffered by his death ; for where
shall we find so laborious and keen-sighted a worker ?'

These words were written of Mr. Jukes after his death
by his old master and faithful friend Professor Sedgwick,
whose commendation was throughout life more grateful to his
pupil than the praise of any other.

West-hill, Highgate, October 1869.

EARLY LIFE AND LETTERS.

1811–1839.

EARLY LIFE. 1811–1839.

Parentage—Boyhood—Early education and pursuits—Matricu-
lation at Cambridge—Study of geology under Professor Sedgwick — Walking tours and lectures — Letters from Derby,
Nottingham, Brailsford, Seaham, London—Surveying work
in Leicestershire.

JOSEPH BEETE JUKES was the eldest child and only son
of John and Sophia Jukes, and was born at Summer-
hill, near Birmingham, on October 10th, 1811. His
father and grandfather were manufacturers of that town
—the latter then residing chiefly at Bordesley, a village in the neighbourhood, where he carried on a little
amateur farming, as a recreation from business. This
place is now absorbed into the great mass of smoky
chimneys, which meets the eye of the railway traveller
on arriving at the Midland metropolis. Joseph Jukes
built at Bordesley a substantial house, with oaken beams
and floors, doubtless cherishing the idea that his de-
scendants, for many generations, would appreciate its

solidity, and sun themselves in the pleasant gardens
and fields surrounding the house—the property being
entailed in the male line. House and garden, fields
and farm, now form part of the unlovely buildings and
sheds belonging to the goods department of the London and North-Western Railway.

Joseph Beete Jukes was christened after his mater-
nal grandfather, Joseph Beete of Demerara. He had
the irreparable misfortune of losing his father before he
had completed his eighth year; and his mother having three daughters to support on very slender means,
and having been herself highly educated, commenced a
school for young ladies in the neighbouring town of
Wolverhampton. In a few years she left it for the
pleasant village of Penn in that vicinity, where she
lived for ten years, greatly respected and beloved by all
her pupils. The house being then required for other
purposes, she removed to the village of Pattingham,
six miles on the Shropshire side of Wolverhampton.
Her son's maintenance and education were provided for,
he being heir to the property at Bordesley above mentioned. He was sent to the endowed grammar-school
in Wolverhampton, where the tuition was of the unsatisfactory kind prevalent at that time in many minor
foundation-schools. The teaching was confined entirely
to one branch of learning, namely Classics; and the
discipline was of one kind only—the cane—which was

administered in the undiscriminating manner frequent in those days.* He used to describe the astonishment of some new-comer, who had perhaps never heard of Homer and Virgil, and who, stumbling upon such a name as Thucydides, would pronounce it as a trisyllable with a long *i*, and receive in consequence such a sharp corrective as would doubtless impress the true sound on his memory, and at the same time greatly confuse his sense of right and wrong. But if the teaching was defective and the treatment tyrannical, the system was not of the prematurely forcing kind that now so often prevails for the mental and bodily faculties, and which, although it might have made of him a kind of 'Admirable Crichton' among boys and youths before he was twenty years of age, would probably have prevented him from becoming an eminent man among his compeers at thirty or forty. He passed a healthy and happy boyhood, unmarked by precocity, but showing a general intelligence and ability combined with quickness at work when he set about it.

He was an eager follower of all youthful games and sports, very strong and active, devouring mental and bodily food with an omnivorous appetite, but especially delighting in books of travel. It is characteristic of

---

* The above description of the Wolverhampton school has long ceased to be applicable, it having participated in the educational reforms which have since taken place.

his tastes, that he spent his first pocket-money—of sufficient amount—in purchasing *Cook's Voyages*, which
were read and re-read with the untiring interest that the
favourite books of childhood commanded in the times
when they were not so abundant as now.   His interest
in geology commenced, as he himself said, 'from the
time he first saw a boulder-stone, and wondered how it
got there ;' and was fostered and strengthened by an
aunt in Birmingham, who possessed an excellent and
admirably arranged cabinet of silurian fossils, in which
reposed among other treasures that known as *Illænus
Barriensis,* since beautifully figured and described in
the *Memoirs of the Geological Survey,* vol. ii.   But the
study of geology did not commence until he was at
Cambridge, when the spirit-stirring lectures of Professor
Sedgwick excited that love and enthusiasm for the
science which afterwards distinguished him.

From the Wolverhampton school, after a short time
spent at that of Mr. Thursfield, Vicar of Pattingham,
Mr. Jukes went to King Edward's School in Birmingham, then under the head-mastership of the Rev. John
Cook, the second-master being the learned and eccentric
Rann Kennedy, in whose house he was boarder.   Here
the tuition was of a much higher order than at Wolverhampton.   Mr. Jukes was too young at the period of
his father's death to feel the trial acutely, or to know
all that he had lost ; but the death of the two sisters

next to himself in age, and to whom he was warmly attached, was a severe blow to him. The annexed are fragments of some of those 'mortal lullabies of pain,' by which he probably sought to soothe his griefs to rest after the death of his second sister in 1831 :

> ' Thou art gone, and long, long years may roll
>   In sullen silence o'er my head ;
> And Time may seem to blot the scroll
>   Of memory where thy name is read.
> The smile again may curl my lip,
>   And joy again may light my eye ;
> But till I of Death's waters sip,
>   Thy sweet remembrance shall not die.
> Yes, thou art gone ! Thy youthful years
>   Were like an April morning's beam—
> They rose in smiles, soon set in tears—
>   A bright, but, ah ! too transient gleam.'

> ' They are gone, and o'er them now
>   Gently the grass is waving ;
> Cold, cold is each brow,
>   Not a sigh their bosoms heaving.
>   But there they lie
>   So peacefully,
> And heed not their loved one that here sits weeping by.'

It may in this place be appropriately stated why, from amongst many verses and fragments of poems, all but these few lines are withheld. Some years later in life, Mr. Jukes, in reply to a letter from a friend requesting his opinion on the advisability of publishing some poems, gave his views in the following words :

'In early youth poetry is a passion, and we naturally imitate what we so highly esteem. Even in manhood, all who have any natural sensibility, or romance,
or ideality, in their constitution have their moments of
wayward fancy, in which the feelings are open to any
impression, and the soul broods over them and nurses
them into a condition of excitement. One of the natural outlets and escape-valves for this feverishness
seems to be *verse*—good, bad, or indifferent, according
to the powers of the individual. So far, so good; let
every one write for his own gratification, or that of his
immediate friends or confidants. But when it comes
to publishing, poetry is an art for which the life of an
individual and the devotion of all his time as well as his
talents is requisite; and even then, unless his powers
are of the highest and rarest order, he must now be
disappointed. Poetry, so far as good versification and
mere elegance, or even beauty, is concerned, is abundant. A poet to succeed now must be of a far higher
standard than formerly. It is no longer a question of
style or words; *thoughts* and *things* are the sole foundation—simple, plain, condensed language, " thoughts
that breathe and words that burn"—in a word, grand
intellectual power, far above that of the common race
of mortals, must now be the attribute of one who as-
pires to the name of poet. For myself, then, I should
as soon think of publishing verses, even if I had given

them all the study and polish of which I was capable,
as I should of attempting a novel system of astronomy,
or endeavouring to found a new empire.

" Mediocribus esse poetis
Non homines, non dii, non concessere columnæ,"

is far truer now than even in Horace's time.'

As may be judged from this extract, Mr. Jukes had
a high standard and a full appreciation of poetry. Indeed, it might be said that he inherited the love and
admiration of our greatest poet from his father, who
was a thorough Shakespearian.

It was the great desire of his mother that he should
become a clergyman; and it was chiefly with this view
that he was sent to Cambridge, where he matriculated
at St. John's, with an exhibition from the Birmingham
school in 1830. Here he did not evince such devotion
to study as might have been supposed from his after
career, being far more inclined to outdoor recreations
than indoor work. Open-hearted and generous, with
abounding health and youthful spirits, without a father's guiding and restraining hand, no wonder that, as
he himself stated in his address to the Geological Section of the British Association at Cambridge in 1862,
he 'displayed but a truant disposition to study, and too
often hurried from the tutor's lecture-room to the river
or the field,' and 'that, had it not been for the teaching of Professor Sedgwick in geology, his time might

have been altogether wasted.' But that ' old man eloquent,' who was then in the height of his powers, had
a most beneficial influence on the young undergraduate,
and gave direction to the future course of his life. Great
was his reverence and affection for his old master; and
when he presided at the Geological Section as above
referred to, he avowed that his ' chief claim to occupy
the post was, that he was a pupil of Professor Sedgwick.'
In the study of this science his talents found their fit
exercise, combining as it does both indoor and outdoor
work, mastery of details and broad generalisation. No
illustration is now needed, or his case might furnish
one, to show the great desirability of including natural
science in the university course, and of allotting some
honours and rewards to successful attainments therein.
More than ten years after this time (in 1844) the late
Professor E. Forbes speaks of natural history as being
greatly discouraged at Cambridge, and regarded as
' idle trifling;' now so great an advance has been made,
that the study of the works of God is being gradually
placed upon a level with the study of the works of man,
and the college of St. John's stands prominent in the
encouragement it gives to the students of natural
science.

Mr. Jukes was a fair classic, and had a good knowledge of geometry; but neither classics nor mathema-
tics had for him that powerful attraction which

'doth the spirit raise
To scorn delights and live laborious days.'

He was writing from experience in the *Quarterly* in
July 1859 when he said : ' The Book of Nature has this
advantage over the stories composed by men, that it
has no end, and its interest grows with every fresh
perusal. To read it, we must breathe the free air and
live for a time in the open field, with not only the mind
amused, but with the muscles invigorated, the nerves
braced, and the blood coursing through the veins with
that pleasurable glow that makes every breath we draw
a pleasure in itself, while good digestion waits on appetite; and health on both.'

Mr. Jukes had given up all intention of entering
the ministry, being far too conscientious to become a
member of any profession, especially one so sacred, for
which he felt himself unfitted. The attitude of the
clergy at that time towards geological science was little
calculated to smooth any difficulties which might arise
in a young man's mind. Indeed, so hostile were some
members of that body to such pursuits, that, looking
back through a vista of more than thirty years, and recalling many warnings of the dangers which were supposed to be incident to geology, it appears as if the
creation of the world and the laws by which it is governed might have been attributed to some evil power,
the study of whose works must necessarily be harmful,

instead of to an all-wise and beneficent Creator, who
gave the earth unto the children of men that they might
not only replenish but subdue it.   There were few to
speak such words as those uttered lately by Dr.
Temple at Plymouth : ' I have a very real conviction that all
this study of science and art comes from the providence
of God, and that it is in accordance with His will that
we should study His works; and that, as He has given
us a spiritual revelation in His word, so also He has
given us a natural revelation in His creation. . . I am
convinced that all light, of whatever kind, is good and
comes from God; that all knowledge comes from Him,
and can be used in His service.'   Happily the spirit of
these words is fast spreading in the Church of England,
and clergymen are more and more inclined to leave
warnings against the study of science to the Œcumeni-
cal Councils and Encyclical Letters of the Romish
Church.

   Mr. Jukes's determination to give up the profession
of a clergyman was a great disappointment to his mother, and unsettling to his own prospects ; and on leaving Cambridge it was not easy for him to decide upon
his future course, there being but few openings into
the world of science by which a young man could tread
its paths and at the same time earn even a small income.   But it was impossible for one so active in mind
and body long to lead an idle life ; and while his future

career remained uncertain he employed himself, and endeavoured to arouse an interest in his favourite science by delivering lectures upon it at several towns in the middle and north of England, by which also he extended his knowledge of his native country and examined for himself geological phenomena, quite satisfied if he could make his lectures pay his travelling expenses. As specimens of these early lectures, the following passages are taken from the first and last of a course delivered in the Mechanics' Institution in Wolverhampton in the early part of 1837:

'To us—the insects of a day, who creep upon this mighty globe, and are not of more relative importance to it than the dust is to the surface of this floor—nothing is accustomed to seem so firm and stable as the ground on which we tread. We walk upon the plain, and regard it as the base on which the heavens themselves could rest; we stand in the valley, and look up to the hills as the very emblems of eternal strength and duration; or when seated on the cliff we mock the raging of the ocean, and laugh in our security as its billows burst beneath our feet. Upon the solid bulwarks and buttresses of our globe the elements seem to waste their strength in vain. We live and die without seeing either the hills lowered or the valleys altered or filled up.

'There is therefore, at first sight, something which seems absurd in the idea of supposing them to have

been once otherwise than they now are ; but the changes
which affect them are not less certain and constant because we do not regard them. Nature is not limited in
her operations to the time of threescore years and ten ;
and though during our brief existence the amount of
her work may be imperceptible, yet when multiplied by
centuries, it assumes a vast and most important appearance. The river glides through the plain, silently and
imperceptibly eating away its banks, sweeping off a
patch of sand here, carrying away a little heap of mud
there : trifles in themselves, but continued through the
lapse of ages, we behold their result in the extensive and
fertile deltas of the Nile, the Ganges, or the Mississippi
—tracts of land containing hundreds, nay thousands,
of square miles, where the corn-fields spread forth their
abundance, or the trees of the forest wave their heads
over what was once a barren waste of waters ! If we go
to the cliffs upon the shore of the ocean, we shall behold the same vastness in the amount of seemingly
petty action—in the masses of rock lying ruined at
their feet, or in the sands and shoals, the gulfs and bays
that attest the extent of the territories which the sea
has wrested from the dominion of the land. Even the
peaked summits of the granite hills themselves are not ex-
empt from the universal rule of change ; the jagged ruins
which lie scattered on their sides, or piled in heaps
around their base, giving evident and convincing proof

of the silent power of the rain and frost, or the destructive force of the thunderbolt. Everything sets
forth the universal truth : that the operations of nature
(which means but the action of those laws which God
has impressed upon matter) are carried on upon a scale
utterly beyond the attainment of our feeble faculties,
and embrace periods and powers, of which the periods
of our lives, or the powers of our minds and bodies, are
but as dust in the balance. It behoves us, then, when
entering upon the study of any of these operations, to
have this truth well impressed upon our minds, and to
have a due distrust of our preconceived opinions on the
subject.'

'We have, in these lectures, been reading the history of the earth. Human histories are but too often a
tissue of falsehoods and misrepresentations, or at best
are but uncertain and subject to mistake. On the contrary, the history which we have been studying contains
no falsehood and is liable to no error. If we do but
make ourselves masters of its language, and interpret
its lessons aright, we are sure that every word it contains is absolute and unqualified truth. And though
some portions of it are yet obscure, though no one has
yet arrived at a clear understanding of the whole; yet
many of its grand truths and some of its fundamental
principles have by the labours of its expounders been

clearly set before us in a manner too plain to be misunderstood, and too certain and convincing to be rejected or
despised. We find in this history the record of times
whose annals we shall in vain look for elsewhere.
Periods of time, compared to which that of the existence of the human race is but a point; when millions
of creatures lived, moved, and had their being upon this
globe, whose living forms were never given to the eye
of man. We can trace back the history of the world
through age after age and period after period; every
age and every period giving evidence of the existence of
organic beings still more and more remote in the details of their structure and their parts from those which
are now our fellow-inhabitants of the earth ; till at last
we arrive at times when every plant and every animal
were utterly and widely distinct from ours. But throughout this lengthened series of changes unfolded to us
in the pages of geology we cannot fail to be struck
with the grand and consistent uniformity which reigns
throughout the plan. Not a change has taken place in
the structure of the minutest animal or plant—not a
single convulsion has affected the mountain masses of
the globe—but it may be traced to the necessary and appointed order, the fixed and immutable laws which have
been stamped upon every atom of the universe. The
Master-mind, if I may be allowed so to speak, is visible
throughout; working indeed by means, but yet as the

Author of those very means themselves, and having
therefore given them their unchangeable and perfect
character.   This is indeed the very triumph of Science,
when by unfolding to our view the grandeur and sublimity of the works of Creative Power, she becomes the
handmaid of Religion, and disregarding all petty dis-
tinctions of sect and creed, she leads the whole human
race to worship in one common temple—the temple of
the world—the common God and Creator both of their
temple and of themselves.'

Mr. Jukes delighted in walking with knapsack and
fossil-bag through wild and lonely places, and made
his way chiefly on foot to the meeting of the British
Association at Newcastle in August 1838, where he was
gratified by becoming acquainted with men famous in
all branches of science.

The following are extracts from letters written prior
to and during this excursion :

Liverpool, June 5, 1837.

My DEAR MOTHER,—I have just finished my last
lecture here, but have been arranging the geological
cabinet of the Royal Institution, which I shall not
finish till to-morrow evening; and then it will take me
two days to pack up my things, and call upon several
kind people here with whom I have become acquainted.
On Friday morning, then, I intend to start for Man-

chester per railway, from whence I shall proceed into
Derbyshire, and walk homewards; so that you must
not expect to see me before the end of next week.
After having seen you transferred to Wolverhampton,
I shall in all probability have an engagement with
Sedgwick, and shall then set out on another lecturing
tour, and have promised to be in Liverpool again in
September. I have altogether spent a very pleasant
time here, having had my hands full of business during
the whole of it; but am now longing for a little country,
for Liverpool is almost as bad as London. I had a
day on the coast with the Rev. T. Devy, a Cambridge
man, and visited the remains of an ancient forest, now
covered by the sea at high water, in which are large
tracts of peat, with stumps of trees still standing erect
from their roots, and their trunks lying prostrate by
the side of them, almost as soft as chalk, and many
of them full of the shells which bore into wood and
stone.

J. Beete Jukes.

Derby, February 6, 1838.

My dear Mother,—I did not set out for this place
till Saturday, and got here in a semi-frozen condition.
Since then I have been hard at work writing a paper on
the geology of Leicestershire, reading conchology, and

making many acquaintances. I have now made all arrangements, published a syllabus, and shall commence next Tuesday. I am not at all sanguine as to great success here; but I hope, at all events, to pay my expenses. I shall not finish till the 13th of March, after which I hope to get an engagement in Manchester; and then shall either come back and work at the South-Staffordshire coal-fields, or shall go on and lecture my way up to Newcastle, so as to be at the meeting of the British Association in August.

J. BEETE JUKES.

Derby, March 17, 1838.

MY DEAR A.,—I have just finished my course here, and have succeeded better than I expected—have made some friends and many pleasant acquaintances. There was a grand bazaar here on the 8th and 9th; consequently I put off my lecture and went to Matlock, and thence across the country to the Dove, down Dovedale. The scenery is most beautiful, and it would excessively astonish you to find such on the borders of your own county. One Staffordshire river, the Manifold, runs in a deep valley with precipitous sides of five or six hundred feet high, covered in some places with wood, in others the bare rock standing out, in one of which there is a large and remarkable cavern called

Thor's House, from which you look down three or four hundred feet, and see the river flashing among the woods and rocks below you. It was a splendid day when I was there; and though loaded with fossils, we rambled without anything to eat from nine in the morning to six in the evening on the banks of this valley. The next day we walked down Dovedale, which is still finer, the hills being more lofty, and the scenery, if anything, more wild and magnificent. I am going to stay a week at Mr. Strutt's of Belper, about eight miles north of Derby, where I am going to give three lectures to the Belperites. Mr. Strutt lends me his schoolroom. The Strutts are great cotton-spinners, and one of them is member for Derby. I stayed a day with them before, and went over their factory. It is an immense place, with twelve or fourteen hundred people at work; it took me two hours merely to see the different processes the cotton went through, from the time it came out of the bales till it was ready to make into stockings or to sew with. Immense rooms full of machinery, all going fiz, whiz, rattle, clash, hum, spin, whirl, buz, twist, twirl, clitter, clatter, all kinds of movements going in all kinds of ways, from great slow-moving majestic-looking wheels, to little fizmagig things spinning faster than your eye could follow them; girls cutting or joining threads, their fingers moving as if they were playing the piano. In another place I saw two men take a *tree*, as much as they could lift; by pressing it against a circular rotating saw, it lay in blocks three inches thick at my feet in a little under [piece of letter here torn off], and then put into a machine, when they came out as cotton reels—those little things you have in your work-box. A tree is brought in with its bark on, and in less than half an hour is all converted into little reels ready to wind cotton on; and in less than another half-hour they may all be covered with cotton, and packed up in small parcels to be sent out. Everything, too, is done under the same roof: the machines are made, the cotton is cleaned and twisted, even the tickets are printed, in the different rooms of the same immense building. After leaving Belper, I shall probably go either to Chesterfield or Sheffield.

J. Beete Jukes.

Nottingham, May 22, 1838.

Dear Aunt,—I received your last at Whatstandwell-bridge, so called because the man who built it said it was one 'what would stand well.' The inn is just at one end of it—a pleasant little house, with a beautiful sitting-room and bedroom over, each having a bow at the end, with a window looking up and down the valley, and admitting the sound of the Derwent as it ripples and foams and flashes over the shallows and rocks below

the bridge. The valley here is most beautiful. The
river is bounded by lovely rich meadows lying in the
bosom of the 'many folding hills,' sometimes only two
or three hundred yards across, giving just room for the
road, the river and the *canal* (here not a disagreeable object) sometimes receding in gentle curves to the width
of half a mile or so, with rocky ridges running out
into the meadows. The hills are three or four hundred
feet high, very various in their outline, and clothed
with magnificent woods ; while at the back of these rise
higher and barer eminences, covered with dark heather,
or showing white masses of limestone among the light-
green turf. At one place the valley narrows and makes
a semicircular curve round a fine eminence, with a narrow transverse valley at one side of it, down which a
brook flows to join the Derwent. The canal here crosses
the river by a fine aqueduct of one arch, and the road
which runs along the outside curve of the valley is cut
out of the rock on one side and walled on the other.
Over this wall you look down about a hundred feet,
through the trees, on to the narrow meadows, the river,
and the aqueduct, the latter backed by the beautiful
little transverse valley, and the whole embosomed in the
most glorious woods. After a climb of about two miles
from Whatstandwell and east of the river, you may get
to the top of Crich Cliff, 700 feet above the Derwent,
and 1250 feet above the sea. Here you see the whole of

this lovely valley north and south, from the rocky gorge
by which it passes through the massive hills around
Matlock, till it widens and opens and gradually slopes
down to the level country about Derby, with the fine
old tower of All Saints just visible in the distance. In
this direction the horizon is closed by the hills of Charnwood Forest. To the east you look over the Derbyshire
coal-fields, to the fine escarpment of the magnesian
limestone, on the projecting promontories of which
stands Bolsover Castle, Hardwick Hall, Annesley, &c.;
while north and west your view is confined by high
dark moors of gritstone or green hills of limestone.
But the beauties of Crich Cliff are not confined to the
view from its summit; it contains things worth seeing
in its interior. This is got at in the following way:
you get to a hut on the hill-side, where you denude
yourself of all your own clothes, and indue those of a
miner, composed of flannel and bed-ticking, the texture
of which is dimly seen through the clay with which
they are imbued. With a flannel cap on your head and
a tallow candle in your hand (stuck in a piece of clay),
you declare yourself ready for your descent. The miner
accompanying you walks to a trap-door at a corner of
the shed, opens it, and very deliberately disappears,
and leaves you to follow him. On opening the door,
you perceive it covers the mouth of a large shaft like
a coal-pit, and by the help of the miner's candle below

you, and your own, you discover some bars of wood stuck
in the sides of the shaft. These are arranged one
under another, about a yard apart, all down the shaft;
and by resting one foot on one side and the other foot
on the other side (it is a good straddle across), and
holding by your right hand and so much of the other
as is not employed in holding your candle, you gradually descend. Thirty or forty yards down you come
to a landing - place, then perhaps go down a perpendicular ladder, then crawl through a hole and go along
a level, and then to another hole with the bars of wood
(called stemples), and so on for five or six hundred
feet of perpendicular descent. The ladders and stemples are all slippery with clay, of all sizes, at irregular
distances, some of them nearly worn through, some
feeling loose and rotten ; so that a novice is inclined to
chant *Te Deum* on his safe arrival at the top again.
Besides, it is such hard work, that it makes your arms
stiff for a week. . . .

On Friday, when I went thus down the Pearson's venture shaft, I went into the sough which drains the mines
and comes out just above the Derwent. By leaving our
candles behind, and wading a few yards in the dark, we
could see daylight at the other end, at the distance of
a full mile, like a large star. You could have come up
the sough and joined me here in a little boat or canoe
made of one hollowed tree. The lead-veins are very

curious, and many of them remarkably rich. At Crich they cross each other in all directions, and have occasional chambers and caverns full of beautiful spar, and some of them full of magnificent masses of lead-ore. Altogether, Whatstandwell is a very interesting and delightful place to stay a week in. The charges are reasonable in the extreme, and coaches by the door; and I prefer it to Matlock, which, moreover, is within a walk.

May 26. This letter never will go. Instead of finishing it, I mounted on horseback at half-past six o'clock this morning, and had a most delightful ride with two friends to see the lias of Granby, near Belvoir Castle. It is a very out-of-the-way place, with three or four quarries; and the men talk of having got fish and 'horses' backs' and other bones — 'horses' backbones with all the ribs and everything;' but as no one ever asked for them, they ' only admired them themselves, and then broke them up.' They have promised to save all they can for me. I rode a beautiful chestnut horse, who took quite a delight in leaping the lias ditches.

I suppose Combe is beginning his course in Birmingham. I shall not finish mine till the 22d of June, when I don't exactly know what will become of me; but I must go again across Derbyshire, and then make for the North, so as to lecture somewhere within reach

of Newcastle. What would you not give for a ramble
with me (in a pony-carriage) through Derbyshire ?   I
know some quarries in which you would spend a week.

J. Beete Jukes.

Nottingham, June 19, 1838.

My dear A.,—I finish here on Friday, but am engaged to conduct two geological excursions into Derbyshire next Monday and Wednesday.   Then I have
some work to do between here and Ashbourne, which
will take me a fortnight ; and I want to walk through
Yorkshire and be at Newcastle by the middle of August.
Now if I come to Shrewsbury, I shall be spending both
time and money which might be better employed.   To
put against these considerations, I could be arranging
a course of lectures at Shrewsbury, and I should have
the pleasure of introducing you to some of the Shropshire hills, as I could drive you all over to Church
Stretton,—which are certainly great inducements.   I
have had a very good class here—never less than two
or three hundred, and frequently four or five hundred ;
and I have been wandering about the country a little.
I spent a day at Lincoln, and went to the top of the
cathedral—a noble building, with a most commanding
view ; and am going presently with a gentleman, on one
of his hunters, to a spot at the junction of the three

counties—Notts, Derby, Leicester—where there is a
very beautiful prospect over the three.

J. BEETE JUKES.

Brailsford, July 28, 1838.

DEAR A.,—* * * Since I wrote to you I have left
Nottingham, and went with W., a brother of my college friend, into the north of Derbyshire, and one day
walked over Kinderscout, the highest hill in the county.
It is a place in which dreariness becomes grand from
its extent. The top of the hill is several miles across,
perfectly flat, and occupied by a great black bog or peat
moss three or four yards thick, furrowed in every direction by gullies, from the sides of which oozes some
deep brown water that stagnates in their bottoms. The
whole looks like a great mass of black sponge full of
water, with patches of green occasionally about it to
heighten its effect by contrast. The bog is bounded on
every side by deep valleys, with precipitous sides furrowed by dark ravines. At the north end of the hill
there is a line of great crags of dark gritstone—many
of the stones the size of your parlour—piled one upon
another into the most fantastic shapes; and by selecting
openings through these, in which we were a little sheltered from the fury of the wind (which was blowing a
hurricane), we could look down into the valley below.

The crags among which we stood formed the summit
of a precipice of 200 or 300 feet, quite perpendicular,
from the base of which a steep slope descended about
800 feet more to the little river Ashop. This steep
slope was strewed over with fragments of gritstone that
had fallen from the heights above. This magnificent
escarpment of 1000 feet in height stretched away for
several miles on either hand ; and on the opposite side
of the narrow valley the eye wandered over moor after
moor, rising and falling in regular undulations to the
horizon. After a minute investigation, we discovered
two trees that looked like little bushes in one of the
ravines below ; all the rest was brown heather, black
bog, or bare gritstone. We had to cross three miles
of this bog, by the way, on our road to Castleton, and
preciously difficult work we found it ; hop, skip, and
jump, or walking as if treading on eggs, with an occasional stick-fast that required another's assistance to
extricate you. I have been staying hereabouts since
Tuesday, finishing my next paper for the *Analyst.* I
mean next week to walk through the north of Staffordshire into Cheshire, and eventually get to Liverpool,
and so on by sea to Carlisle, and by railway to Newcastle, where the British Association meet on August 20.

I am becoming quite a well-known character in
Derbyshire ; not a place do I go to, but I meet some
one that knows me, by report at least, if not by sight ;

and I have a general invitation to people's houses in all
parts of the country, some of which I accept when they
least expect it.

J. BEETE JUKES.

Seaham Harbour, Durham,
August 30, 1838.

MY DEAR A.,—As it is probable you may expect to
see me about this time, I write to tell you that in all
probability I shall be a month longer, as I am going to
see a friend at Darlington, and shall perhaps lecture
there. We have had a glorious week at Newcastle,
where Sedgwick, Buckland, Lyell, Phillips, Murchison,
&c. &c. all met. I was on the committee of the Geological Section, and read a short paper of very little interest; but it was merely by way of introducing myself.
I send you above a sketch of one of the new streets;
the new part of the town is in the middle of the old,
and it looks like a city of palaces surrounded by hovels.
The street above is finer than any in or out of London.
It is all a private speculation of one man, who builds a
palace in a few weeks, or, if necessary, can change it
into a theatre or a market by a wave of his hand. I
got acquainted with Dr. Pye Smith, and others known
to fame. I spent Monday and Tuesday in Allendale,
looking at Mr. Beaumont's lead-mines; and this morn-

ing walked here from Shields. The weather was delightful, and my path principally lay on a fine sandy
beach, with bold projecting and overhanging cliffs of
magnesian limestone on the one hand, and the foaming
waves of the German Ocean, like sea-horses with white
manes chasing one another to shore, on the other. At
one point, called Marsden Rocks, pillars, towers, and
pinnacles of limestone stand out in the sea, which has
cut them off from the land; and perpendicular ridges
of rock occasionally run out into it, with narrow bays
and creeks full of immense masses that have fallen
from the cliffs above. Great caverns are scooped out in
the cliffs, and sometimes two or three of these join, and
leave merely natural pillars to support the rocks above.
There is one portion of the cliffs that once existed far
out, where the sea now is about two or three hundred
yards beyond the present shore. It is about a hundred
feet high, and is worn into caverns all round its base;
about the centre the sea has breached a magnificent
arch thirty or forty feet high right through. When I
was there it was nearly high tide, and a fine breeze
blowing right upon the land. The waves came in quick
successive lines of long rolling swell, each of which as
it approached the shore rose higher and higher, till the
upper part at length curled over, and the whole was
shivered, as it were, into a magnificent burst of foam,
that surged over the rocks, dashed clouds of spray up

into the air, and at last spent itself in spreading sheets
of dazzling white foam on the level sands. The rock
with the arch in it looked as if it lay in a sea of boiling
milk; and every now and then a strong wave would
dash into the arch, and fill it up to its very top with
foam and spray; then, as it fell, you saw the green
ocean beyond, with a ship or two in the distance; so
that the arch looked like a frame to a fairy picture,
where even motion was delineated. I sat on a projecting rock, watching this scene for more than an hour,
where I could see the coast and the surf for miles on
either hand, and an occasional splash of spray came up
into my face, with the sun overhead, and the whole sea
studded with white sails. I just went in and dined at
Sunderland, and continued my walk along the shore till
dusk, when I stopped here, which is a new bathing
village. The edge of the cliff is about eighty yards
from the door; and as I sit here the dull heavy roar of
the surf sounds on the ear with a regular rise and fall
that, though monotonous, is not wearisome, for it is
the sound of motion and of mighty power. It is impossible to mistake any other noise for the sound of the
sea.   J. Beete Jukes.

It was about this period of his life that Mr. Jukes
made himself acquainted with practical surveying, and

joined a gentleman, who was working with a staff of assistants, in the neighbourhood of Leicester. He afterwards wrote for the *Analyst* a sketch of the geology of Leicestershire and Derbyshire. He also surveyed geologically the district of Charnwood Forest, the results of which were published in a separate memoir annexed to the history of that forest by Mr. T. R. Potter. By these means he rendered himself competent to accept the post of Geological Surveyor of Newfoundland, which was offered to him in 1839. His life had for the few preceding years been too desultory, and he gladly availed himself of the opportunity thus afforded him of accomplishing more definite work. The only drawback to his pleasure in being thus enabled to see more of the world, arose from his mother's pain at the parting, her dread of the dangers she thought he would encounter, and the loneliness of the life he would lead; but he assured her that he was not likely to undergo any great hardships, and felt no misgiving as to his health and strength being equal to the undertaking. She was too unselfish in her love to throw any impediment in his way.

The following letter was written by Mr. Jukes whilst in London preparing for his Newfoundland expedition, and is noteworthy for its reference to the dawn of photography, and the neglect with which the anticipatory discoveries of Niepce were treated :

14 *Cecil-street, Strand, March* 14, 1839.

MY DEAR AUNT,—Sedgwick has just taken tea with me, and I have been with him to the Royal Society to-night, where, among other things, I heard a notice from Sir J. Herschel read on the new process of drawing by sun-light, or, as he calls it, photography. A book of specimens was exhibited, in which he had caused several prints to copy themselves. These were mostly a dull brown, and looked something like old sepia drawings. He said his experiments were at present imperfect, and he showed these merely as first attempts. The effect in some was very good. Afterwards Professor Wheatstone showed Sedgwick and myself some specimens of what had been done by Niepce fourteen years ago and more. Niepce was then in London, and offered a paper to the Royal Society, with specimens; which, as the process was kept secret, they could not entertain. After many efforts to bring it into notice, he at length desisted. His views are on metal, apparently slabs of lead. One view, which was reduced in the process from an engraving of two feet to one of six inches, is beautiful, like a very highly finished and elaborate pencil drawing. Another, the first ever taken by the camera from his chamber, was very plain and distinct, in a similar style; something like a lead-pencil drawing, where the pencil has afterwards been rubbed to produce a soft effect. But the most remarkable thing is, that he

D

hit upon a process of making the light continually act upon the same lines, so that at length it burnt into the metal, as it were, and etched the picture, the impression from which had all the character of a very spirited etching. The neglect and, apparently, contempt with which Niepce was treated at the time is wonderful.

I have promised to stay over the next meeting at the Geological Society, as Lyell has a paper on some curious facts concerning North America to read then. I have seen Owen once or twice, and promised to collect some *beasts* for him.

J. BEETE JUKES.

NEWFOUNDLAND SURVEY.

1839–1842.

NEWFOUNDLAND SURVEY. 1839–1842.

Explanatory notes—Grant by House of Assembly—Newfoundland—Nature of the country and work—Mr. Jukes as a
master—Letter of an attached servant—Duties and qualifi-
cations of a geological surveyor—Letters at sea—St. John's—
Bays and creeks visited during coasting voyages—Inland expeditions—Professor Stüwitz—Dr. Stabb—Lectures at St.
John's—Expedition to the ice—Seal-catching—Preparations
for and return to England—Wolverhampton—Address to
the branch of Dudley Geological Society—Appointment as
naturalist to the expedition for surveying Torres Straits.

In choosing letters for publication, the desire has been
to give those which most clearly bring out the character
of the writer without intruding upon the privacy of
domestic life. When, however, they seemed to present a fuller view of a nature that was at the same time
both strong and tender—careless of self, yet consider-
ate for others—the editor has not shrunk from inserting
some letters which she might otherwise have withheld.
Her object has been to convey as complete a picture as
possible of a man

'Who battled for the true, the just;'

who sought the pure gold of wisdom and knowledge for
its own sake, without the alloy of power and wealth,
which is so generally needed to make that gold pass
current among us.

Mr. Jukes sailed for Newfoundland in April 1839;
and his letters form a complete series, by means of
which we can accompany him through the work of his
first year. Soon after his arrival in St. John's, he made
several short excursions in the neighbourhood, one especially to Harbour Grace, where it was reported that
coal had been found. These journeys showed him the
impracticable nature of the interior, and the impossi-
bility of obtaining any knowledge of the structure of the
island, by traversing the woods and marshes with which
it abounded. He therefore requested to be furnished
with a small coasting vessel, in order that he might
examine the cliffs in detail; and to enable him to follow
this plan, the original grant of 350*l.* was increased to
600*l.* During the second summer of the survey, his
correspondence was not so full, as he had then no vessel of his own in which to spend any leisure he could
command, but was obliged to live in open boats, or to
encamp in the woods with his men. The House of
Assembly was unfortunately seized with a fit of short-
sighted economy, and refused to continue the grant
which had been voted the preceding year. The services
which were being rendered in so efficient and yet inex-

pensive a manner do not appear to have been understood or appreciated, except by the governor (Captain
Prescott), the surveyor-general, and some few of the
better-informed among the members, by whose strenuous efforts a second grant was at length obtained.
This was, however, on too small a scale to allow of his
hiring a vessel, the want of which subjected him to
great hardship and much inconvenience. He nevertheless performed his work with just the same energy and
cheerfulness as if he had been receiving adequate com-
pensation. He thus notices an incident which had
amused him, and which shows the view taken by some
of the 'honourable members' of the work of the survey :
'After the vote was passed, a member called on me
privately, to recommend me not to take a vessel, since
I would then *make as much money* out of this grant as
out of the former one; as if my object had been to make
money rather than a geological survey.' As the hire of
the vessel, and its attendant expenses for the first year's
survey, amounted to more than 400*l.*, it may readily be
supposed that Mr. Jukes was more likely to have lost
than gained, in a pecuniary point of view, especially as
it was quite contrary to his disposition ever to do anything on a small or economical scale. Throughout this
survey, he shared in all the labour and lived on the
same fare as the men who accompanied him, and who
became much attached to him. A letter from one of

them, named Simon Grant, bears such a simple testimony to the power Mr. Jukes possessed of winning the confidence and affection of those who worked with him, that no better proof could be given than the letter itself, which was received during his subsequent absence in Australia, and is here given in its own orthographical integrity :

London June 1845

DEAR MAM—I take this oppertunity of writing to you hoaping you and family are in good health you will be pleased to let me know when you expect Mr Dukes home and how he is I am most anxious to heer finding him the kindest of masters whilst serving him in Newfoundland its nearly twelve months since I heard from him in a letter to his friend Dr Stab pleas to answer this note as soone as possible as I am leaving London on Monday next. Direct to the Stoord on bord of the barque Margaret Daviss wharf toley street   I remain your obbedient sevt   SIMON GRANT

This power of attracting those who served him was gained by very simple means. It was a part of his creed, that all men had more good than evil in them, and that to the better part of their nature an appeal could always be made ; and he ever acted on this belief, giving credit for truth and honesty, the trust that begets trust, and draws out the latent sparks of good feeling and good faith, which a different mode of treatment so frequently smothers and hides. Added to this charity which ' hopeth all things,' there was that self-possession and courage in danger, and that energy and perseverance in difficulty, which never failed to inspire with confidence those who depended on his leadership.

In order to judge of the difficulty of this survey, it will be well to give an extract from his report:

' A geological surveyor, in entering upon a country with which he has no previous acquaintance, has a task of some difficulty, and often of great perplexity. He must make himself acquainted with the characters of the various beds constituting the series of stratified rocks of which the country is formed. He must discover the natural order of this series, and ascertain what are the general characters which are common to certain large portions of it, in order to divide it into groups or formations. He has to accustom his eye to these characters, in order to detect them where they may be obscure; and many other little points of detail must be worked out before he is thoroughly qualified to enter on his task of describing the solid geometry of the district. For the more perfect fulfilment of this, too, it is necessary that the country should be accessible in every direction; that the surface should not be too much covered up by vegetable soil, or hidden by a thick growth of wood; and that the cliffs by the sea-shore, the natural sections exhibited in the beds of rivers or the banks which enclose them, and the appearances on the hill-sides, or, in their absence, artificial cuttings, sinkings, or borings, should be attainable. The very first requisite is, that a good map of the country should be placed in his hands, showing the general ranges and heights of the hills, and the courses of the rivers and valleys; and that this map should be on a sufficiently large scale to enable him to trace his daily routes, and mark down the phenomena observed. Furnished with this knowledge and these materials, it is the business of the geological surveyor to trace on the map the surface boundaries between the igneous and aqueous rocks, as also between the different formations or subdivisions of the latter. In doing this, he must also collect such information from the continued observation of the dip, strike, faults, veins, and other phenomena, as shall enable him to show the positions of the rocks below the surface, to as great a depth as he is able. This is done by means of sections, or supposed perpendicular cuttings through the earth in various directions, by which its internal structure may be best exposed. By means of these coloured maps and sections, a cabinet of specimens, and written descriptions, the geologist is enabled to convey an intelligible idea of the physical structure of the country, and the qualities and properties of the materials of which it is composed. . . .

'Nor must it be forgotten by the geological surveyor that, while he is bound by every means in his
power to discover and lay open the natural resources of
the country under examination for what are commonly
called practical purposes, he must not neglect the theoretical portion of the science of geology. A knowledge
of the mode of formation of the rocks, and the cause of
their being placed in their present positions, will constantly help him in difficult or obscure points, and will
more especially be of the greatest possible assistance
to him in mining or other operations in which direct
observation is impracticable. He is bound, therefore,
to store up every fact, and even every speculation that
may arise in his mind from the observation of fact,
bearing on theoretical geology, and to contribute, as
far as he is able, to the great mass of knowledge which
is gradually working out for us the history of the formation of the present crust of the globe.

'Those who are acquainted with the island of Newfoundland will be best able to appreciate the difficulty
of applying the general principles and rules laid down
in the foregoing observations to its particular case.
The interior being trackless, uninhabited, and obscured
by woods and morasses, the coast affords the only means
of continued observation. Here, though the cliffs are
bold, they are frequently inaccessible, and often either
too perpendicular, or two well guarded by surf to render

landing practicable. Not only is there no map of the interior, but no general knowledge of it exists. No guide can be found who knows more of the country than a few miles round his own dwelling, or a particular path to a neighbouring settlement. Much time was therefore necessarily devoted to gathering materials for a rough map, and acquiring some information on the physical geography of the country; and the present report can only be looked on, as a collection of so much preliminary information as would have been of use in the commencing a detailed survey, had the nature of the country rendered such an undertaking advisable.'

Long. 32°, *lat.* 48° 52', *April* 26, 1839.

MY DEAR MOTHER,—To relieve a little the monotony of sea life, I think of commencing here a small journal of my proceedings.

[After giving particulars of the first few days at sea,] *Monday* 22. Stiff S.W. breeze, but cleared up in the evening. The sea, running in great ridges, left us every now and then buried, as it were, in a valley of water; and then our good ship, mounting the back of the wave, dashed the crest of it into clouds of glittering spray, washing the decks and sails, and giving us a view of the wide waste of waters, green flecked with white, which was again lost to sight and hidden by the masses of water alongside as we descended into the opposite hollow. Porpoises were showing their brown backs, and strange sea-birds called ' boatswains' wheeled about us, while now and then a bigger wave than usual would dash against the sides of the vessel as she heeled over, and treat us with an unexpected shower-bath as we held on by the rigging. Our admiration was unmixed by the slightest particle of fear, our confidence in the ship and captain being perfect; and the laugh and the joke were mingled with the word of command, and the ' yeo-heave-yeo' of the seamen, the whistling of the wind, and the hoarse sound and dash of the sea. There's a picture of the Atlantic after a gale, as near as I can go.

May 4. Strong breeze from W.S.W., thick and foggy. While drinking our first glass of port-wine after dinner, the cry of ' ice ahead' brought us all on deck. Fog very thick. Many pieces of jagged white ice whirl-ing about, hardly to be distinguished from foam. All hands to tack the ship. Just as we got her about, the edge of a great field of ice just visible through the fog. Had it been night, we should have gone plump into it, which would certainly have delayed us, and might have damaged the vessel. Stood to S.E., away from ice, and having thus fallen in with it, and weather moderate, passengers' watch dispensed with. . . .

May 8, High-street, St. John's, Newfoundland: 9 at night. At length we are here, with some prospect of peace and quietness, a broad bed, and a steady floor. Coast fine, bold, and picturesque; bare red rocks, in form something like Great Orme's Head, beds dipping at a great angle, and cliffs rising to six or eight hundred feet straight from the sea, with deep water close at the foot of them, and no beach whatever. H. and I stayed and packed our things till eleven, when Mr. J. Stuart, one of the partners of the house to whom the vessel belongs, came out for us in a boat, took us to his house, lunched us, walked us over the town, introduced us to at least fifty people, brought us to these lodgings, which H. and I have taken together, and which are very comfortable. He then went with us up the high hill on S. side the harbour, a pretty stiff climb over rocks of red grit and quartzose conglomerate, covered with short spruce trees (the interior looks wild and barren, like parts of Wales and Scotland), gave us a good dinner and tea, got our baggage out of the vessel, and sent it up here in a cart, and has now taken H. to the theatre while I write this; thus doing the honours of the island in prime style. Our arrival has caused quite a sensation. I walked about in a five days' beard, and old hat, pilot cloth dress; H.—a fine fellow, exactly six feet five inches in his stockings, and well-made withal—clean, and in the highest style of London fashion. In one hour it was all over the town that I was come to look for gold mines; but H. puzzled them amazingly, since

they could not think in what he could possibly *trade*.
I go to call on the governor to-morrow; but may perhaps
send off this letter before, as a vessel is about to sail for
Portugal. The next vessel for England is in about ten
days, by which I shall also write. I find there is no
regular post-office here, as letters go round by Halifax. . . .

May 11. The vessel by which I intended to send
this sailed while I was writing it. I have met with a
very kind and friendly reception, both from the governor and the colonial secretary, Mr. Crowdy. The
governor took me all over his establishment, and after
an hour's conversation, he, his little boy, Mr. Crowdy,
and myself mounted and rode six or seven miles into
the interior, to visit the farm of a Captain Pearl, which
the owner confidently affirmed to contain coal. Nothing
was visible except slate-rocks in the neighbourhood;
but some man who lived near, and who had been a
miner, pointed out to the captain two small grass-covered hills as the receptacle of coal. We did not see
this man, but the governor sent an order to him to be
at Government House this morning at ten o'clock, and
he will send him to me. His excellency invited me
to dine with him, which I was obliged to decline, as H.
and I had invited the captain of the Diana to dine with
us, upon which he fixed to-day, and said he would ask
two or three of the most scientific people to meet me.
He says he expects a man-of-war shortly, when he shall

get her to take me round to St. George's Bay, on the W. side of the island, where there is known to be coal.

May 15. I went with the man mentioned above to the summit of Branscombe Hill, about seven miles, and found its height above the harbour to be nine hundred feet. It is clothed with wood and brushwood, making its ascent, joined with the bogs, very toilsome. In the evening I dined at Government House, with Colonel Saul the commandant, Mr. Crowdy the colonial secretary, Mr. Noad the surveyor-general, Mr. Browne the chief-justice, Drs. C. and S., and their ladies. On Sunday the most remarkable occurrence was a fire, which broke out about six in the evening, in the centre of the town, and raged with the greatest violence till three or four houses were consumed; and it was only got under by great exertions, and in consequence of the little wind, and the occurrence of a passage separating the burning houses from the next row. Where everything around is dry wood, the appearance of a fire is terrific, and the heat, even at thirty or forty yards off, was quite intense; all the neighbouring houses smoked like a furnace when the engines played on them. The military and fire-companies were soon in attendance, but the people generally seemed very stupid and apathetic.

May 13. Rode with the governor and Mr. Noad to Portugal Cove, a small harbour on Conception Bay, about ten miles off. Crossed high bleak moors, and

woods of spruce, firs, and birch, containing many beautiful lakes, one in particular called Windsor or Twenty-
mile Pond, four or five miles long, and from half to a
mile broad. The last two miles descended into a most
lovely and romantic little valley, full of woods, rocks,
and waterfalls, and bounded by fine peaked and serrated
hills, and opening out upon the sea by Conception Bay,
with a beautiful island called Bell Island a few miles
off right opposite, backed by the distant hills on the
other side of the bay, whose sides and summits were
still white with snow. One of the most beautiful scenes
I ever saw. The rocks are various, almost all schistose,
with some granwacke and conglomerate, and the coast
of the boldest and most precipitous description, not a
trace of sand or mud, and the sea water as clear as the
air above it.

May 14. Walked with the governor and Mr. Noad
to Quiddy Viddy Pond—a very pretty little lake two
miles hence, out of which a small stream runs into a
little cove hidden among the bare lofty rocks, through
which there is a passage just wide enough for boats
into the sea. This passage they are now widening by
blasting the rocks. Found a basaltic dyke running
from top of the cove to St. John's. . . . I have had an
offer from the Hon. ——, Esq., member of the House
of Assembly, to act as my paid servant, carry my kettle,
and make my fire, on my expeditions into the interior.

E

This may astonish you, but he is a hunter and fisherman, sent by one of the outposts. The country, even
within ten miles of St. John's, is still absolutely un-
traversed, except by new roads in one or two directions,
and in a complete state of wild nature, full of impenetrable woods and almost impassable morasses. The
view, however, from Branscombe Hill is really magnificent, and if the country is reclaimed, it will be a very
fine one; hill, wood, and water are most abundant.

May 18. I believe a vessel called the Queen sails
this morning for London, so I have risen early to finish
my letters. Yesterday was one of the most beautiful
days I ever saw. I ascended the south side hill, and
found it burning hot, but the air as clear as crystal. I
could see for many miles homeward over the blue Atlantic, whose colour was certainly most lovely, dotted
here and there with white sails. I found the height of
the hill to be 790 feet above the sea. The House
of Assembly opened in the middle of the day, and
salutes were being fired, and drums, &c. beaten, as I
looked down upon the assemblage of wooden boxes
which constitute the town. It was curious to see the
little jet of flame and puff of smoke from the cannon,
and then wait half a minute before the silence around
me was broken by the distant report booming up the
hill. H. and I dined with the governor last night, and
his excellency said, if I had attended the opening of the

House, I should have heard myself *puffed*. . . . On Monday I start for a short tour round Cape St. Francis, and the week following steer for the south round to Placentia Bay. This latter excursion may take me two or three weeks. Having come to the end of my two sheets, I now can only wish that you may all be in as good health and spirits as I am.—Yours ever, though distant now,

J. BEETE JUKES.

St. John's, Newfoundland,
June 11, 1839.

My dear mother, Amelia, grandmother, aunts and uncles, as also cousins, individually and collectively, and all other my friends and relatives, and to whomsoever these presents may come, be it known that in me, J. Beete Jukes, you behold not only the ' Geological Surveyor of the Province of Newfoundland,' but ' Commander of the ketch Beaufort, forty-tons burden ; Garden, Master.' But let us do everything in order. I believe I dispatched my former letter with about twelve others on May 16.

On May 18 I rode with the governor and Mr. Noad, surveyor-general, to Logie Bay, a small indentation of coast, rocks, cliffs, and surf, very bold and romantic ; thence to Torbay, a very fine bay indeed, something more English about it, but still very bold precipitous cliffs ; the roads are mere tracks marked on the rocks and stones by frequent passing, but cut through the woods. On my

return I engaged a man of the name of Kelly for the
summer, for 28*l.* and his keep (wages are very high here).

On May 20 Dr. Stabb and I, with Kelly, started on
foot with knapsack, barometer, rifle, &c. Walked to
Torbay, took a boat and examined the cliffs nine miles,
then dined on tea, eggs, and bread-and-butter, and
walked, partly on coast and partly through woods of
firs, over a very rugged road to Flat Rock, four miles.
Thence right through the thick woods, seven miles, to
Pouche Cove; this path was just opened, the stumps
were not cleared, and small poles merely laid down in
the *softest* places, by help of which, and roots of trees,
you kept from sinking; some small trees laid side by
side helped us to cross the rivers. In one part the
wood had been burnt, and the black ruins rose around
us, with ashes scattered on every side, worse than Bilston. The lakes numerous; from one near the track a
fine trout stream sprung, and rushed over the rocks and
stones through the woods towards the sea. Sat for
some time on the rude stumps in Canadian solitude—
no sound but the river; no sight but it, the lakes, and
the pathless impenetrable woods. At Pouche Cove at
night, tea and eggs again. No inns, but the best house
in the place receives strangers.

May 21. Stabb returned; Kelly and I walked through
woods to Biskia Cove at Cape St. Francis, twenty-four
miles north of St. John's. Kelly is a Newfoundland

Irishman. Passing the small schoolhouse at Pouche
Cove, he remarked very gravely, 'What a fine thing
it is, sir, to have them schoolmasthers erected in every
part of the island!' Cape St. Francis wild in the extreme; a fisherman gave us tea, eggs, and fish for dinner.
We then boated back to Pouche Cove, and set off immediately on our yesterday's track; but before I reached Flat
Rock, I felt so intolerably faint from hunger, that the
first house I came to I went in, got a quart of milk, and
then *gorged* on hard (*very hard*) biscuit and red herrings,
and the constant beverage *tea.* Reached Torbay after
dark, and got a comfortable bed in a private room.

May 22. Rose at four, and returned to St. John's to
breakfast ; very glad to get some meat ; enjoyed a meat
dinner at night, and found my English stomach, in
fact, very inconvenient when no meat was to be had to
eat, and nothing to drink but tea or water. Found tobacco to be in these circumstances an absolute necessary,
and being unprovided with either that or cigars, for I had
not smoked for a long time, borrowed a dirty little pipe
and strong tobacco of a fisherman in Torbay, and found it
quite a luxury, as it allayed the gnawing of the stomach.

May 23. Rain and fog ; cleared in the afternoon,
and Kelly and I started south for Petty Harbour. On
the road, Kelly's tongue running as usual, asked me
about my family and my mother, whether I had written
to her. On my telling him I had, 'I'll warrant the ould

woman would be mighty glad to hear you were safe
landed, sir ; she'd think no harum could happen to ye
afther that.' The walk was a fine one, over bare broad
eminences full of beautiful little tarns, winding among
the rocks and woods. On approaching the sea you
suddenly drop down about 500 feet, by a winding road
into a little narrow creek, running in among the rocky
hills. Round this the houses of the village called Petty
Harbour are built. Here we got decent accommodation,
though Kelly slept by the kitchen-fire.

May 24. Breakfast on tea and eggs, and at six
started in a boat with four stout hands down the coast
to Shoal Bay (nine miles). Coast equally bold in places
as towards the north, but more bare and barren-looking.
In Shoal Bay found the vestiges of the old copper-mine
worked in 1750, and got some specimens of carbonate
and sulphuret of copper. After staying an hour or two
of a most lovely morning, returned to Petty Harbour
just time enough before the breeze arose ; then walked
back to St. John's, and in the evening attended Mrs.
Prescott's 'At home' on the Queen's birthday.

On May 29 started in a boat to Bell Island, landed
and measured cliff, found it gritstone and shale. Then
pulled right round it in a thick fog and a breeze of
wind. Cliffs high and unbroken precipices. As I sat
in the stern of the little boat dancing on the waves, it
was very fine to see headland after headland looming

through the fog, and then frowning on us as we passed
its foot, and gradually fading again into the mist as we
proceeded. The island is nine miles long, and four
across : one great rock called the Bell, like Marsden
rocks on the coast of Durham. Next day went across
Conception Bay in a small packet-boat to Harbour
Grace (twenty-one miles); heard of coal being found
there, but all slate. A very pretty little town, with
neat stone church, all like England. The harbour
magnificent ; eight miles long, three quarters of a mile
across. Walked over a hill called Peak of Teneriffe
(560 feet) to a beautiful lake called Lady's Pond. The
only inn in Newfoundland is at Harbour Grace, pretty
comfortable. . . .

June 5 (St. John's). Made some arrangements
about hiring a craft, as Dr. Carson (speaker of House
of Assembly) told me they had raised the sum voted to
600*l.* to enable me to do so.

June 7. Received tenders for vessel, and accepted
Garden, Master of Beaufort, a ketch of forty tons (built
as a cutter by Captain Bayfield for surveying service),
he to find four men and a boy, a good boat, &c., provisions for himself and crew and everything complete,
to fit up cabin for me, &c., for 55*l.* per month. Since
then I have been engaged in selecting men, charts,
provisions, stores, &c., till I am bothered to death. The
vessel is being cleaned, fitted up, and painted, and will

be ready for sea to-morrow morning. It will take me
two days to get my own things on board, so I hoist my
flag (a red pennant with a black hammer) on Saturday
morning. .... My first trip will be to the northward, round
Conception and Trinity Bays; this, I expect, will occupy a fortnight or three weeks. I shall then return
to St. John's, and start again for the S. and W., and run
along to St. George's Bay. This will occupy me the
remainder of the summer till the latter end of October,
after which it would be dangerous to stay out on coasting service. I am victualled with a bag of flour, ditto
biscuit, ditto potatoes, ditto pork, eggs, tea, coffee, &c.,
a case of hock at thirty shillings per dozen, and other
wines, spirits, &c., for my own stock. Then fish is to
be had at a moment's warning, ducks ditto for the
shooting; and in the season curlew, plover, ptarmigan,
deer, &c. The account of my first trip you shall have
by the first opportunity. .... Everybody says I shall
have a very pleasant trip during the summer, barring
the mosquitoes when I land, the receipt for which is to
smear yourself over with *tar* and *oil*. Having now filled
my sheet, I can only say how glad I shall be to see you
all again and give you a more particular account of my
adventures than these few hints, and that I shall often
think of you on the seas, or hills, or lakes of Newfoundland; and with kind love to all, you must believe me
ever your affectionate   BEETE.

Middle of Conception Bay, Newfoundland,
June 17, 1839.

I sent off a despatch last Thursday, but begin another now, that I may fill it up at *quatres d'heures.* On
June 15, at eleven o'clock at night, having got all my
luggage, books, instruments, guns, &c. stowed away in
the cabin of the Beaufort, two friends took a glass of
grog and a cigar with me, and left me to my repose in
my berth.

June 16. Awoke by noise on deck. My captain having been in search of one of the crew all night, had just
caught him drunk in the street, and he was being
brought on board. Gave orders for sailing as soon as
possible; and at half-past eight we tripped our anchor,
hoisted the topsail, and glided down the harbour with
a gentle breeze, a red pennant flying at mast-head, and
union-jack at peak, being on government service. Wind
died away, and obliged to tow through the magnificent
pass of the ' Narrows,' the entrance to the harbour being
a mere crevice in the great sea-wall of that part of the
island. . . . At eleven at night, a red streak of sunset
just visible in the west, with a beautiful arch of aurora
over it, from which fine streamers of flickering yellow
light shot up into the stars. The young moon, and
Venus, clear and lovely, and all the host of heaven,
looking on the broad expanse of the bay, and the fine
shores and hills that bound it. . . .

June 20. Lovely morning, with a good stiff breeze from W.S.W.; went round Spaniard's Bay, walked over neck of land to Bay Roberts, examined singular brick-red slate with white stripes, then stood out again, and are now spanking away across mouth of next harbour, called Port-au-Grave. The sea a beautiful blue, sky perfectly clear, and a fresh breeze making our ship jump across the waves; while every imaginable shape of rock, bay, and headland, with mountain-like hills, surround us on one side; and on the other the blue ridge of Cape St. Francis and the beautiful cliffs of Bell Isle encircle us at from ten to fifteen miles. To the north and north-east only is there clear sea; all the rest is land as beautiful in its aspect as it is wild, uncultivated, and difficult of access when upon it. What should you think of me in a check shirt, sealskin cap, and Indian boots coming up to my knee, over some brown velveteen trousers, pacing the deck, a sea-geologist ! . . .

June 25. Mr. Green, surveyor, came and break-fasted with me; and at seven we set out in boat for head of next bay, called Collier's; thence struck into woods for some fine three-peaked hills that looked close by. In three hours and a half we arrived at the foot of them, having travelled about that number of miles through marshes, up brooks, but above all, through trees, on trees, under trees, and among trees. We sometimes were obliged to travel like squirrels, from

one fallen tree to another, for thirty or forty yards, it
not being possible to force our way down to the ground
for branches and fallen trunks; and a tree having happened to fall in the direction we were going was a
grand thing, as we could walk along its trunk. Climbing, creeping, crawling, twisting, pushing, pulling, tug-
ging, tearing,—every possible method of progression
except flying, swimming, or riding. Then the earth,
covered with boulders and great rocks, piled with moss
up to your knees, concealing slippery stones, deep holes,
sharp stakes, &c. &c.; and, to crown all, the *mosquitoes.*
Mr. Green's face and neck and ears were one mass of
trickling blood; you literally would have thought he
had been fired at with small shot. He was the worst
off; but even now (July 1) I am covered with pustules
from bites received that day. But when, after our
climb, we arrived at the top of the hills, our toils were
amply repaid. The whole of Conception Bay stretched
at our feet on one side; while, on the other, hill, dale,
valley, and lake stretched to the blue hills that bounded
the view on every side, in every possible variety of shape.
Walked back again in two hours, being down hill; and
finding a wood path, got back to our boat by dark, and
got some molasses tea in a ' tilt.' Molasses tea is tea
and molasses *boiled* together in a tea-kettle; and a tilt
is a temporary wooden hut, made of poles stuck in the
ground. . . .

July 1. Ran down along the coast, landing at one place where Indians used to come for red ochre. In afternoon put into a place called Northern Bay for shelter, as it seemed likely to blow hard. Thunder in evening. Walked about on sea-shore. Hundreds and thousands of beautiful little fish called 'capelin,' all green and silver and burnished red, rolled in by tide, so that the beach is covered with them, every wave seeming but a mass of fish. They are about eight inches long, and most delicate eating, and we catch them by simply picking them up alive on the shore. . . .

June 5. Beat up as far as Buonaventura Head, and anchored in harbour close by. Saw whales chasing the capelin in the bays, blowing up their spouts of water into the air, and heaving their huge tails up and down, making the seas foam under them. Saw a thrasher— that is, some cetaceous animal, which attacks the great whale above, while the sword-fish is said to be thrusting him below. The thrasher kept his tail out of the water—and a huge one it was—and lashed it backwards and forwards, making the waves boil. The men told me the whale is sometimes so hard bestead as to be heard to roar like thunder at the distance of three miles.

July 8. Fitted out the boat with provisions, and I took two men in her, leaving the captain with two more to return and take the vessel round to meet us at Hick-

man's Harbour. We rowed away, landing occasionally.
. . In the evening, seeing no signs of the vessel, we proceeded; but night coming on, we put into a small cove,
where was a brook, a beach, *and mosquitoes.* We set
to work to cut and collect wood, made a fire, set our
kettle to boil, then collected moss and sods to put on the
fire and make a smoke, by which we got rid of some of
the mosquitoes. Then we cut a lot of branches of
young fir for a bed, cleared away all the big stones,
and strewed the branches on the shingle. Over these
we spread one sail, and kept another for coverlid. We
then had tea and supper all in one, smoked our pipes,
took our grog, wrapped our faces in our handkerchiefs.
I rolled myself in my cloak, and put my knapsack for
a pillow. The men crept under the sail, and we were
soon snoring. During the night, I was awoke twice by
heavy rain on our fire, and clapped on another tree or
two; but otherwise slept soundly.

July 6. Woke at six, got up, bathed in the sea—water
very cold, but so clear (as it is all round the island)
that I could see lots of lobsters, crabs, fish, sea-weed,
zoophytes, and hundreds of green echini, below fifteen
or twenty feet of water as plain as in air. . . . We
pulled to examine some islands, then rowed along the
coast to the opposite headland; and when the fog
cleared, saw our ship, and in an hour or two got aboard
of her and sailed for the S.W. arm, another great inlet

of the sea. At night it came nearly calm, and in the dark we could not distinguish the entrance of Fox Harbour, where we meant to anchor, so I went to examine it in the boat. After half a mile rowing, found a narrow entrance, sounded and got it all right, and returned. There was not a cloud in the heavens, ' studded with stars unutterably bright,' Jupiter especially casting a strong gleam across the still water. This might have been taken for a lake, but for the gentle heave of the ground-swell, which told of the stirring of old ocean. The air was balmy; and in the N.W., where the latest streak of twilight yet lingered, a beautiful aurora was flickering with yellow light; while every dash of the oar and surge of the boat made the sea flame with phosphorescent sparks and flashes. Around this brilliant scene the dark headlands lay on every side in stern repose, a fitting frame for the picture. All was still except the boom of a whale as he rose occasionally to blow (making a noise like the puff of a steam-engine) and the surge of the waters as they closed around his descending carcass.

July 14. Started with four hands for the other side of the Bay of Bulls to climb a hill called Centre Hill, about four miles. In half mile found myself nearly knocked up. Roused by fresh track of deer, but could not get sight of one. With much difficulty got to foot of hill; lunched on biscuit soaked in whisky-and-water.

Set to work again, and found myself at length on the summit, but too giddy and tired to enjoy it. Counted, however, one hundred and fifty-two small lakes within eight miles. . . Got back to ship faint and weary. The fact was, I had had nothing but salt meat (of no very good quality or curing), and my English stomach had acquired a distaste for it; so that, as the weather got hotter, I lived principally on tea and biscuit, the latter sometimes mouldy. This agrees very well with the people here, but I wanted something more substantial. Before night, between the showers, I took a walk in the woods, and on returning found Kelly had turned one sail into a hammock for me, under which Joe had spread boughs for himself, while Kelly and Dick lay on boughs and the other sail. We then had supper, grog, and pipes, and spun yarns; among which Kelly's were conspicuous, partly for the humour with which they were told, and partly for the astounding lies they contained, himself the witness and voucher for every one of them.

July 16. Sailed to Tickle Head, then to Tickle Point, where I landed and shot five brace of sea-pigeons, a small kind of teal or widgeon, with red legs and black-and-white feathers. Thence into Collier's Bay, and anchored in a small cove remarkable for nothing but tolerable water (the first since we left Trinity that you would think drinkable, as we could generally study botany and entomology in our glasses) and the vivid bites of its mosquitoes, as also for the abundance of its sand-flies—an infinitesimal creature, like a needle's point in size, but much sharper and more stinging. . . .

July 18. Another famous day. Found a few faint impressions of shells in a red rock, the first fossils I have seen, something like small orthes of Dudley. The men thought I must have discovered a silver mine at least. I saw a paper, by the bye, at New Harbour, professing to give some account of my proceedings, and saying I had discovered a copper mine, gypsum, and I don't know what. I suppose I was asleep when I did it, for I don't recollect it. . . .

July 21. I landed at Quiddy Viddy, and walked to St. John's; found letters from you. . . . In answer to your letter, my dear mother, I have only to thank you, and to assure you that, though I hope I shall not shrink from any danger in discharge of my duty, I am by no means of that chivalric temperament to seek danger for its own sake. . . . I am specially obliged for the copy of Sedgwick's letter; his approbation is indeed a spur.*

* The letter here referred to was one from Professor Sedgwick to Mr. Alfred Jukes, surgeon, of Birmingham, from which the following words are an extract:

'Your nephew, whom I saw in London about a month since, is now, I presume, on his way to Newfoundland. I am delighted with his boldness in undertaking the laborious task that is before him. But he has chosen a line for himself, and there is only one good way for him—to follow it in such a manner as may tell for

I sincerely hope that my fears may not be fulfilled, and the Birmingham meeting may prove a good one. . . . I am afraid the postage of this letter will be very heavy. . . . I have been much more particular now than I can be in future. I shall not return from my next expedition till the latter part of October, soon after which we shall be frozen-up. I shall not be able to send any letter after this till November, when it is a chance if there will be a vessel going. You must not, then, be uneasy if you do not hear from me after this until next spring. At the same time, it is very probable I shall be able to send a letter in the autumn or winter. . . . By dint of great resolution and keeping indoors this most lovely day, I have finished letters and copied part of my large chart of Newfoundland, of which Avalon forms but a small corner. I have also traced my track in red ink, as you will perceive. Mr. Noad, the surveyor-general, has copied it for my use with his pentagraph, on six times the scale, or thirty-six times the superficies. But only think—if it takes me five weeks to examine these two small bays, what it will take for the

his own honour and the good of his country. His health is excellent, and he will stand in need of it. His zeal is unbounded. A new world is before him, from which, I trust, he will bring back a rich harvest. The work, if his life be spared, will put him in the first class of geological observers, and be, I hope, the means of eventually securing for him some good position as a man of science.'

F

whole island ! . . . Finis. I shall be too busy to-morrow to add a word, so now farewell. My best love at-
tends you all. I wish you all health and happiness ;
but why need I say this when I who say it am your
affectionate

BEETE.

St. John's, July 30.

Got the last of my things on board at nine o'clock ;
but detained to get clear of Custom House, which does
not open till ten. Then getting my men together, so
did not set out till twelve, taking a Will instead of a
Joe, who was too drunk. Having gone three miles, two
guns were fired from Fort William. Captain thought it
was a signal for us to put back ; would not attend to it,
so cracked on ; no time to be lost.

Aug. 1. In afternoon close in to St. Pierre, but too
thick to venture among rocks. Wind shifted, and blew
very hard ; nearly driven on lee-shore ; in the morning
made land, and found ourselves fifteen miles back again
away from St. Pierre ; by direction of fishermen, steered
for small harbour called Lamelin. Got in in the fog.

Aug. 2. When fog cleared, saw we had passed through
a narrow passage, seventy yards wide, surrounded by
shoals and rocks. Country here a dreary waste.

Aug. 5. Got into St. Pierre in evening. Captain of

the guard-ship came on board to know who we were.*
Offered his services in any way. He spoke a little English, which I answered in as little French. Sent in my
name to French governor. Midshipman came on board
to say governor would be happy to make my 'honourable acquaintance at *midi.*' Invited me to dine at five
o'clock, when I met two Frenchmen and an Englishman
settled there. Complete French dinner; and at nine
o'clock, instead of tea, a glass of bottled ale handed
round.

Aug. 7. Went in boat to neighbouring islands of
Langley and Miquelon, connected by a beach five
miles long, through which there was a passage fifty
years ago. Governor has a country house here, at which
were two gendarmes, one of whom was married to a
pretty Irishwoman. She left Ireland with her father
and brother in the time of the cholera, and went to
Quebec, where, finding it very bad, they went to Mon-
treal. Here it was worse; and both her father and brother sickened and died. She embarked for home, and

---

* ' The islands of St. Pierre Miquelon and Langley are the
only territorial possessions left to the French in this part of the
world.

' By treaty, the French are not allowed to erect fortifications,
nor to have more than fifty soldiers on the island at one time.
They have very strict regulations in the port. No English boats
or vessels having fish on board are allowed to come in, on penalty
of being seized.'—*Jukes's Excursions in Newfoundland.*

was wrecked on the beach here; and being very young
and utterly destitute, she shortly after married this gen-
darme.

Aug. 10. Found ourselves in Gulf of St. Lawrence,
with west shore of Newfoundland about ten miles off.   In
evening a splendid aurora—a diffused yellow light all
over north, with beautifully brilliant primrose-coloured
streamers, dotted with stars, and dabbled with black
clouds.   The next morning weather splendid ; the Beaufort bending under wind, and leaping from wave to wave.
Got into Bay of islands, and anchored at dark.

Aug. 12. Sailed up an arm of the sea to a fine river
called the Humber.   Made preparations to go up it. At
three miles came to first rapid, and obliged to put out
three men and a line to tow her up—difficult navigation,
great rocks, shallows, deep holes, and whirlpools, and
current like a mill-race ; rapids half a mile long. About
ten miles up another rapid, more dangerous and difficult than the first. Took us a long time to get up. One
mile above this came upon a fine lake.   Began to rain ;
so encamped in the woods, having brought the square sail
of the Beaufort, which made a fine tent, and kept us
dry, though it rained very hard all night.   Started at
seven o'clock, explored a small river, rowed and sailed
up lake, which was fifteen miles long, and got into continuation of river, and encamped at five o'clock under
some fine beech-trees.

Aug. 15. Early in the morning heard roar of falling water; mist cleared, and saw great rapids half a mile ahead; despaired of getting boat up—great rocks and breakers. Examined provisions; only for two days more. Obliged to go back. A little lower down found another river sixty yards broad; turned up it, but in half a mile came to similar rapids; and back again. Saw Indian wigwam, and found old arrow-head of red Indians. Sailed back to lake, and encamped on a very pleasant spot. Rain at night; but tent dry.

Aug. 16. Rowed down lake; eat last of our prog on small island; finished dinner on berries of various kinds —some like bilberries. Got back to the Beaufort at dark.

Aug. 17. Examined limestone hills; no fossils. In woods got gooseberries, currants, raspberries, a black berry intermediate between black currant and rough gooseberry, bunch berries, squash berries, maidenhair berries; and saw three kinds of poison berries: one a beautiful cerulean blue, growing on leaf and stalk like lily of valley. Woods much loftier and better to walk in than in east of the island. Went to Indian wigwam; saw three Indian women; bought a pair of mocassins— white deerskin ornamented with coloured cloth. These Indians are not aborigines, but came from Cape Breton originally.

Aug. 24. In evening got out a fishing line, and

in an hour and a half we caught fifty-five cod-fish, several of which weighed thirty pounds. The men split them, and salted them down in two old beef-barrels.

Aug. 27, St. George's Bay. I have heard of an opportunity of sending to St. John's, so am just finishing off this sheet. I hear of coal, and 'plaster-of-paris,' and lead-ore at different points of the bay, which will be the most interesting spot I have seen. Meanwhile I have just agreed with an Indian to go with us into interior. Grand Pond, which I left unreached from Bay of islands, comes to within twenty miles of here, and we can reach it in a day and a half. Then he has a flat-bottomed boat on it, by means of which I can get to within five miles of where we were obliged to turn back. We shall be away a week, as there is thirteen miles to be walked, which will take a day and a half each way. My future movements will be hence in about a fortnight; then cruise along south shore of Newfoundland to St. John's, which I should reach about the latter end of October. . . .

On August 27 I finished my last despatch, and made it into a packet to go to St. John's, to be sent to England by the first ship. On August 28 we set out, after several heavy showers and strong winds, and got to the mouth of the brook, whence we were to set out inland. Here we slept in a small cabin ten feet long, six feet wide, and four and a half high. It just held a fire at one end and the four men, the Indian Sulleon and myself lying heads and points at the other, my dog Bell creeping under a corner of my blanket, and lying along my back. Up at daylight, boiled the breakfast, arranged our packages, and started. I carried my hammer-bag, my new knapsack—chockful — cape, leggings, and blanket. One man a bag of bread, the others beef, pork, &c.—provisions for a week. The day was fine, and we trudged on, following a kind of track which to Indian eyes was visible, but scarcely to less practised ones. We passed through several small woods and long marshes, and toiled on with our loads until four o'clock. We had then come eight miles, according to Sulleon; but all were sufficiently tired. I believe we had come at least ten, and I had rather walk forty in any part of England. We slept that night in an old tilt which happened to be near, and managed pretty well, except that the men made such a fire, that we were near burning it down.

Aug. 30. Left some provisions till our return, and again set forward over similar ground, but more hilly. The wet moss sinks under you at every step, making every movement a labour, and obliging you to watch your footing. The woods are more lofty and open

than any I have seen; but the path still impeded by
roots, stumps, and fallen timber. It would, however,
be much easier to clear than that on the east of the
island, and contains much finer timber. At three in
the afternoon we stood on an eminence of about 100
feet, from whence we could see the Grand Pond lying
below us. Descending the hill, we came to the bed of
a brook running into it, and half a mile farther we
came on the flat-bottomed boat which was to help us
forward. We all laid down our loads with considerable
satisfaction, and shortly had a fire blazing and the
kettle boiled for dinner. Having put the things in the
boat, S. and I left the men to bring it on, and walked
to look for game, and among some swamps we shot
four or five wild ducks. The first part of the pond is
from one to two miles wide, and about seven long,
lying between precipitous hills of six or seven hundred
feet high, all gneiss and mica slate. At the south side
of the lake was a wigwam, which had been built by
Sulleon; and as we had to make another oar, a mast,
and repair the boat, we stopped here for the evening.
A wigwam is a lot of sticks stuck round a circle and
meeting at top, covered round the side with birch-bark,
and the top open for smoke. The fire is made in the
middle, and the small boughs of fir are spread neatly
round in a circular fan-shape for the couch. If you
stand up, your head is over the fire in the chimney;

but you can sit or lie very conveniently at the circum-
ference of the circle, with your feet to the fire. Six of
us lay in this way, each man's head behind the other's
legs. It forms a tolerable shelter from rain, but is, in
fact, only a great fire-place, with the draught drawing
under the bark all the way round. Wrapped in my
blanket, however, and with my knapsack for a pillow, I
slept soundly enough. Up early; but Sulleon and I
both unwell, from wading after wild ducks while glowing with perspiration, and then sitting in wet clothes.
After proceeding seven miles with much difficulty,
against a head - wind and high waves, found another
wigwam, in which we took shelter. During that night
the wind blew so violently that we were all afraid the
wigwam would be blown over, or some of the trees
fall on it. At daylight the Pond was a sheet of foaming
waves, like snow. The Pond, after the first seven
miles, divides into two, enclosing an island twenty-two
miles long. We sailed up the north arm between lofty
precipices. At length such a squall came driving
through the narrow pass, and such a rolling swell, as
I never saw except at sea. We kept as much as possible under lee of the land; but had it not been for
Sulleon's careful steering, we should certainly have
been wrecked; and the banks on either hand so steep,
as to make it uncertain whether we could have got
ashore at all. Arriving at the end of the island, we

landed for a few minutes to recruit. . . . By five o'clock in the afternoon, we had sailed twenty more miles, making upwards of forty that day. We landed in a little cove, and slept in a fine spacious wigwam. At night there was the most splendid aurora I ever saw. A long band of brilliant yellow light, with a well-defined base, but shading into glancing rays upwards, proceeded with a rapid but stately march in three bold curves from north-east to north-west, passing through both Bears. It was like a host of heavenly spearmen marching in battle array; on each side a fainter light, the reflection of the principal body. At first the north-east was a mass of light, from which the array seemed gradually to unfold itself, and march onward to the north-west, till at length it all passed away in that direction.

Sept. 2. Up with the dawn, and sailed eight miles to the head of the Pond, which is here five miles broad; then proceeded up small brook to look for a bed of coal, said, by old Indian I saw at St. George, to be there. The place where he saw one three feet thick was covered with fallen rubbish; but we found another six inches thick in a newly-fallen bank. As this is on the borders of the flat country, and as this flat country is of considerable extent, and the rocks dip in that direction, I hope this is the beginning of a large coal-field, stretching into the country. Many pieces of coal in the brook above this.* Our provisions were beginning to run short, so we made a scanty supper.

Sept. 3. Wind lulled about midnight, so at three in the morning got into the boat and rowed back. Dreadfully cold till an hour after sunrise. By eight o'clock, having gone fifteen miles, we stopped to breakfast; then proceeded, and kept under shelter of the land until one o'clock, when we were within two miles of the island. Here hoisted sail, and crossed over to go down the other side of island from that we came up. Landed at one point to look for game, but found only squash berries, on which we dined. In a few miles more came to a narrow pass, where deer or cariboo were accustomed to swim across from and to the island. Stopped to see if we could get one. Shot a couple of divers, on which we supped, and before dark had the good luck to kill five more, which secured us till we could get back to the end of the Pond, where we had left some provisions.

---

* The beds of coal here alluded to are thus mentioned in the Report annexed to the *Excursions in Newfoundland:*

' The beds of coal on the south side of St. George's Bay, as well as in the country north of the Grand Pond, do not seem to be of any great thickness. It is perfectly possible, however, that more important beds may be found, should the districts ever be thought worth working. This can only become the case, either from the exhaustion of the present mines of Cape Breton, or from the settlement and increased population of the districts themselves.'

Sept. 4. Off at daylight, but wind still against us, and blowing hard at eight o'clock. Landed to breakfast, having only got four miles. Here, being uncertain whether we could proceed, determined to try and get a deer. Accordingly I divided my remaining stock of provisions equally among all hands. Sulleon and I and two men then started for a marsh famous for cariboo, in the centre of the island. After a hard climb through thick woods, we got on marsh at top of high land. Very hot, and for six hours no sign of deer. I then went to look for partridges with Bell, but could not find one. Meanwhile Sulleon had got a fresh track of a cariboo, and soon after, I heard a gun and a shout, and on going down, found Sulleon had shot a fat buck. We immediately set to work to cut him up. He weighed about 400 pounds, so we had a decent load; but by dusk we reached our camp, and great was the boiling, roasting, frizzling, and frying. Unluckily, we had no salt and but little bread; but we all fared gloriously, more especially poor Bell, who, being a pointer, would not eat birds, and had therefore nothing for two days.

Sept. 5. Off at daybreak; quite calm. The island principally granite and chlorite slate. By two o'clock reached camp at the end of the Pond. Here Sulleon cut and hung up to dry half the venison, till he and his family came this way, which they intended to do shortly.

Sept. 6. Getting our things together, and hauling

the boat up the river, detained us till half-past seven before we could make a fair start. Our first plunge into the bushes wet us to the skin. The trees and roots were all slippery; the brooks rising and the woods so bad, that it was eleven o'clock before we got out on to the marshes. I kept urging on the men. The wind was very cold, and pelted the rain and fog upon our backs most unmercifully. By one o'clock we reached the tilt where we slept on the 29th. The roof was bad, and it afforded little shelter. Accordingly I allowed but one hour for dinner. The marshes got wetter and more sodden, and our burdens heavier, every step. Every time we went into the woods, we came out chilled, and my hands purple with cold. As my load was equal to any of the men's, they could not grumble, and I pushed on. Sulleon himself began to flag and complain of pain in his stomach; but just as the last light of day faded away, we emerged from the last wood to the little brook where we had left our boat. Every one's spirit rose as he laid down his burden; and we quickly dragged our boat into deep water, hoisted our sail, and actually began singing to keep ourselves warm, as we drifted before the blast. I wrapped myself in my blanket, for the wind was bitter; and our singing soon sunk into talking, and that into silence, and, on my part, an eager looking out into the dark night for the first signs of the harbour. After two hours' rapid sailing, we gained the side of our vessel at half-past ten. I ordered a glass of brandy all round, took one myself, stripped off my wet things (the first time except once for ten days), got the captain to rub me down with a coarse towel, pulled the sheets out of my berth, and jumped in between the blankets. Some fried venison and a glass of porter soon raised the vital warmth a little; and sending down a tumbler of hot grog by way of a ' raker'* to keep in the fire for the night, I was soon wrapped in the arms of ' Morbus.' . .

Sept. 13. We got round Cape Anguille, and shortly after into a small harbour between an island and the mainland near Codwy River, and ever since it has been blowing a hurricane with driving rain and fog. This gale increased, when, soon after dark, a small schooner near us snapped her cable, and nearly ran foul of us, but brought up with her other anchors just in time. Presently after another small schooner began to drift. Gale increasing about nine o'clock, we also began to drift. Jumped on deck; the sea a sheet of white foam; the wind itself white with spray, obliging us to cling to the ropes, and shout in each other's ears, and our little vessel drifting backwards towards the entrance of the small harbour in which we were. Giv-

---

* A ' raker' is the term used in Staffordshire for a large lump of coal, put on the fire at night with a ' backing' of slack, to keep it in until morning.

ing her all the chain and anchors we had, she brought
up within fifty yards of the point forming the mouth of
the harbour, and close inside the bar. Here, luckily,
she held on, but struck once or twice rather heavily on
the bar. Had she gone on, we determined to have
taken to the boat, rather than accompany her out to
sea on such a night. About twelve o'clock the wind
shifted to N.W., and blew us back into the harbour.

Sept. 14. The wind having blown hard all night,
such a sea was rolling in as I never saw. Even the men
and the captain kept exclaiming at it, and thanking
their stars they were not blown out on such a night.
Long waving ridges of sweltering water, with flickering
tops like mountain peaks, came rolling in, and on approaching the shore, fell over in broad cataracts of foam
like the falls of Niagara. At one time I counted seven
of these ridges one behind the other, occupying a space
of three miles long by half a mile broad, in any part of
which a vessel would have been knocked to pieces as
easily as a china tea-pot.

Sept. 15. Heard rumours of much damage; many
boats lost; a large vessel ashore a few miles from us,
sails set, and nobody in her.

Sept. 16. Small schooner came in that was out
in the gale, her deck boat smashed to pieces. Quite
tired of this spot. To-day I have been mending my
own clothes, having sewn up a great rent in my shoot-

ing-jacket, darned some smaller ones with black silk, put some buttons on my waistcoat. My general costume is a pair of canvas trousers (washed occasionally), with a pair of Indian boots tied over them above the knee, a canvas jacket or frock, with pockets inside and out, white when fresh washed, at other times the colour of the last rocks it has been among, a blue shirt, sealskin cap, and a bushy beard, no razor having violated the integrity of my face since I left St. John's. My hands and face vary from a raw blue to a red brown, according to the temperature. . . .

Sept. 18. Towed the vessel some miles into Codwy River, where are some English and some Indian inhabitants. All the men gone to the wreck near Cape Ray. One fine intelligent young Indian, lying in his wigwam, is crippled, having nearly broken his back two years ago. He gave me much information about the country.

Sept. 23. Detained in this place, partly by south winds, partly by promises of Indians to show me a bed of coal up the river. This they at last refused to do. I bought of one a fine black-and-white dog of the right Newfoundland breed, called Camouge, which is Indian for 'stick.'

Sept. 24. Sailed to Cape Ray, where we saw the wreck, a fine ship of 700 tons, called the Onondago, lying on the rocks under the shore, several boats at work, as well as men ashore, getting out her cargo, which is only timber. She looks like the carcass of a racehorse being devoured by a pack of wolves. Followed a boat by a narrow channel through horrid-looking rocks and reefs into Port aux Basques, one place not more than twenty yards wide, with high black rocks of mica slate.

Sept. 25. Hammering the rocks—such crystals of quartz, mica, feldspar, horneblend, and garnets! Made sail against the captain's inclination; but we have lost so much time. Wind shifted to S., and it began to look dirty. Got a fishing-boat to pilot us into the harbour at the back of the Dead Islands. These are a cluster of small rocky islets which form a snug harbour, but very dangerous and difficult to strangers. Four families live on them. Visited one man, George Harvey, a fine jolly old cock, a regular sailor, with a large family, full of hospitality in his small cabin. He has been the means of saving more than 200 lives in his time. From one wreck he brought off 160 men, women, and children, assisted at first only by his son of twelve and his daughter of seventeen years. For this he had a gold medal and 100*l.* from Government. Last September he saved twenty-five from another wreck, being the crew of the Rankin. They were clinging to a piece of the wreck which was driving out to sea, and he brought them off six at a time, in a small boat,

through a tremendous surf. Nothing but wrecks and
rumours of wrecks along this coast. There were four
small vessels wrecked the other night, but no lives lost.
I shall be glad when we get off this coast, which is
a bleak, bare, desolate mass of rocks, with only a few
stunted bushes here and there, and is the terror of seamen going to Canada. . . .

Oct. 10, St. Peter's Bay. My work along this shore
has been entirely cut up by the unusually bad weather.
My contract with the captain expires at the end of this
month, and I have got 200 miles of coast to visit. All
I can do now will be to push on from harbour to harbour, as the winds will permit, and work my way to St.
John's. There is here a large ship which was dismasted
in the first gale, and put in here in distress. She is
bound to Liverpool, and I shall take the opportunity of
sending this packet by her. Two ladies, whom I met
at Mr. Thorne's, with their children, were on board of
her in the storm, expecting every minute to go to the
bottom, the skylight dashed to pieces, and water pouring down into the cabin; and yet they never cried out,
nor worried the captain with any questions, although
they were left alone in the dark for three hours. I did
not intend to send off this packet—as it is full of such
disastrous accounts—before I arrived safe in St. John's.
From this place, however, the coast is both far less dangerous, and my captain and men are acquainted with it.

The equinoctial gales are past; there is therefore no more, indeed less, fear of our meeting with an accident than if all these had never happened. Mr. ' Camouge' is now resting his great head on the table, examining my writing with great gravity, and occasionally turning up the whites of his eyes as he looks in my face. He would no doubt send his love and respects if he had been used to civilised life, but being an Indian you must excuse him. I hope there is no need of my composing an elaborate sentence to assure you of my continued love to all. In my adventures, travels, and voyages, though occupied at the time, when my thoughts have time to repose, they return to the recollections of home, and wander from scene to scene till you are all included; and I hope the only effect of my absence will be to render me more worthy of being ever your affectionate

J. Beete Jukes.

Trepassée Harbour, October 26, 1839.

On the 11th of this month I sent from St. Peter's a transcript of my journal up to that time. On that evening I sailed for St. Lawrence, having Mr. R. T. on board as a passenger. We got in during the night. The next morning I went with him to his house or his brother's. We then went ptarmigan shooting, and on my return I slept at Mr. T.'s. Their house is a large

and pleasant one, on a very pretty island near the head
of the harbour of Little St. Lawrence, and they gave me
a very snugly-furnished bedroom. It was the first time
I had slept in a bed since July. The winds detained
us at St. Lawrence till the 17th, which time I employed
in seeing what was to be seen, and shooting what was
to be shot, and found it quite a treat to have some con-
versable companions. On the 17th we sailed to Mortier
Head, where I was as hospitably received by Mr. H.,
the magistrate of the district. On the 19th to Audience
Island, and thence to Great Placentia, which was once
the capital. I was there entertained by Mr. B., the
surgeon, and the next day I walked with him across to
Little Placentia. There is a road from Great to Little
Placentia, five miles; and as it was frozen hard, it was
quite a treat to walk on it. On 25th the wind shifted
round to N.; and after waiting all day to see whether it
would stick there, we sailed at twelve o'clock at night.
The wind blew a gale, and the pitching of the vessel
capsized my stove and chimney, and knocked all my
cabin furniture about into an utter wreck. This morning it moderated; and after a beautiful day we anchored
here just before sunset. It is a noble harbour, but the
country about is low, bare, and bleak-looking. I have
got my cabin to rights again; and while my men are all
snoring after last night's work, I am sitting by the stove
drinking a glass of grog in a very sailor-like manner.

How little I thought, this time last year, that I should now be addressing you from such a place! I should have then looked upon it as quite an adventure to have spent even a night in a little vessel like this on the coast of England, much more on that of Newfoundland; and now I feel as comfortable as in a London hotel. I believe I was intended by nature for a wandering life, and only wish I could have begun it ten years ago in some capacity. As my fire is going out, I shall now say good-night.

Oct. 27 *to* 30. Still at anchor in Trepassée, not able to stir, wind E. Heavy rain outside; stove inside is a smoke manufactory on a small scale. Every now and then open the window, in comes the rain; shut the window, out comes the smoke; and if the wind were but fair, I might have been in St. John's, lolling on a sofa reading your letters, or going to dine with the governor. A most humiliating condition of humanity to be dependent on the direction of a puff of wind. Waiting for a coach is bad enough; but only fancy waiting four days at one time, without even room to walk about in—getting out of your berth to sit down on a chair, getting off the chair to lie down in the berth, or occasionally going two steps up the companion ladder to look at the rain, and be satisfied that it is going to last. Well, it is a long wind that has no turning.

Nov. 2. On this day wind blew a gale from north-east, drove us from our anchorage, and hove us ashore, wrecking us in a gradual and deliberate manner delightful to behold. Luckily, the shore was there a beach of pebbles; and though we got a hard pounding and thumping, and the sea broke over us and washed our decks, and gave us all a thorough soaking, pouring down into my cabin, no farther harm was done. High and dry at low water, and the vessel half buried in pebbles.

Nov. 4. Got all the ballast out to lighten her; tried to get her off, but tide did not rise high enough.

Nov. 5. Got vessel off; then all day getting in ballast and putting to rights. Shall be in St. John's in eighteen hours. Such a fine breeze from west! But just as we were heaving up the anchor, pop comes the wind again from the north-east; and now, Nov. 7, here we are still, thick and heavy rain; and when we are to get out of this confounded hole, heaven knows. What makes it worse, the only gentleman who resides here, Mr. G. S., is clerk of the court, and is now making the circuit with the judge. Mrs. G. S., however, has lent me some books; she is likewise waiting for a wind to go to St. John's. What would an English lady think of being obliged to go a hundred miles in a small boat, with a little cabin on deck, in the stormy month of November, along a rugged and dangerous coast—taking a baby too —in order to buy provisions for the winter? She has gone twice a year ever since she was married—sometimes with her husband, sometimes without. She was wrecked once with a baby and a maid-servant, and out in an open boat all night.

Nov. 9. Fine S.W. wind. Started at eight o'clock; and by twelve beat out of the bay, and left Trepassée at last. By night we got to Ferryland, and anchored. This was the first settlement in the island, and its name is, I think, a corruption of Fairyland, as the first settlers, under Sir Humphrey Gilbert, give rapturous accounts of it in the sixteenth century, and speak of corn and vines. The climate must have altered very much, if their accounts are true, as no corn will grow here now. The country around is very picturesque, with many fine harbours. I am afraid the wind will get east again, and catch us here.

Nov. 11. As I feared, a furious storm has been blowing from the north-east, with thick fog and heavy rain, snow and hail. We are snug enough in our little natural dock, in a corner of the harbour, behind a pebble beach, and moored to a wharf. I can hear the wind raging and blowing and whistling in our rigging with tearing malicious violence, and the pauses of its shrill treble filled up by the hoarse deep roar of the sea breaking on a reef of rocks at the mouth of the harbour. I should set out for St. John's by land to-morrow, but the country will be impassable for a day or two, and even with a fair wind we must wait to let the sea run down. This is the sixth heavy gale since September ; all have done more or less damage on the coast. Two or three vessels were lost last week, bringing coals from Sydney to St. John's. 'Blow, winds, and crack your cheeks.' I wish the north-east would crack his cheek, and let the west have his turn.

Nov. 12, 13, 14. Still at Ferryland, but a sharp frost has hardened the marshes, and made the walking tolerable. The hills in the interior are all white with snow. If we don't start to-night in the vessel, I shall go to-morrow without her.

Nov. 15. Wind shifted to S.W. Sailed from Ferryland ; and at eight o'clock entered the narrows of St. John's, and anchored in the harbour. Immediately went to my lodgings ; found H. still there, and looked upon this year's work as finished. The next day I called on the governor, who received me kindly, and congratulated me on my safe arrival. Every day since, I have been invited out to dinner. There is a Dr. Stüwitz here, a Norwegian naturalist, sent out by his government to examine the natural history of North America. He is a capital fellow, and an excellent naturalist. He means to winter in Fortune Bay, and return here time enough in the spring to go to the ice, and see the seal-catching. I daresay I shall go with him. . . . I sent my barometer and thermometer in the summer to be mended ; they

are both come out again, but broken much worse than before. Here I end my journal, and commence answering
letters. Farewell.

J. BEETE JUKES.

St. John's, Newfoundland,
Nov. 20, 1839.

MY DEAR A.,—I have received your letters of July,
August, and September, for all which I am greatly obliged to you. The last came in a parcel brought by
Ellen Highfield. She arrived here the day before yesterday. Who is Ellen Highfield? you say. She is one
of that class of ladies ' that walks the waters like a thing
of life.' . . . . I more than ever regret the exaggerated
accounts I received, and sent you, of the difficulties of
communication here in the winter. Vessels are coming
and going from and to all parts of the world as regularly
as at any time. . . . . At this present moment the sun
is shining, and a greatcoat an encumbrance except in
a sleigh. By the bye, I like this method of progression
uncommonly, it is so safe, swift, and silent; the silence
at least is only broken by the jingle of the bells round
the horse's neck.

Jan. 7. The lower orders keep up the old custom of
mumming from Christmas-day to Twelfth-day. Forty
or fifty fellows, in fantastic dresses and masks, dance

about the streets with flags and music, and thump everybody with bladders tied at the end of whips, cows' tails,
&c.; jump up into the footman's place behind the sleighs;
and take any other liberties that come into their heads.
All the shops too are shut up on New-year's-day and
Twelfth-day.   The session of the legislature was opened
the other day by the governor, who mentioned ' my arduous and important labours' in his speech.   My report
is being printed, and I mean to send you some copies. .
Last Saturday I went with five more in three sleighs to
the Bay of Bulls, about twenty miles off.   The road is
not finished, being full of drains, ditches, great rocks,
and stumps of trees, and the snow not above three or
four inches deep.   Accordingly, after the first six miles
it became rather exciting, and the motion more like a
boat in a heavy sea than a sleigh.   Two bridges tilted
over with us.   (N.B. The bridges consist of two long
poles across a ditch, with some short ones laid across
them.)   All the sleighs upset: shafts broken, seats
coming off, and a long catalogue of misfortunes.   The
weather, however, was most splendid ; and J. S. and I,
who walked the most, arrived in a bath of perspiration
and with a marvellous appetite.   We returned next
day, walking the greater part of the way, having hired
men to bring our sleighs, and dined in a tilt with some
wood-cutters on potatoes and tea, flavoured with salt
fish.   You talk of my wonderful hardships and priva-

tions. I assure you, I begin to look upon myself as a
monster of idleness and luxury. Here is Professor
Stüwitz set off at the beginning of December in a boat
with a little cuddy to which my cabin was a palace.
He is gone to see the winter fishing in Fortune Bay,
and make other observations, with the chance of being
frozen up on his return, and having to get ashore, and
come through the woods and snow. He is returning to
go to the ice in March to see the seal-fishery. In that,
however, I mean to accompany him. H., too, was out
shooting for four days in the woods, lying on the snow
at night, and his clothes stiff with ice—having fallen
into a hole of water—and had nothing to eat for two
days but some spruce-boughs boiled in snow-water.
Had they not struck a path when they did, he would
have been obliged to shoot a dog for supper. The people here think nothing of sleeping on the snow by a
great fire, burnt on one side, frozen on the other. Pray
don't talk of my hardships and privations and courage.
Everybody here envies me, and looks upon my journeys
as mere pleasure excursions. . . .

Feb. 18. We have had the finest and mildest winter
that has been known for many years. The harbour has
never been frozen over, and last week we had the thermometer up at fifty-two degrees. Since then it has
frozen again, and a little snow is coming down, for an
increase of which we are all anxiously looking, as all

the sleighing-parties are broken up. In these parties
every gentleman takes a lady in his sleigh, and away
we go, twelve sleighs one after the other, into the country. Prog, wine, &c. are stored under the seats; and
picking a nice smooth place, as on a pond, for instance,
where it is sheltered from the wind, we cut down some
trees, make a great fire, warm up the soup, &c., and
picnic on the snow. Returning to town by dark, we
adjourn to the house of some one of the party, and finish
off with a quadrille. Dinner-parties are likewise abundant, in very good style and taste, and of these I generally have to refuse one or two a week from preëngage-
ments. In order to contribute to the stock of amusement, Dr. Stabb and I are now giving some lectures on
chemistry and geology. We were obliged to take a
room which had previously only been used for the public
meetings of the Catholic party (as they are called), and
some persons would not enter it. After all, however,
we were ' most numerously and respectably attended.'
I am writing a condensed report of my lectures to be
published in a pamphlet form. At the end of the first
lecture, a High-Church clergyman got up and attacked
me in a very harsh and loud tone of voice; but the
next day he wrote me a note to say he misunderstood
one of my expressions, so we are on good terms again.
Mr. Bridge (the Evangelical clergyman, with whom I
am very intimate) called on me next day to say he heard

nothing in what I said to which any one could reason-
ably object. It is a fatality here, however, that you can
do nothing without having to quarrel about it with
somebody. . . . Stüwitz is not yet returned, which I
can't account for. I only hope he is not lost. If he
does not come back, I don't think I shall go to the ice
by myself, though it is a very favourable season. The
vessels are now preparing for the sealing-voyage; seventy
sail leave this port, seventy that of Harbour Grace, and
so on. Give my best love to my mother and all other
friends, and believe me ever your very affectionate
brother,

J. BEETE JUKES.

I must work hard next summer, but mean to return
earlier, in order to get home before Christmas.

St. John's, Newfoundland,
Jan. 20, 1840.

MY DEAR AUNT,—I am going with Stüwitz to the
ice in March with the annual sealing expedition, an
undertaking that is looked on even here as the last extremity of quixotism, not so much on account of the
danger as the discomfort and the stench. I have not
yet seen Fitzroy's travels, as the governor, who supplies
me with books, has not finished it. He did, however,

lend me Darwin's journal, with which I was greatly
pleased. . . . The missionaries appear very different
people when you meet them in a savage land from what
they are at home.   They seem to me in New Zealand
like respectable settlers, getting good farms, and showing the people the comforts of civilisation.   At all events,
while other classes of people are too indolent to make
exertions to civilise the savages (their only chance of
escaping future extermination), they have no right to
cry out on the religious class for attempting it in their
own fashion, though that fashion may be a bad one.   As
for leaving them to the tender mercies of chance settlers, the fate of the aborigines of this country is an
example of that, of whom whole families have been shot
down by the old settlers for the sake of the skins they
had about them, until the whole race have now vanished
utterly from the face of the earth. . . . Dr. Stabb and
I are attempting to get up a course of lectures here on
chemistry and geology.   There are, however, here, as
elsewhere, two parties ('confound their politics'), the
Catholic, or low party, and the Protestant, or high party,
as they please to call themselves.   The governor, being
impartial, has offended them both, and I, as I receive
money voted by the House of Assembly, am suspected
to belong to the Catholic party, notwithstanding my
Cambridge degree.   There are only two suitable rooms,
both belonging to charitable institutions, one of each

party. We applied first for the Protestant, and got it;
but meeting with no great cordiality from some of the
members of the committee, and some obstacles being
thrown in our way, I have now given it up, and applied
for the other. How it will end, heaven knows; but I
will try what a good laugh at their folly will do.

This winter has as yet been the mildest known for
many years. There is not as yet a sign of ice in the
harbour, and the snow is not deep. Indeed, up to
January there was none at all.

Extracts from Journal of an Expedition to the Ice,
March 1840.

After some trouble and consultation, Professor Stüwitz and I at length agreed with Captain Furneaux, of
the brigantine Topaz, to make room in his cabin for us
and my man Simon. The vessel was of 130 tons burden, and the cabin very neat and clean; the captain too
is the only one in the shape of a gentleman that ' goes
to the ice;' in other vessels crew and captain are all of
the same stamp, and all ' pig' together. I only finished
my lectures on Friday, but by Tuesday morning, March 3,
had our stores and everything on board, and only waited
a fair wind. The captain said he should not sail for
two hours, and I strolled with MacBride, at whose
wharf the vessel lay, into the billiard-room. We had

played a rubber or two, when happening to look out of
a back window, we saw the topsails of a vessel under
sail, and his flag at the mast-head. Guessing it was
the Topaz, we rushed out of the house down to the
wharf, found they had been waiting an hour, jumped
into a punt without stopping to speak to friends assembled to take leave, up the side of the vessel, cast off the
rope from the wharf-head. They gave us three cheers,
called all hands aft for three cheers, huzza! one cheer
more from the wharf, and away we go, pushing through
the thin ice of the harbour, amongst which vessels were
working in all directions. Passing out of the narrows
we found the sea clear of ice, and steered straight away
for Cape St. Francis. We sailed at four o'clock and
reached the Cape at sunset. The sea was smooth, and
frequently its half-frozen surface looked like skimmings
of broth, being of the consistence of pea-soup, and having a greasy aspect as it rose and fell in long undulations. In the night awoke by a great noise caused by
the grinding, rushing, and roaring of thin ice against
the sides of the vessel as she cut through it.

March 4. Up at seven, a few miles north of Bacca-
lieu Island, standing across Trinity Bay. Many vessels
in sight. North of us, in a field of ice near two vessels,
saw some little black dots—the crew cutting and breaking the ice. Got close in to Cape Bonavista, near
enough to see the bedding of the rocks. Here we

found the edge of the ice, and several vessels tacking to and fro on it. Taking advantage of a small channel near shore, we pushed through, doubled the Cape, and then followed a channel in the direction of some vessels ahead. In a few miles this likewise closed, and for the rest of the evening we sailed about some smooth lakes of water, that kept slowly opening and closing about us. The day was beautiful, and the dreariness of the scene relieved by the number of vessels about. The ice was about a foot thick, more tough and spongy than fresh-water ice; it was much cracked and broken in every direction, but a small piece would bear the weight of a man. Nothing like a smooth plane of any extent existed, the whole consisting of pans jammed together, with broken angular pieces sticking up along the cracks, and openings of greater or less width filled with 'lolly' or frozen snow. The thermometer was not below 40° during the day. At night stuck the vessel's nose into the edge of some ice, and lay secure and motionless as by a wharf.

March 5. Morning overcast. At seven o'clock, after tacking about a lake to find an opening, dashed into ice in company with two other vessels on a N.W. course. Soon got stuck. All hands overboard breaking the ice with axes, poles, gaffs, and handspikes, sawing through the harder parts. A crowd of men get about the bows of the vessel, which shung round with ropes,

H

and break the ice to pieces; then crowding all sail, and putting out a great claw with a tow-line, they warp the vessel slowly on. When the vessel makes way, they hold on by the ropes and stamp and bear down the ice under the vessel, pulling her over it and being often over their knees in water upon the bending and broken ice. In attempting to cross before the vessel, I trod on a part that seemed sound, but had been smashed up, and in I went to my middle before I could lay hold of the pan behind me; one of the crew helped me out. As it was my first dip, I had to pay my footing in rum to the crew (for having lost my footing). This cutting, pushing, and shoving, relieved by an occasional smooth lake of water, occupied the whole day, during which she made about fifteen or twenty miles.

March 7. Ice gets heavier and heavier, great blocks and slabs piled on one another. In afternoon ice opened a little and became broken into pans, or rather the pans previously jammed together separated. Through these we cracked along with a fresh breeze, running 100 yards or so through clear water and then, crashing into the ice, brought up all standing, sometimes top-gallants set. The motion at this time is like that of a steamer, the vessel constantly trembling and thumping against ice; the scraping of the pieces along her sides is like the roar of machinery.

March 8. Quite calm. Brought up by thin ice

with large pans. Saw a great seal on the ice a quarter
of a mile off; took a rifle and a man and went after him.
Many channels in ice full of lolly, too wide to leap;
slipped in jumping from one pan to another and fell
into lolly up to my arm-pits. Luckily the gaff stuck on
the two pans, by means of which I got across to the
man and he helped me out. Returned on board in a
semi-frozen condition. It appeared, however, to do me
good, and cure a cold I had coming on before. They
say every man has three dips each season, so I have
only one more to come. The other night, as ship was
passing quickly through broken ice, the phosphorescent
sparks were most brilliant, not only in the water, but
apparently in the small lumps of ice. Stüwitz says he
found infusoria *in ice* in Fortune Bay which, when the
ice was melted, lived uninjured. The sparks in the
small lumps of ice were very pretty; their pale light
shining through the crystal might have formed lamps
for the Cave of the Naiads.

March 9. Passed through some heavy ice broken up
and piled in slabs one on the other. Many pans tilted
over and standing upright, jammed in among others,
large pinnacles, some higher than the sail.

March 10. At dawn a gale from S. After dark the
clouds gradually cleared off, and unfolded a most lovely
sky, brilliant with stars and the young moon, and
adorned by the yellow flickering streamers of a fine au-

rora in the N. The ice, too, opened, and we sailed
with a gentle W. wind through calm water, among
numberless floating pieces of ice, with broken fragments
piled in fantastic forms, 'quietly shining to the quiet
moon.' Everything was still, and even the men hushed
their most sweet voices as we stood on the deck looking
at this most beautiful scene. The hoarse voice of the
master of the watch, as he sang out from the foretop
brief orders to the helmsman for avoiding the lumps of
ice, was not out of harmony with the feelings of the
time. The lonely voice sounding at intervals served
but to make the utter stillness around us more deep
and solemn. We stood in the bows gazing quietly till
the aurora faded away, and the vessel, passing from
among these fairy islets, struck again into a field of
close-packed ice-pans, and the ground-thumping and
grinding became accompanied with the cries of the men
to shove this way and haul that, and the enchantment
was dissolved by 'gaffs' and 'pokers.'

March 12. Getting into a lake of water, we sent out
three punts to collect seals from the scattered pans
about; but as several vessels about us had pretty well
cleared that quarter, we bore away in another direction.
In passing through a skirt of thin ice, a man picked up
a young seal with a gaff, and its cries were precisely
like those of a child in the extremity of fright, agony,
and distress; something between shrieks and convulsive

sobbings. It at first thrilled one's nerves; but when I recollected that their sole employment when alone on the ice is uttering these cries, and that the sounds differ but little when meant to express intelligence, enjoyment, or defiance from those which appear to be the utterance of pain or fear, I became more reconciled. We soon after passed through a lot of loose ice on which they were scattered, and all hands were overboard slaying, skinning, and hauling. We then got into another lake, and sent out five punts. The crews of these dispersed over the ice, and dragged the seals or their pelts to the edge of the water, whence collecting them in the punts, they brought them on board. In this way we had 300 seals on board by dark, and the deck was one great shambles. When piled in a heap together, they looked just like a flock of slaughtered lambs; and occasionally from out the mass one poor wretch still alive would heave up its bloody face and flounder about. I employed myself in knocking these on the head with a handspike to put them out of their misery.

March 13. As soon as it was light, all hands overboard, and the whole of the day was employed by the men in slaughtering young seals and hauling their pelts (the skins with the fat attached) on board, three, six, and seven at a time. They cut holes in the sides of the skins, and passing a rope through the nose of the under one, they lace them through one another; so that when the cord is pulled tight, the lowest skin partly envelopes the rest, and the whole forms a compact bundle. Fastening the gaff into this, they haul all together by a rope over the shoulder, bringing them into the vessel on all sides, sometimes from one or two miles' distance. The ice was very slippery and much broken, and as my boots were not sufficiently fitted with 'sparables' and 'chisels,' I stayed on board to assist in management of vessel and hauling the seals in. By twelve o'clock my arms and hands ached sufficiently with hauling ropes and tackles. Some of the men brought in sixty skins in the course of the day. As they came on board, they occasionally snatched a hasty moment to drink a bowl of tea and eat biscuit and butter. Hundreds of old seals were popping up their heads in the lakes of water, anxiously looking for their young; and occasionally one would hurry across a pan in search of the snow-white darling she had left, and which she could not recognise in the bloody and broken carcass stripped of its warm covering, which was all of it that remained. The sun set most gloriously, gleaming across the glittering white expanse, stained, alas! with many a bloody spot and long ensanguined trail, marking the footsteps of the unhallowed intruders on the peace and beauty of the scene. Notwithstanding all the slaughter, the air was still resonant as night closed in with the cries of the young seals on every side. As the sunlight faded in the west, the quiet moon looked down from the zenith, ' shepparding flocks of stars,' and a brilliant arch of aurora crossed the heavens. The men were busy all night heaving out ballast and stowing away the seals. There were 1380 brought in to-day.

March 16. We brought one young seal unhurt on board by my request. He lay on deck very quiet, opening and shutting his curious nostrils, puffing out his breath, and occasionally lifting up his large, dark, expressive eyes at the strange scene about him. On patting his nose, he drew his face in under his fur, knit his brows, and shut his eyes, looking very round and funny. His fur, when clean, is beautifully white, and the creature very pretty; so round, and warm, and comfortable. When teased, though so young, they are very fierce, biting and scratching everything about them; but this one became immediately quiet on my patting and stroking it. As the men and dog, however, would tease it until it became exhausted and seemed dying, I passed my knife into its heart.

Some very thin ice was made up of scales about the size of the palm of one's hand, frozen together into larger flakes about a yard in diameter. These overlapping each other, and being very thin and shining, looked like the scales of a fish.

March 20. Breakfasted on seals' hearts and kidneys —very like pigs' fry, but rather more tender and deli-

cate. Pushed into ice on a N.W. course. Captain shot
an old hooded seal, and Stüwitz and I went to measure
and cut him up. He was six feet long, with a great
pulpy mass on his nose, and very like a rhinoceros in
the face, with a savage-looking ghastly eye.

March 21. Foggy. Scarcely stirred all day. Frequently heard guns and signals from vessels about us
for lost men. Got two or three young seals, one of
which was shedding his white coat, disclosing a smooth
elegantly-spotted skin beneath. At night moored to a
large pan; heard a noise as of breakers ahead, which
we rightly concluded to be a 'rolling pan,' or oscillat-
ing iceberg. One small schooner passed close by us;
she is six men short, and cannot find them. The small
iceberg we heard last night came down on us at break-
fast-time. It passed close under our lee, and got foul
of our tow-line, obliging us to cast off and make sail.

March 26. A silver thaw, that is, rain freezing as
it falls, and covering everything with a coat of thin ice.
A young hooded seal brought on board. Stüwitz measured and drew him, and we put his carcass by in a
punt to dissect in the morning; but the men ate him
during the night. Dined on a young seal. It was par-
boiled first, then fried in onions. It was very good, and
tasted like a tender wild duck.

April 1. Saw a singular island of ice a few miles
off, with an arch through it some distance above the

water; height about sixty feet. Brought a very pretty young seal on board alive. He had lost his white coat, and had a shining skin of short hair, silver gray, with dark spots. Tied a rope to his hind flipper and let him swim about, which he did quite at his ease. Nestor went out to him; but he showed no fear of the dog, standing upright in the water, attacking him, when he came within reach, tooth and nail, and driving him off. His motion in swimming beautifully swift and easy.

April 7, 8, 9, 10. Heavy gale; seals beginning to *run;* can smell the fat in the cabin: fine cure for sea-sickness.

April 12. At 3 A.M. saw Cape Spear light, and about the middle of the day, got once more into St. John's harbour. Altogether, one of the most interesting excursions I ever had, and no disagreeables until last week.

St. John's, Newfoundland,
April 22, 1840.

MY DEAR A.,—I returned from the ice all safe last Sunday week, having had a very interesting excursion. I found your letters of January and February. By thus receiving them all together, I was spared in some measure the anxiety about my mother's health, since you say she is now better. . . . The year 1840 does not seem to open auspiciously for us, as I believe my situation here is at an end. In the beginning of the session, the House of Assembly voted 600*l.* in the supply bill for the geological survey; they also voted a large sum for road-making. The latter item the council thought too great for the revenue, and in consequence rejected the supply bill. The House of Assembly, always at variance with the council, then said, if the country was too poor to make roads, it was too poor to do many other things, and among them to make geological surveys, and accordingly voted me 100*l.* to pay my passage home. The governor and council were excessively angry at this, and the governor sent down a special message on the subject, and there it rests for the present. The attorney-general told me the other day he thinks they will be shamed into continuing it for one year more, in order to give me sufficient notice. . . . In any case I shall not return home immediately, as I must see the other side of Avalon, and shall probably visit Halifax and Nova Scotia on my road home. I write in a great hurry by a vessel just going to Cork; by the next that goes to Liverpool, I will send you some journal of my ice-doings. Stüwitz is a splendid fellow in every way; I am not sure but I shall join him during the summer, if I can anyhow manage it. . . . . Since writing the above, they have voted me 350*l.* for myself, and 100*l.* for expenses, recommending me not to take a vessel. I shall accordingly set out immediately on foot round Avalon, and in June or July take my chance of going in the man-of-war to the northward, and then make my bow to the House, and return home. . . .

———————

St. John's, Newfoundland,
May 10, 1840.

MY DEAR MOTHER,—I am much concerned to find that you have suffered from one of your old attacks, but I sincerely hope that by the time you receive this you will have recovered the same health you had when I left you. I am very glad that you are going to give up the school. You will then have no uncertainties to trouble yourself with. . . . . For myself, while I have a head and a hand, and money is to be earned, I do not fear but I can earn it, if it becomes absolutely necessary. What I have wanted all along was a little roughing and knock-ing about the world. It sharpens a man's wits and re-solution wonderfully. Though I could not reconcile myself to settle down and vegetate in a provincial town, yet I am satisfied that I can at the same time gratify my love of travel, and ultimately gain by it. . . . . Pray take every opportunity of cheerful exercise and amuse-ment that you possibly can. Nothing strengthens the mind so much as occasionally putting aside all feelings of care and anxiety, and letting the blood circulate

freely through the veins. I regret that I cannot be at
home to assist you in the labour of removing at mid-
summer; but this is my first outset in a new and uncertain profession, and unless I carry it out with something like success, and show that I am capable of greater
things, were the opportunity offered, I may never get
another appointment. The temptation will be very great
to me to accompany Professor Stuwitz, whom I like and
admire more and more every day, in his tour through
Canada and the United States, when he would return
with me, and winter in England in 1841. He has re-
commendations from the Secretary of State to all the
authorities. I could get them from the governor here, so I
should never have such another opportunity. It will,
however, greatly depend on how you both are. . . . Stuwitz takes my old vessel, and my new charts and baro-
meters, quite stepping into my shoes. I go with him
in a day or two to St. Mary's Bay, where we stay a week
or so.   He then goes to the Banks to catch cod-fish, and
I on foot through the woods to Placentia. I then return how I best can, and shall get back here about June
10, I hope. By that time the bishop and the man-of-
war will have arrived, and we shall be quite gay; and I
hope I shall get a large packet of letters. I shall then
push off to Bay of Exploits, where I am to meet Stüwitz at the latter end of August—errors, &c. excepted—
and we go into the Red Indian Lake together. I shall

get back here in October, I hope; and in a month or six weeks after, shall have finished my work, and quit Newfoundland for good, as they say. . . .

With best love and hopes for your speedy and permanent recovery, believe me, my dear mother, ever your affectionate son,

J. B. JUKES.

The other day the Diana arrived from London. I have been to-night to see her and the captain, and was astonished at the different appearance she had to my eyes. When I came out in her, she seemed small, confined, and dirty; now, having been accustomed to the craft here, she looks large, roomy, and delightfully comfortable—quite luxurious, in fact. I believe, when I get back, Wolverhampton will seem a city of palaces.

St. John's, July 6, 1840.

MY DEAR A.,—I am very glad to find my mother so much better; and hope sincerely her amendment has continued. I have no spare time now to copy my journal, not having had a cabin, as before, in which I could spend my leisure time. I will, however, give you an outline: I set out on May 20 with Stüwitz, who had engaged my old vessel the Beaufort for the summer. Went first to Aquafort, and made an excursion to the top of a hill

called the Butter-pots, where I sprained my knee and
broke my barometer. Then round to Trepassée, and
overland to St. Mary's, very lame. In a boat to the head
of St. Mary's Bay — very pretty country. Overland to
Placentia, where I found Stüwitz again. Across with
him to an island called Merasheen, whence we made excursions to Island of Valen, Ragged Islands, &c., and
finally left Stüwitz there, he intending to proceed to the
Banks, and round the island to the Labradore. Out three
days in a little boat, in order to get twenty-five miles.
Thence into the country, and walked across from Placentia Bay to Trinity Bay, in thick fog and rain, heavy
marshes, thick woods, and no road. Got to Chapel
Arm in the evening. Walked across from Conception
Bay to Trinity Bay, and came to Carbonear, and A—n
accompanied me to St. John's, where the bishop had
lately arrived. I liked the bishop very much; he is
now round in Conception Bay on a visitation. A day
or two after I accompanied the governor and the surveyor-general to Shoal Bay, about fifteen miles from
here, where there is an attempt at a copper mine going
on. We rode nine miles, then got a guide, who led us
astray into the country, and were walking four hours
through woods and marshes to get five miles. The old
man was nearly kilt entirely, being obliged to lie down
on the moss. We got back at dark, having been thirteen hours. The governor never had such a tramp be-

fore. . . . . I have been anxiously looking for vessels
from Liverpool, several of which are expected, but have
been disappointed. I am going in about four days to
set out again for Bonavista, and the northern part of the
islands; and unless letters come time enough to over-
take me in Conception Bay, I must give up all hopes of
hearing from you till that time. Should I meet with
any opportunity from the northward, I will not fail to
take advantage of it; but you must not be surprised if you
do not hear till I write to announce my own coming. . . .
I miss my little vessel very much; and as we have often
to walk, our supply of clothes is very scanty. Luckily
I have a most excellent servant, who can do everything
and anything, and who is much attached to me. My
other man also is a good honest fellow, and we get on
in the woods very well as long as the weather is fine.

July 12. A vessel has just come in from Liverpool,
bringing your letters of May 11 and June 7. In the
first place I am most provoked about the parcel I sent
in January. It contained a coloured map and sections
which cost me a week's work, and a whole host of letters, which cost me another. . . Concerning my return,
I think it is highly probable that I shall come home in
November; for though the journey with Stüwitz dazzled
me at first, business calls me to England, and I have a
kind of a slight yearning just to see you all again, which
grows stronger every time it returns; and then, after

all, I could join Stüwitz in New York in the spring.*
I have been very busy map-making, with which the
governor is so pleased that he intends to have them
lithographed for the public service. Indeed, I had no
idea myself that I could shade a map so well.

Carbonear, July 21. I left St. John's several days
ago, but brought my letters with me, as I thought it
highly probable the mail would come in from Halifax, and bring me more news and letters to answer. I
am now very busy—buying tea, sugar, and hard biscuit ;
packing up clothes, powder, shot, matches, thread,
needles, fishing-lines ; sharpening knives, hatchets, to-
mahawks ; cleaning guns ; and all the multiplicity of
little details necessary for a backwoodsman's existence

---

* The friendship between Mr. Jukes and Professor Stüwitz
was most cordial, but was soon sadly terminated by the death of
the enthusiastic Norwegian naturalist. The following conclud-
ing paragraph of a letter, written by Stüwitz to Mr. Jukes after
the return of the latter to England, is charming in its expression
of simple affection :

'By the next packet, I shall send you a letter; if it will be
long or short I cannot know; but howsoever it will be, it can
never tell how often you are wanted and wished by your true and
sincere Norwegian friend STUWITZ.'

Mr. Jukes thus writes from Australia respecting the loss of his
friend :

'Captain Prescott has sent me an affecting account of the
calmness with which he met his end, regretting only that the
fruit of his scientific labours would be lost to the world. You
may imagine I was much pained by it.'

during the next three months. . . I do not know how
I shall be able to manage when I get back without my
excellent servant, Simon Grant; for he takes all manner
of trouble off my hands, and can do nearly everything.
According to my usual luck in such matters, the man-
of-war came into St. John's just two days after I left,
otherwise I should have got a passage in her with the
bishop down to the northward. . .

Toulinquet, Newfoundland,
Sept. 1, 1840.

MY DEAR A.,—I dispatched letters to you from Carbonear in July, and I hope by this time you have received them. I set out from Carbonear in a boat belonging to Mr. P. of that place, accompanied by a Mr.
B. into Bonavista Bay. We went up into a fine arm of
the sea, called Clode Sound, where we remained nine
days, while the men were cutting timber to build a vessel. We then had a most beautiful sail through a great
number of picturesque islands, down to a place called
Greenspond, on N. side of the bay. Here I found a
very pleasant gentlemanly man, named W., with whom
I stayed a week. I then hired a fishing-boat, and went
back among the islands and arms I had left, visiting
Freshwater and Bloody Bay, and making an expedition
of two days up the country. The weather the whole

I

time was most beautiful, but desperately hot. During this time my bed consisted of a lot of boughs laid in the centre of the boat, and a blanket. I then went to Cape Freels and Cat Harbour, where we were detained by the first N.E. gale we have had these two months, and then proceeded to the Island of Fogo. There I stayed two days, and then came on here to Toulinquet, which is another of the cluster of islands with which the N.E. coast is fringed. I am staying in the house of a young merchant, and he is now fitting out his yacht to take me up the Bay and River of Exploits, and round Green Bay; and as he is a very pleasant fellow, and has seen much of France and Spain, I expect a very pleasant time of it. Mr. P. an old resident here, who knows more of the interior of the country than any one else, accompanies us to the Exploits. Since June we have had the most uninterrupted succession of glorious weather I ever remember, but it has been too dry for the gardens, which are burnt up; and the woods and country are as dry as touchwood. The consequence has been many extensive fires. The country is on fire in four or five places in this bay, the whole air is thick with smoke, and the clouds glowing at night in several parts of the heavens. Thousands of acres of wood, some of it fine timber, are thus being destroyed. I had appointed to meet Stüwitz here, but have heard nothing of him yet. . . .

St. John's, Oct. 24, 1840.

Returned from my northern tour on the 18th of this month, and found your letter of August 31st. . . . Owing to my not having had a vessel of my own, I have been obliged to waste much of this beautiful weather. I have, however, done enough to enable me to close the survey. I have now a vast number of specimens to label, arrange, and pack up, and a short report to draw out. This will take me a fortnight longer, and then I shall be off by the first good opportunity. I can assure you, the picture you draw of home-comforts has all the greater charm for me from my knocking about here in all sorts of discomforts. I never had so wretched a voyage as from Fogo here. Living four or five days in a coal hole—for our cabin was no better—eleven people screwed into it with barely space to lie down, the sea sweeping our decks every wave, and nothing to eat but biscuit and rancid butter. Sprung our main mast, &c. &c. I have no doubt I shall like —— and ——, because you like them. I heard from Sedgwick the other day; he has been an invalid at Cheltenham. He says in a P.S., ' I yesterday met with Mr. Lycett of Minchinhampton. He is a good fellow; and no wonder, for he tells me he is a friend of yours.' By similar rule, I am bound to believe Miss —— and —— to be ' good fellows.' As I can't tell you all that has befallen me this summer, I will reserve it till we meet.   I can only say I was never better. . . .

After reading the preceding letters, the following passage from his *Excursions* may be appropriately quoted :

'Thus ended my last excursion in Newfoundland, and I can only give my advice to any one who wishes to lead the life of a traveller, to commence with this country in order to get well accustomed to rough living, rough fare, and rough travelling ; and to get rid of all delicate and fastidious notions of comfort, convenience, and accommodation he may have acquired by journeying in England.

'I must add, however, that so far as the inhabitants are concerned, under a rough exterior, he will meet with sterling kindness and hospitality.   In November I sailed from St. John's in H. M. Steamship Spitfire, which I mention as she was the first steamer ever seen in a Newfoundland port.   She happened to touch here in order to bring a few troops from Halifax ; and great was the wonder and admiration she caused among the population of St. John's.   Some boats and schooners outside were so astonished as she approached, that they had scarcely presence of mind to get out of the way, and she had very nearly run them down.'

Mr. Jukes returned to England at the end of 1840, and spent some time at home. His mother's health had been seriously affected during his absence, and she had resigned her school. In 1841 he accompanied Professor Sedgwick into Devonshire and Cornwall; but no letters of interest, relating to their journey, have been preserved. He was also engaged in preparing from his journals the *Excursions in Newfoundland,* which was published by Murray in 1842. Great interest and an active part were always taken by Mr. Jukes in any movement made in Wolverhampton for the intellectual advancement of the town; and in letters written to local papers, he urged his fellow-townsmen to give more of their time and money to the support of a literary and philosophical society, which had been established on a small scale.

'It is not,' he tells them, 'a mere question of obtaining amusement or even information for themselves —though this alone would amply reward them—but it is a question whether they will keep pace with the age in which they live, whether they will join hand and heart in the great work of intellectual reformation and improvement which is now just beginning to regenerate our country, or quietly suffer the mighty stream of knowledge to flow around them and above them. . . . . The dullest eye cannot fail, I think, to see signs of the approach of a mighty change; the most fearful and fore-

boding mind cannot but be conscious that, to those who
will join in it, it must be a change for good.'

  After enumerating some of the intellectual advantages to be derived from such institutions, he proceeds :
' But there are, if possible, still greater benefits to be
derived from this society to the moral feelings. The
numerous sects and parties of religion and politics into
which mankind, in a free country, are necessarily divided,
are so very apt to become imbued with rancorous and
embittered feelings, that surely it behoves us to omit no
opportunity of alleviating their intensity, and of retaining as much as possible of that spirit of sociality and
mutual kindness, which so greatly heightens and adorns
the pleasures of existence. By nothing can this be so
well and so effectually accomplished as by meeting together with a common object, and engaging in the same
pursuits, more especially when those pursuits are the
most worthy. . . The study of external Nature in all her
boundless magnificence and infinite detail, or the examination of our own moral and intellectual natures, is
most peculiarly adapted for hushing to silence the petty
quarrels which disturb mankind. In the calm regions
of philosophy and science, or the delightful walks of
poetry and letters, the jarring notes of discord are over-
awed or soothed to peace ; and fierce and unrelenting
indeed must be the feelings of that man who can draw
but one breath of their atmosphere, and not feel himself

the wiser and better for the draught. . . . . Let me also
hint to those who hesitate because they see in the society a religious or political opponent, they are by the
very act acknowledging themselves his inferior in Christian charity and anxiety for the well-being and improvement of themselves and their fellow-men.'

A branch of the Dudley and Midland Geological Society having been formed at Wolverhampton, Mr. Jukes
was requested to give the opening address, which contains the following passages :

‘ We must never forget that geology consists of two
branches — palæontology, or the study of the forms,
habits, and history of the ancient beings, the animals,
and plants that in old time lived upon our globe; and
that part which may be called physical geology, or geology proper—the investigation of the solid structure of
the earth, and the causes by which that structure has
been produced or modified. This is the high and important branch of our science ; and fossil animals and
plants are only so far valuable to the real geologist as
they throw light upon his investigations. We must
never forget that the object of the Geological Society of
London, which we must always look up to as our model,
is declared in its charter to be the investigating the
mineral structure of the earth. The collecting a museum of fossils, therefore, is not our sole or even our

principal object. We must investigate the structure of our neighbourhood ; we must make ourselves acquainted with the character, thickness, and extent of every bed of rock or other mass of mineral matter ; we must study their natural order and arrangement, their actual and relative positions. In short, we must work out the *solid geometry* of our district. To do this properly would require us first of all to procure perfect physical maps on a large scale, and upon them to mark the out-crop of every important bed, the course of every fault, and the boundaries of every formation. We then ought to know the thickness of every bed in every pit, in every cutting, and in every hillside. We should thus have materials for making accurate sections in any direction we pleased; we should then be able to state at once the condition of any of the known parts of our district ; we should be able to trace the different beds in all their changes of thickness and character; and we should know the precise relations of every one fault with all the others by which it is surrounded. Suppose us to have once acquired and registered this knowledge, what would be the practical result ? There need be no longer any blind working in the dark, no longer any doubt about cutting through one fault or leaving it alone for fear of letting in the water; the miner's work underground would be as plain and evident as his course about the streets and roads on the surface. But we could do much more than this ; having a thorough general knowledge of the strata and the faults in all the *known* parts of the field, we might work confidently towards the unknown. The direction of powerful faults at a distance being ascertained, their farther extension could be guessed at with great probability. The law and the *a priori* probability of their affecting any particular district or not would be at once perceived ; and thus the most favourable spots selected for experiments, if any farther experiments should be necessary. But it is not simply a knowledge of actual facts that would result from such extended and minute investigation. We should rise higher. The general knowledge of phenomena invariably conducts us by the surest and safest road, either to the clear perception of the causes by which the phenomena were produced, or a close approximation to it, and therefore to a highly probable speculation upon them. We should reduce the theoretical possibilities of the case within very narrow limits. In geology this is often of the greatest practical utility. When a man knows the causes by which the things he deals with were produced, he knows what is their *possible state,* and what will probably be their circumstances under any given conditions, or, at all events, he knows what circumstances connected with them *are impossible.* In sinking for coal, to mention one instance, in situations where the merest tyro in geology could have at once decided that it was impossible to find it, thousands of pounds have been thrown away during the last few years.'

After briefly relating the geological history of the immediate neighbourhood, he concludes as follows :

'Some of you, I have no doubt, will withhold your belief of the truth of the facts I have stated to you. It is natural that it should be so. I do not wish you to receive them on my bare assertion; but I am ready to stake my credit, not merely on the truth of the facts, but that, if you will only take the trouble to examine into the evidence, you yourselves shall be as firmly convinced of them as I am. The evidence lies everywhere around you; you tread upon it daily; you cannot stir from this town without its lying open before you, and it only requires your attention to be once directed to it in order to force itself on your conviction with a daily and hourly increasing strength.

'It has been well said by one of our first philosophers, the great Herschel, that geology yields only to astronomy in the greatness and sublimity of its objects and its views. They are indeed similar, as opening out to our contemplation, the one an infinity of space, and the other an eternity of time. Great and simple general laws, regulating and producing various and complex results, are common to both. They both appeal through the evidence of the senses to the highest powers of our intellect; both exalt the imagination and

act in two ways on our affections: they show us realms
of truth, compared with whose extent the most daring
flights of fancy are but feeble essays; and while they
impress on our hearts the humble consciousness of our
individual littleness and insignificance, they yet raise
and strengthen us by the thought that of the vast
scheme of creation we still form a part; they place us
in communion with the Eternal and the Infinite; we
feel the high consciousness of immortality glowing in
our souls; we look down upon the earth as our present
wondrous residence; we gaze from it into the wide regions of space above, around, and beneath us, and claim
the universe as our dwelling-place for ever.'

SURVEYING EXPEDITION TO TORRES STRAIT,
GREAT BARRIER REEF, &c.

1842–1846.

SURVEYING EXPEDITION TO TORRES STRAIT, GREAT BARRIER REEF, &c.

FIRST CRUISE. 1842, 1843.

Appointment of Mr. Jukes as naturalist—Abstract of the voyage
—Letters and preparations—On board the Fly—Voyage out
—Madeira—Funchal—The Corral—Ascent of the Peak of
Teneriffe—Cape Verde Islands—Crossing the Line—Cape of
Good Hope—Island of St. Paul—Van Dieman's Land—Ho-
barton—Norfolk Island—Sydney—Cape Upstart—Explana-
tory notes—Journal of work, explorations, and adventures on
Barrier Reef and north-east coast of Australia—Interviews
with natives—Murray's Islands—North coast—Port Essington
—Timor—Swan River.

THE office of naturalist to the expedition for surveying Torres Strait, New Guinea, &c., was offered to Mr.
Jukes early in 1842, and gladly accepted. Indeed no
position could have been more to his taste at this
period, when the thirst for new scenes and a life of
adventure was almost as strong as his love of science,
and this appointment opened a prospect of gratifying
both. There was no little of the sailor in his dis-

position, as he himself perceived during this voyage, and had he spent his boyhood in a sea-side town, doubtless the sea would have been his destination. The ample details given in his letters of this important expedition have formed one of the chief grounds for this memorial of him; few travellers have time or inclination to write so fully, and it will be seen that he neglected no opportunity of sending letters home. They form a series complete in themselves, and the greater variety of life and scenery which they portray renders them much more interesting than those from Newfoundland, the island of fogs and swamps. Another farewell to his mother, whose health was much impaired, was again his chief trial in quitting England; she was, however, spared to welcome him home, and lived for a few years after his return.

In order to connect and render the following series of letters more intelligible, an 'Abstract of the Voyage,' taken from the appendix to Mr. Jukes' 'Narrative,' is here given. This may be divided into two parts: the first cruise comprising the voyage out to Hobarton, the work of surveying the Barrier Reef and N.E. coast, and the circumnavigation of Australia until arriving again at Hobarton; the second cruise including the building of the beacon-tower on Raine's Islet, the visit to Java and New Guinea, the voyage home, and arrival at Spithead.

Abstract of the Voyage.

We sailed from Falmouth in H.M.S. Fly, in company with her tender, the Bramble schooner, on Sunday afternoon, April 11, 1842, and anchored at Funchal, in the Island of Madeira, on the following Sunday morning, April 18. We remained here a few days to rate the chronometers, and then sailed to Teneriffe, where a party of us ascended the celebrated Pic de Teyde. We carried up a mountain barometer, and the mean of our observations gave 12,080 feet for its height above the sea. We left Teneriffe on May 3, and on May 9 touched for a few hours at Porto Praya, in St. Jago, one of the Cape Verde Islands. We crossed the Equator on May 18, and on the 23d hove-to for a few hours, and landed on the little island of Trinidad. Thence we sailed to the Cape of Good Hope, where we anchored in Simon's Bay on June 19. We stayed here some time, to refit and refresh, and also to compare our magnetic instruments with those of the observatory at Cape Town. We again sailed on July 14, and on August 5 anchored under the little island of St. Paul's, and visited the interior of the crater in our boats. On August 27, we entered Storm Bay in Van Dieman's Land, and remained at Hobarton till October 6. Then, calling for a day or two at Port Arthur, we proceeded to Sydney, where we arrived on October 15, and remained till November 24. On No-

K

vember 26, we entered Port Stephens, and having then
completed our preparations, and collected all the preliminary information we could acquire, we sailed thence
on December 17 to commence the survey. This was
begun at Sandy Cape by the examination of Breaksea
Spit. From this point the survey was regularly carried
on, through the Capricorn group of islands and Swain's
Reefs, up to lat. 21°. We were then obliged to go to
Port Bowen to repair some damages. Here we remained
till February 28, 1843, during which time a detailed
survey of the Port was completed. Thence we sailed
through the Percy Islands to West Hill, a little north
of Broad Sound, where we found a supply of water, of
which we were beginning to run short. The Bramble
had been dispatched to look for water, and found it in
abundance a little north of Cape Hillsborough. The
coast from West Hill to the northern part of Whit-
Sunday Passage was then surveyed, a part that had
been only hastily sketched in by previous expeditions.
On March 30, we anchored under Cape Upstart, where
we remained the rest of that month, repairing the boats,
and raising and decking the pinnace, to enable her to
keep the sea during the surveying operations.

We were joined at Cape Upstart by the barque
William, with a fresh stock of provisions and stores
sent up from Sydney by previous agreement. We made
several boat-excursions during this time. On May 17,

we sailed from Cape Upstart with the Bramble and the pinnace (now called the Midge) in company, and the next day anchored in Rockingham Bay, of which an accurate survey was completed by the end of the month.

On June 1, we sailed to the northward; and after heaving-to for an hour to look at Endeavour River on June 4, we anchored under Lizard Island. Here was commenced the survey of the outer edge of the northern part of the Great Barrier Reef, which was completed up to Murray Islands. On August 14, we left Torres Strait, touched for a few days at Port Essington and at Coupang in the island of Timor, to procure water and refreshments; and on September 30 anchored at Swan River, where we remained a month, and then returned to Hobarton.

16 *George-street, Euston-square,*
Jan. 26, 1842.

DEAR A.,—I have packed up the greatest portion of my baggage, amounting altogether to 5300 boxes—that is, including pill-boxes and chip-boxes—besides two or three hundred bottles. I have just got my instruments, with such a beautiful little microscope, the highest power of which makes squares of the five-hundredth part of an inch the size of pin's heads. I spent a very pleasant evening at Murray's yesterday, and I go to a soirée at

Murchison's to-morrow night. . . . London was illuminated last night, and some of the club-houses were very splendid. The King of Prussia is here, but he has not called upon me. I hope to see Humboldt next week at the Geological. I was at the Zoological last night, when mad Lieutenant A— addressed us at the close of the meeting, and nearly suffocated us with laughter. . . . My baggage altogether would about fill the study up to the ceiling.

Best love to my mother and yourself, from your ever affectionate brother,

J. BEETE JUKES.

I am getting more valuable books, and I like Blackwood more the more I see of him. He is very anxious I should be fitted out completely, and seems to enter into the natural-history part with great eagerness. I hope we shall have a plant-collector, after all, and have written down to a man I heard of near Bridgenorth.

Between science and socialities, my time is fully occupied.

H.M.S. Fly, Devonport, March 28, 1842.

MY DEAR MOTHER,— . . . We are all very busy on board this morning, as the admiral is coming off shortly to inspect us. All my companions are getting out their cocked-hats and swords; but as I do not sport a uniform, I shall receive him in a dress-coat and round hat merely. After our inspection, we shall make our first voyage of a mile or two from our moorings off the dockyard, down into the Sound, behind the Breakwater. We shall then be put into perfect sea-going order, the men will have three months' pay in advance, and about Saturday or Sunday next we shall sail. We shall, however, I believe, put into Falmouth for a day or two, to take some magnetic instruments on board. We shall then sail for Madeira, where, if we get a fair wind, we shall arrive in ten days, and there we shall stop for two or three. You may therefore calculate on hearing from me in about a month after we sail. We shall then be six or seven weeks going to the Cape, where we shall stop three weeks. From thence we shall sail direct to Hobarton ; and how long we shall be, I can hardly tell you. Everything goes on very smoothly hitherto, and I think we have every prospect of a pleasant voyage. The first part of it will be the worst, as we are lumbered with an immensity of stores, and are obliged to have barrels of flour, &c. alongside our dinner-table in the gun-room, besides a great locker under the table, to hold tea and sugar for eight people for six months. We do not go to Rio after all. There is such a bustle about, that I can write no longer.—Your ever affectionate son,

J. B. JUKES.

Murray has made me a present of Byron's works, in one beautiful volume.

H.M.S. Fly, Falmouth, April 10, 1842.

MY DEAR A.,—This, I expect, will be my last letter from any part of England.   Since I wrote last, I have been into the great mining district, and yesterday took a party of our officers into the bowels of the earth, down the East Wheal Crofty mine.   We had much laughing at each other in our miners' dresses, with candles hanging at our button - holes.   I have received all your letters, the last containing the music and the address. It seems very fair, barring a little Wernerian geology, which is *rayther out of date*, and some incomprehensible chemistry, in which he seems to treat carbon as a compound of hydrogen and nitrogen.   Why does not somebody get up and say 'No' to what somebody else says at your discussions, to smarten them a little ?   If the present easterly wind continues to blow at this time to-morrow, we shall be raising the anchor, and in the afternoon put to sea.   We shall be at Madeira in eight days, if it continues to blow fresh, which I hope it will ; but as all letters from Madeira come through Portugal, you must not expect one for some time longer.

April 11. I was interrupted yesterday, and now all is bustle.   I am just going on shore for the last time in England, to buy a few things. You and my mother must now, therefore, accept my best love and farewells. . . . There are several things I meant to say to you, but I shall have plenty of time to think them over between here and Madeira.—Your ever affectionate brother,

BEETE.

Funchal, Madeira, April 18, 1842.

MY DEAR A.,—Exactly a week ago we left Falmouth, with a piercingly cold N.E. wind. We are now luxuriating in the rich warm air of this lovely island. We had a splendid passage; a strong wind right after us the whole way, with a clear sky overhead, the air and water getting gradually warmer and more serene as we approached the south. We made the 1200 miles in rather less than the week, and the first two days we were delayed by the slow movements of our tender, the Bramble, which we then left behind, and expect here to-morrow or the next day. . . . This morning, when we made Madeira, a strong wind was blowing from the N.W., and the island was enveloped in clouds and rain. On rounding the western point, however, we found the sun shining on the southern side, and a smooth sea, with light breeze that took us gradually to our anchorage. The island is most picturesque. It is about thirty miles long, and its highest point in the interior (which we have not yet seen for clouds) is upwards of 5000 feet. From this point it slopes gradually down towards the sea, ending in abrupt and lofty precipitous cliffs. On the north side, indeed, is the loftiest precipice in the world, being nearly 2000 feet. The cliffs and the whole surface of the island are worn and furrowed by numerous steep narrow winding crevices and ravines, each with its silver thread of water leaping along its bottom from cliff to cliff. Sharp ridges run consequently on all sides down to the sea, with here and there a solitary peak, the sides frequently showing precipices of bare columnar basalt or tabular trap. The tops of the hills and ridges are brown and bare. Lower down they are covered with pines. Below these the slopes are lined with trellises for the vines; white houses or chapels being sprinkled here and there, and many small thatched huts. In the bottoms of the valleys and near the sea grow bananas, oranges, lemons, and other tropical fruits in the greatest profusion. We have been feasting to-day on bananas and oranges. There is some recent book of travels in the Wolverhampton Library containing a panoramic view of Funchal. Pray get it, and imagine the Fly at anchor a little to the east or right-hand of Loo Rock (on which is a fort), and a five-oared gig pulling off with the purser, the doctor, myself, and a midshipman in the stern-sheets, going ashore to take (for me) the first look at a Portuguese

town. The houses are all white, the streets narrow
and irregular, but paved with small pebbles and perfectly clean, with squares and open places surrounded
with sycamore-trees, all in full leaf. The gullies and
banks and walls, and more especially the gardens, are
full of large plants of tropical forms, graceful curling
and pendent leaves, or broad and umbrageous ones;
and there is a brightness in the air and a warm rich
glow, something like that in a hothouse, but without
its oppressiveness, that is totally different from anything I ever felt before. Add to all these the quaint
dresses of the Portuguese, their pretty little oxen drag-
ging small sledges or drays, with the strange cries of
their drivers and the totally foreign air of everything
about, and you will have a tolerable idea of my first
impressions of Madeira. Having given you these, I
shall now retire to rest. . . .

Tuesday, April 22. The day after we landed Lieutenant Ince (a jolly fellow, the picture of good-humour),
Downes, and myself went a trip along shore, lunching
in a sugar-shed (where they grind the canes) on brown
bread and Malmsey wine for a shilling a bottle, and
then bathing in the clear blue water below. The next
day a party of thirteen of us went on horseback (beautiful Spanish ponies) to the Corrál. We galloped up
the paved streets and steep roads near the town, each
with a Portuguese guide holding on by the tails of our

horses, with long poles in their hands. After crossing
several ridges, we came to a steep hill, up which winded
a narrow path, a yard broad and full of broken stones
and masses of lava. After holding on for some time,
at considerable risk of slipping over my horse's tail, we
dismounted and clambered up, the horses climbing like
so many cats. We then came to the brink of a precipice looking down into a most magnificent gorge. Along
the sides of this we rode for some miles, winding among
the folds of the mountains, galloping whenever we got
a few yards of level road, then clambering, walking, or
climbing over places that would astonish and utterly
baffle any English horse, till we came to a ridge two or
three thousand feet above the sea and looking down
into the head of the gorge. I can't describe to you the
magnificence of the scenery here. Peaks and precipices
of black rock rose on all sides of us, their tops covered
with curling mist, and the most savage grandeur mingled with exquisite beauty. We dismounted, and giving
our horses to our guides, descended by a path narrow,
steep, craggy, and slippery, overhanging precipices, or
winding with short turns down their sides in a way that
made me very glad when we got to the bottom. Here
there was a small chapel and a priest's house, with
some cottages. It now began to rain heavily, and un-
packing our provisions, we repaired to the priest's house,
who supplied us with wine. After remaining an hour

or two, we again mounted, and galloped up into the mists and clouds that had by this time settled down upon the hills. It was a most picturesque sight to lag behind and look up and see our cavalcade dimly through the fog, winding up the face of the cliff above me, the spirited little horses clambering up bare sheets of rock, with nothing to save them from a plunge into the valley below except their own sure footsteps. The cries and shouts of the guides in their drawling Portuguese added excitement to the scene. After many minor adventures, such as my horse losing a shoe and the guide putting it on again, we all arrived safe on comparatively level ground; and having mustered at a venda, or small wine-shop, we formed into close order two and two, and marched into town in a slow trot, greatly to the aston-ishment of the people and our own amusement; for we made as much noise clattering over the pavement as a troop of cavalry. It was altogether an excursion never to be forgotten. . . . We shall sail, I expect, on Sun-day, but the letter-bag is to be made-up to-morrow morning. We go hence to Porto Praya in the Cape Verde Islands; but we only stop there a few hours, so you may expect to hear from me next from the Cape. . . . You would be delighted with the flowers here. Geraniums, roses, fuchsias, and hundreds of large and beautiful flowers, known only in hothouses with us, flourish here among the rocks by the waysides, with beautiful ferns and mosses in the wet gullies. The geology, too, would interest my aunt J., as the whole island is a mass of volcanic materials, partly ejected in a molten state, partly blown out and stratified under water.

Cape Verde Islands, May 9, 1842.

MY DEAR MOTHER AND A.,—After writing to you from Madeira, I spent another pleasant day there, and on the Sunday we sailed. Before going, however, we had a wedding on board, an English gentleman and Portuguese lady coming off to be married, as the ceremony could not be performed ashore between a Catholic and Protestant. We then had very light winds for two or three days; but on April 28 came in sight of the Peak of Teneriffe, patched on his summit with snow, and towering high above the clouds. On the 29th, we anchored at Santa Cruz, and made inquiries as to the possibility of ascending the Peak. The English consul and all the authorities said it was impossible at that season of the year, and that no guides would go with us. This had the effect of determining Captain Blackwood to stop and try. We accordingly made a party of eight: Captain, Shadwell, Ince, M‘Gillivray and Dr. Sibbald, Burgoyne and I.

April 30. We set off on horseback, each with a guide, carrying a knapsack and blanket strapped on our saddles. We rode up causeways and streams of lava to Laguna, about 2000 feet above Santa Cruz. Here we found a large and handsome town, but almost deserted, and beyond, a plain surrounded by hills, and covered with ripe wheat, barley, potatoes, lupins, maize, &c. Riding across this, we came on the slope of the other side of the island, and had the most glorious view I ever beheld of the coast; the hill slopes covered with vines, palms, fig-trees, cactus, &c., rising gradually into a level stratum of clouds, above which rose the cone of the Peak in all his majesty. About 3 P.M. we arrived at Porto Orotava, a considerable town delightfully situated, and about twenty-one miles from Santa Cruz. Here again Mr. Carpenter, the vice-consul, and most other people, said it was impossible to ascend; but at last we found a fine dark-whiskered tall Spaniard, named Aguida, who had been up nineteen times, and he made no difficulty about the matter, but promised to have eight saddle horses, and five pack-horses for provisions and water, ready the next morning. We then strolled out for a bathe; but there being no shelter, I got my legs and arms cut against the rocks in the surf, and the next morning again fell and hurt my knee on the slippery lava blocks with which the town is paved. I also got a wicked horse with a horrid Spanish saddle, something like a pack-saddle in miniature. However, at 6 A.M., May 1, off we set; and riding over one or two lava streams, all bare, galloped along the little rugged paths over the undulating valley of Orotava among vines and tropical plants, greeted on every side by the smiling salutations of a handsome and comfortable-looking peasantry. We then began to ascend by steep rugged little paths alongside a ravine, and in about an hour entered the clouds. The rocks were now covered with beautiful heaths and juniper-bushes, through which we rode for another hour and a half, when we emerged from the clouds into a blazing sunshine, striking on hills of lava and pumice-stone, all brown, bare, and desolate. Over these we rode, climbing occasionally the most breakneck-looking crags on our sure-footed little horses. I should have told you, too, that my beast, kicking out at another, broke the fragile girths of the saddle, and laid me and it on the ground, where he trod on me once or twice, without doing any material injury. At noon we stopped to rest and lunch at a place called the Cañada, and took the height of the barometer; we were now upwards of 6000 feet high, and 2000 feet above the clouds, which stretched away below us in white snow-like plains. We now entered on the great plains of pumice-stone, in small pieces, all dry, dusty, and yellow, with crags and rocks protruding at intervals. On this nothing but a few bushes of broom grew. It was toilsome work crossing these arid and dusty rolling plains, which, though looking smooth, were really hilly. . . . We then arrived at the foot of the great cone, which had, however, a buttress of pumice on one side. Up this we rode, and then, with slow and toilsome steps, scarcely able to get our horses along, we dragged them up a little narrow winding track, up a steep ascent of ashes and cinders, to the Estancia de los Ingleses. Here we found ice, and making a fire under the lee of some great blocks of rock, we ate our dinner about 5 P.M., and prepared to pass the first of the night, which we did very comfortably; we singing English songs at one fire, and our dark wild-looking Spanish guides singing Spanish at another. We were all clothed in caps and blankets, and with our thirteen horses and twenty-two selves must have made a very wild and picturesque bivouac among the black lava and white ashes, lighted up by our fires, and by the clear and brilliant stars. At two in the morning Aguida roused us; and taking each a cup of coffee and a long stick in our hands, we addressed ourselves to the Peak by the light of a half-moon, leaving our other guides and horses at the Estancia. Slowly, and by little and little (*poco a poco*), walking a minute and resting a minute, did we mount the steep mound of ashes and cinders by a little winding track for the first hour. We then came to the foot of a great lava stream, and for another hour and a half we scrambled up round loose rugged stones, tottering, jumping, or

climbing from one to another. At five the sun rose
most magnificently over the hills of the Gran Canaria,
which pierced through the clouds to the west, and we
found ourselves at the foot of the piton, or last and
newest cone of ashes. This last little bit of five or six
hundred feet cost us one hour and a half of the severest
labour; but at length, at half-past six, we stood all on
the summit, looking down into the little crater on one
side, and over a space equal to, or greater than, all
England and Wales on the other. Palma Gomera and
Canaria were visible, and patches of sea, and Teneriffe
below us; all the rest looked like an immense country
covered with snow. It was, however, very magnificent.
Several of the party felt so sick, that we did not stay so
long as I wished, or nearly, on the Peak; and began to
descend at seven o'clock. At nine we reached the
Estancia, and at three P.M. had returned to Orotava.
Our barometrical measurements gave a height of 12,109
feet for the Peak. We were visited by several people,
both English and Spanish, and were invited on all sides
to balls and parties if we would stay. The gentlemen
said all the Spanish ladies there spoke English, and
were longing to see us; but as we had no business to
be in Teneriffe at all, we were obliged to fly temptation,
and early on May 3 we returned to Santa Cruz, and
sailed the same evening. We have had very pleasant
weather ever since, and are now sailing into Porto Prayo

in St. Jago, Cape Verde Islands, where, as there may be an opportunity to send a letter, I have scratched you off this. You will find an account of Porto Praya in the beginning of Darwin's volume, and a panoramic view of Funchal (but not a good one) in Fitzroy. Humboldt describes Teneriffe in his *Personal Narrative,* first volume. I have only room left to assure yourselves and all friends of my frequent recollection of you, either below the clouds or above them. . . .

False Bay, Cape of Good Hope, June 19, 1842.

MY DEAR MOTHER AND A.,—We are now just beating up False Bay, and hope to get to anchor this evening in our harbour—Simon's Bay, about twenty miles from Cape Town. I sent you a letter with an account of our ascent of the Peak of Teneriffe. We only stayed there a few hours; and the day after we left we got among the flying-fish, and into the bright water and lovely skies of the neighbourhood of the Line. As we approached more nearly the Equator, we had occasional heavy showers, it being the rainy season; but everything soon dried again, as the thermometer was at 80° With the awning on the quarter-deck, this weather and temperature was delightful. On May 17, we were very near the Line, with light variable winds, and in the evening a strange sail was reported from the mast-head, bearing

L

right down on us. We all ran on deck, and when the
captain had taken his station on the poop, a voice hailed
us from the sea immediately under our bows, and demanded leave for Neptune to come on board. Aboard
he came accordingly, in strange disguise; and, after a
palaver with the captain, expressed his intention of
boarding us to-morrow with Amphitrite and his son,
and shaving those who were for the first time entering
his dominions. The next morning, May 18, after a
deal of preparation under the forecastle behind some
sails—which was all kept secret from the uninitiated—
a lot of rough fellows, daubed with paint and variously
and hideously disguised, with great staves in their
hands, and denominated ' constables,' came and ordered
all below that were to be shaved, and all on deck that
had crossed the Line before. They then fastened down
the hatches, and arranged their proceedings. Almost
all the mates and midshipmen, more than half the
ship's company and petty officers, and some of the best
men in the ship, were uninitiated. Accordingly, when
all assembled, we called a council, and determined to
show fight for it; and that when Neptune summoned
us one by one on deck, instead of answering to our
names, we would tell him to come and fetch us. No
sooner said than done. We triced the ladders up under
the hatchways to form a barrier, cut off the connection
between the pumps and the hose on deck, and declared

war. After much skirmishing with buckets of water
on both sides, capturing some of the constables' staves,
and passing hard raps about pretty freely, they made
a rush down, took one pump, drove one wing of our
body forward, captured a gunner, and rigged a hose so
as to play upon us (with a snout like a fire-engine).
Some of them slipped a noose over my head, reined the
rope through a tackle, and were hoisting me on deck,
when some of our best men dashed forward, knocked
over the constables, and seized me by the legs. Then
began the thick of the fray, and I was very nearly
pulled in two, but at length got free. We then took
two or three constables, dragged them forward, and
lashed them as prisoners to the mess-tables. In one
great struggle I was nearly drowned, as the first lieutenant kept playing on me with the great hose from the
deck; but at length we drove the constables up, retook
the pump, and cut off their supply of water. After
this, all went against them. Every one of their attacks
were repelled, and at last, as we had more than half
the constables lashed in a row round the bows of the
vessel, Neptune sent to acknowledge himself foiled,
but begged us to comply with ancient custom. Satisfied with our victory, we yielded to these terms and
went on deck; submitted to be blindfolded, pumped
upon, swilled with water, lathered either with soap or
tar, and shaved with the rough or smooth razor (accord-

ing as we stood in the good-books of the barber or
otherwise), lowered into a sail full of sea-water, dragged
through it, and then to scramble by ropes on to the
forecastle under a heavy fire of bucketsful of sea-water.
I was let off pretty easily; but some of them got the
tar-brush and the rough razor. As soon as each one
reached the forecastle, he was at liberty to pitch into
the others; when, providing myself with a bucket and
a wet swab, I proceeded to pay off the first lieutenant
and some other of my friends. By this time, however,
no one cared for water, as, from the captain down to
the loblolly boy, no one had a dry rag on him, and
few had any rags at all except their trousers and shoes.
When all had been shaved, Neptune's procession was
formed, and he was wheeled aft on a gun-carriage, with
his hideous Amphitrite and their daughter (a young
scamp of a boy in an old bonnet, a sail-cloth shawl,
red paint, and oakum for ringlets). He had also his
doctor in tin spectacles, his secretary in an old coat
and a ' shocking bad hat,' and a tribe of other attendants. Having made his report to the captain, and all
hands being promised double allowance of grog, the
word was passed to clean decks and call the watch, and
in a quarter of an hour everything and everybody was
dry, clean, silent, and orderly. It was altogether a good
lark, and our desperate resistance heightened the fun and
formed a subject of conversation for a week afterwards.

I have given you this long account because I have really nothing else to tell you, except that we landed at the little island of Trinidad on May 27. It is a lofty and barren little spot, about five miles long, very difficult of access, and seldom landed upon. We were only there three hours, and I was occupied with getting shells, corals, fish, and rocks. It is of volcanic origin, but contains sandstone with recent shells. A farther cruise on it would have been very interesting.

On May 28, we passed the tropic of Capricorn, having given up all hope of being obliged to go to Bahia or Rio Janeiro. On June 4, the wind, which had been blowing from E. and N.E., suddenly shifted to S., when we were in about 30° S. and 25° W., and blew hard with heavy swell. We then turned our head to the east, and ran along 30° S. till we came to 16° E., which we reached June 17. We were then in sight of the Cape of Good Hope, and hoped to have been in in a few hours. Easterly winds and calms, however, checked us off, and for two or three days we have been idling in sight of port without being able to get in. Notwithstanding its being midwinter, and the hills up the country covered with snow, the weather is most lovely, clear, warm, and pleasant, the thermometer ranging from 58° to 68°. The day after to-morrow is the shortest day with us. Our whole passage has indeed been most favourable and delightful. We expect to stop here three or four weeks before we sail for Van Dieman's Land. I write this off in case a vessel should be on the point of sailing when we get in. . . .

Simon's Bay, June 21. We anchored, as we expected, on the evening of the nineteenth, and yesterday I came ashore to get a place for the twenty-four hours' meteorological observations. I engaged a room at the British Hotel, and last night, having fixed all my instruments, I got into bed very sleepy and tired. Before I could well shut my eyes, however, I was attacked by a host of fleas, and not one regular half-hour's sleep have I had. At half-past four I got up, lighted a candle, and began to observe, as I could lie still no longer; and now the first red tinge of morning is shining across the still waters of False Bay. Simon's Bay is a snug little sandy cove on the west side of False Bay, environed by precipitous hills six or eight hundred feet high, formed of dry brown sandstone resting on granite. The sandy shores are strewed with various kinds of shells—some very beautiful—and as soon as I have finished my meteorological observations, I mean to have a dig at them with my dredge. Although to-day is the shortest day, and the sun does not rise until after seven o'clock, the weather and temperature are delightful. No fires nor even fire-places. There is a small war broken out on the borders of this colony at Port Natal. Some emigrant Dutch boors, who left this place to settle there, have broken out, defeated with some loss the party of military stationed there, and the Southampton (the flag-ship of the station) is gone down with troops. We have no more recent news from India here than when we left England; indeed, the vessel which is to take this letter was detained a month by bad weather off the Cape on her passage from Singapore, and only got in two days before we did. You would delight in the flowers here. Only fancy jonquils growing wild in the middle of winter, besides beautiful heaths and lots of others whose names I know not! . . . Now again farewell, and believe me ever your own dear

BEETE.

Cape Town, July 10, 1842.

MY DEAR MOTHER AND A.,—The very first day I landed at Simon's Bay I was foolish enough to tumble into a hole in an old wooden pier, and bruise and cut my shin. I took no notice of it for three or four days, when it inflamed and laid me up for a fortnight, confining me to the captain's sofa. Three days ago I was able to take the poultice off and run up to Cape Town, on condition I did not walk much. (N.B. I ran up in the mail-cart.) I have accordingly lost all opportunity of seeing much of this extremity of Africa, the first continent I ever trod. Cape Town is really a beautiful place.

Table Mountain and the adjacent hills, rising in magnificent precipices to the height of 3500 feet (the height
of Snowdon), half encircle an *undulating plain* that
slopes to the sea (there's an Irish definition of a plain).
The sea is the wide sweep of Table Bay, opening into
the Atlantic; south of the bay a real plain quite flat
extends for twenty miles, on the other side of which is a
very picturesque range of mountains. The town is well
laid out in spacious streets crossing at right-angles, some
of which have rows of trees with a brook of clear water
down the centre. The houses are of light stone, spacious
and lofty, with handsome fronts and fine rooms. Altogether I know scarcely any town more beautifully situated, or more striking at first sight.

Captains Blackwood and Wickham, Burgoyne, and
myself are staying in this house. It is a boarding-house
kept by a widow lady. . . . They are very pleasant people. We live remarkably well and in good style, din-
ing at six, and having always music and dancing in the
evening. In short, we could spend another month here
very pleasantly, especially when my leg is well enough
to mount on horseback and visit the environs. One of
the most characteristic sights here is a bullock wagon
long, light, and narrow, and drawn by a string of from
fourteen to twenty pairs of bullocks, each of which
knows his name and obeys the voice of the driver.
They sometimes drive, too, eight horses in hand, turning

round the corners with the greatest dexterity. . . . We
leave this place to-morrow, having engaged an eight-
horse wagon to take ourselves and our traps, including sundry cases of schnapps and scheidam, down
to Simon's Town; on Wednesday we expect to sail, or
at all events are sure of doing so within this week.
We shall then calculate on arriving in Hobarton in September, four or five months after which you may expect to hear from me again.   You must not calculate
on seeing me in England till 1846, as the three years of
an ordinary voyage are rather likely to lengthen out in
one like ours.   We shall stay, I believe, six weeks in
Van Dieman's Land, then proceed to Sydney and stay
three or four, and then go on our first cruise, which will
probably last a twelvemonth; so that unless our letters
are forwarded to us (which they probably will be), I
could hardly calculate on receiving an answer even to
this until November 1843.   I suppose it is quite warm
with you. . . . The sun is shining gloriously here on
the gray heights above the town, but the air we consider sharp, the thermometer being as low as 57° in
the house.   How gets on your Geological Society?   Do
the ladies still attend well?   Tell Alfred Browne I
should be delighted to have a letter from him or any of
my male friends and 'fellow socialists.'*   . . . I have
now only room to say best love to you both, and kind

* This refers to the Social Literary Society at Wolverhampton.

remembrance to all friends, and that I am ever your affectionate son and brother,

J. B. JUKES.

Hobarton, Sept. 1, 1842.

My dear A.,—We sailed from the Cape, July 14. About July 26 we had very bad weather and a heavy sea running. The vessel rolled so much as to take in tons of water, first over one hammock netting and then over the other; so that we were obliged to batten down all the skylights and hatchways, and live below by lamp-light. The same thing occurred again on the 31st, making everything very uncomfortable and wet. On August 6, we came in sight of the little island of St. Paul's. This is an old submarine volcano which has got a hoist, the crater being broken down and open to the sea on one side. We anchored and went ashore, rowing into the crater across a shallow bar. The sides of the crater are very steep, rising to a height of 800 feet, and are covered, as well as the whole island, with coarse grass and rushes, no other thing growing there. We found some wild pigs that had been left there, and the captain shot one. Several springs of hot water trickled down through the stones on the beach, one of which was 150°. The island is not more than four miles long and three wide. They caught from the ship nearly a ton weight of very excellent fish in the three or four hours during which we were at anchor. After leaving St. Paul's we had three other gales of wind with very disagreeable weather before reaching Van Dieman's Land. On August 23, a man fell overboard, but, as it was luckily calm and smooth, he got hold of the life-buoy, and we got him in again. These, together with watching and sometimes catching the flocks of beautiful albatrosses and petrels, are the events of our 6000 miles' voyage in the South Indian Ocean.

Aug. 25. We saw land, but did not get to anchor in Storm Bay till the night of the 27th, and anchored at Hobarton on the 28th. The next day I got a letter from Mrs. Meredith, dated July 11, inviting me to see them, and pointing out the best route. I set off on Thursday next, September 8. It is two days' journey, forty miles of which is through the bush. We were all very much pleased with Hobarton. The Derwent river or estuary is two or three miles broad, with hilly land covered with wood on each side of it. In a hollow, at the head of a cove on its western side, stands the town, backed by a broad flat-topped hill called Mount Wellington, 3100 feet high, the top of which is now covered with snow. The town is well laid out, with straight broad streets crossing at right-angles. But houses of all shapes and sizes are yet scattered about; they are building, however, some large and handsome places now, of white stone, of which they

get plenty hard by. . . . A vessel sails from hence this
week, but no other will leave till November. I shall
therefore not be able to send you an account of my
trip to Great Swan Port till we arrive at Sydney in
October. M'Gillivray and I have taken lodgings on
shore while the ship is refitting. It is very cold here
now, so that we were glad to be near a fire and get
some of our things dried, many of which were mouldy
and spoiling.

Sept. 3. We went an excursion yesterday through
the woods on the other side the river. The country is
really beautiful: tall trees of the kinds called Euca-
lyptus, Mimosa, and Banksia with graceful foliage;
pretty little valleys with cleared spots for farms; parrots and beautiful little birds with gaudy plumage.
One lovely little valley, with an excellent road leading
from Richmond in the interior, with a little brook down
it, and steep wall-like faces of rock (something like the
valley at Chesterton, but more picturesque), lead us to a
ferry across the river, there a mile wide with fine bold
hills about it. I could have fancied myself all the
while in some fine part of England. The very mile-
stones and the tracks of the coach-wheels were old
friends. . . .

H.M.S. Fly, Sydney, Oct. 17, 1842.

My dear Mother and A——,—We anchored here yesterday, and I intended to have had a long letter written for you before we arrived; but in coming from Van Dieman's Land we had a tremendous gale of wind, reducing us to scudding under a close-reefed fore top-sail, with a heavy sea washing over the decks, pouring into the cabins, and making everything wet and miserable. I find now that a vessel sails for England this evening. I therefore, in the midst of other occupations, snatch a hasty moment just to say all is well, promising to write a longer despatch before we leave this. . . . . I spent a week with C. and Mrs. Meredith at Spring Vale. I went one day's journey by coach through a beautiful country, partly rocky hills covered with thick woods, partly open grassy plains, with park-like trees here and there, where we often left the road and drove among the trees over the turf, to a little place called Ross. The next day I rode on horseback with a guide over rocky hills and woods to a lofty tract of marshes called Kearney's Bogs, 2000 feet above the sea. Here I slept; and the next day proceeded down the seaward slope of the hills into a fine flat valley, ten miles broad, all covered with woods, except a little clearing here and there. In the middle of this valley, at the junction of two brooks, ten miles from the sea, is Spring Vale. The house is a one-story-high cottage,

barely finished, and very rough. It stands on a slight
elevation, with a fat marsh close alongside, and half-
cleared woods about it. The day after I got there it
began to rain, when down came the floods, nearly
drowned two men, and kept us close prisoners four or
five days. . . . When I left them, I walked back to Hobarton, about eighty miles, but have no space for a description of my journey. On getting back to my ship,
I was involved in a great round of dinner-parties and
balls, and became acquainted with the governor and
Lady Franklin. As the vessel was to touch at Port
Arthur, they in the kindest possible way gave us a
cruise of four days with them round the Probation Station in Norfolk Bay in a colonial steamer. It was a
large and delightful party, and I saw much that to an
ordinary traveller is a sealed book, as these stations are
diligently guarded, and the whole of Tasman's Penin-
sula separated by a military post from the rest of the
island. At Port Arthur, too, the commandant and his
lady put their house at our disposal, and I rarely have
seen a more interesting place. We saw Frost, Williams,
Beaumont Smith, and other celebrated characters.
There were also coal-mines; and you may tell my aunt
Jane that I have got Tasmanian coal fossils, and lots of
spirifers, productæ, &c. &c. from the adjacent rocks,
which I take to be exact representatives of the Devonian
system. I hope we shall return to Van Dieman's Land,

as, though Sydney is more spacious and handsome and like an English city, I greatly prefer the aspect and style of Hobarton. I have not yet seen anybody ashore here, however, and perhaps may alter my opinion. I must not omit to tell you that I have got a new dog, a kangaroo dog, very like a greyhound, of a dark brindled colour; light, active, intelligent, with sharp nose, and fine ears and tail, which (*i.e.* his ears) he cocks up when you speak to him, and looks as knowing as you please. He is called Spring, and whenever I come on deck he almost springs over me. By the bye, I hope Gov. and Bell* are well. . . . I need hardly beg you to give my kindest love, regards, and remembrances to all to whom they are due; and to accept the same yourselves from your affectionate son and brother,

J. B. JUKES.

H.M.S. Fly, New South Wales, Sydney,
Nov. 14, 1842.

MY DEAR A.,—At length, after many arrivals with none, I get a letter from you. . . . Since we arrived here, I have been spending a week with an old college friend, M., who has a parish at Mulgoa, about forty miles from Sydney, and a very nice wife and family. The

---

* Two dogs which he brought with him from Newfoundland. The former, being a parting gift from Captain Prescott, was named ' Governor.'

country around is pretty, with a magnificent river gorge within a few miles. All the grass, however, is thin and brown, like that by the roadside on a dusty day, the trees small and sombre, the woods affording no shade, and everything hot, dry, brown, and burnt-up. It is not to be compared to Van Dieman's Land for beauty and pleasantness, and this entirely from the want of rain. Sydney, looked at as a town of sixty years' growth (a lady is still living who slept in a hammock slung between two trees in what is now the principal street), is a wonderful place.   The governor's new house is a small Windsor Castle ; the spacious streets, large and handsome shops, numerous harbours, wharves, ships, country mansions, and the wide extent of ground it covers, equal or surpass a third-rate town in England, such as Nottingham (it is much larger than Shrewsbury or Wolverhampton). It is, however, badly paved, frequently not at all ; and being filled and surrounded by sand, when the wind blows, which it almost always does, it is most disagreeable. This morning, after a little rain, all was calm and lovely, when about ten o'clock, without any warning, the whole place was enveloped and hidden from sight in a cloud of fine dust, and a squall struck us that brought all hands on deck ; ever since it has been blowing so hard that no boat can go ashore, and fine dust has been falling all over us, though we are at a good distance from shore. This wind the inhabitants call a brick-fielder, as it blows from the brick-fields south of the town. In the summer a hot wind from the interior sometimes brings the thermometer up to 120°, when up comes a brick-fielder, and drops it to 60°.

I have made some pleasant acquaintances here. Rev. W Clarke I knew in England. I have seen here Captain King, Sir T. Mitchell the traveller, Mr. M'Leay the naturalist, and others. . . . I am delighted to find that you have at last had in England a little of that sunshine which glows perpetually here, and that you have enjoyed yourselves in Shropshire. Believe me, I shall be very glad to stand once more on Apewood Castle hill, but suppose before the year 1846 I shall not have a chance. The Beagle was here the other day, and I went all over her. She has been five years and three months out from home. The lamb history has perplexed me much. If Governor becomes a confirmed ovicide or agnicide, I know not what you must do with him. He will never leave it off. I wish I had brought him with me. If, however, he becomes troublesome in any way, the best thing will be to part with him at once. However, do just as you like. Pray be under no alarm about my health. I enjoy heat now and the only illnesses I have had have been slight. When we get to our work, I shall be well enough directly. . . . Have you ever realised (as the Yankees say) our relative positions ? When you are enveloped in

M

fog and snow, we are basking in sunshine and flowers,
parrots and gaudy insects surrounding our walks;
and as we are ten hours before you in longitude, our
daily occupations are very nearly reversed. For instance,
at six in the morning I am just taking my morning
bath, at the same time that you are going to bed; and
now, while I am writing this at eleven o'clock at night,
you are walking out before lunch. Lady Franklin remarked to me how odd your correspondence made you
feel here. For instance, you wrote home about something very particular, but in a month or two entirely
forgot it, when next year comes a letter full of the subject. She said it seemed to her like stopping every now
and then and living a bit of your past life over again.
Your lamb affair will, no doubt, be an example of it, for
you will hardly get this before April. . . . We expect to
sail from this next Tuesday. We take the bishop and
Captain King to Port Stephens, about one hundred
miles, where we remain a few days, and then enter on
our work. In about four months' time, a vessel will be
sent after us with provisions. She will bring on our
letters, and will perhaps take back our answers, if she
return to Sydney. At all events, no vessels will pass
through Torres Straits till April next; so that after
this, unless I write to you from Port Stephens, you
must not expect to hear from me for six months. I
expect it will be thirteen months before we return to

Sydney. We shall then have encompassed Australia; and we shall have to do that three times, besides visiting the neighbouring islands, before we get a chance of being ordered home. As I can only keep one eye open, I shall say good-night for the present.

Nov. 15. Business this morning presses upon me, and the letter-bag closes at noon. I have killed about a hundred golden, silver, and mottled beetles, stowed them away in boxes, partly cleaned and packed some skulls, made a selection of various things to go ashore for this trip; and I have now to go and look over a whole host of traps, and remove them from a private merchant's store to the commissariat department. . . . Tell my aunt Jane that I am writing a geological paper for the *Tasmanian Journal,* a periodical under the patronage of Lady Franklin. And now, my dear A., farewell. With every possible good wish for you and all, believe me ever your own brother,

BEETE.

H.M.S. Fly, Cape Upstart, N.E. Coast of Australia,
April 10, 1843.

MY DEAR A.,—Yesterday, when the morning dawned in clouds and rain, we discovered our expected provision ship at anchor three miles to leeward. About eleven o'clock we got the mail-bags on board, and great was the breaking of seals. I got your October letter,

which I read over in great wonder at several mysterious
allusions to some previous one, but what it was about,
and where it could be, I could not tell; and whilst exclaiming in the gun-room against my hard luck in losing
it, the sergeant brought me three others, which had
got mixed with ship's company's letters. . . .

Of my mother's improved health I am most rejoiced
to hear. The details, too, of your pleasant walks (no
occasion to carry a double-barrelled gun and look sharp
among the bushes) make me almost envy you, did I
not know that my own thirst for the excitement of adventure would soon arise again among turnpike-roads
and hedged fields. As to my book, I am also annoyed
that there is no map. There are maps and sections
sent to Newfoundland with the report, of which I hope
and trust Murray will keep some. for me. The notice
taken of the work in the papers so far exceeds all my
expectations that I am ' as uplifted as a midden-cock
upon pattens,' as Scott says. I really hardly expected
so hastily-written and slight a work to be noticed at all,
except in the Newfoundland papers, or among the short
remarks on ' books published.' Sir R. Bonnycastle, I
see, attacks me; he came to the country just before I
left it. I have a very kind letter from Captain Prescott,
in which he says Bonnycastle's remarks will do me no
harm. . . . I wish whoever stopped the draughts in
your parlour were here to stop the leak overhead in my

cabin. I catch a gallon of water a day from it in this wet weather, besides the quantity absorbed by my bed, &c., and it cannot be cured until we lay the ship up again. We understand there are lady passengers going to India aboard our provision ship, but the weather is so bad she cannot come up to us. As we have had only distant glimpses of ladies, and they have been both black and naked, we are all anxious for the weather to clear, that we may enjoy the sight of a frock or gown again. . . . You will be sorry to hear poor Stuwitz is dead. Captain Prescott has sent me an affecting account of the calmness with which he met his end, regretting only that the fruits of his scientific labours would be lost to the world. You may imagine I was much pained by it. A martyr to scientific zeal, he would, had he lived, have done much and acquired fame, as well as advanced the boundaries of human knowledge. You talk of reading Richardson's *Geology for Beginners.* I never read it, but reading is not the way to begin geology. Get an appetite for collecting fossils. You become interested; you acquire an eye both for organic forms and for the strata of rocks, &c. Your curiosity is excited, and you wish to know something about them; then apply to a book, and you learn. All the natural sciences, by which I mean the knowledge of the animate and inanimate things about you, require the eye and the senses to be used, to be trained, to be educated, as well as the intellect. You positively *do not see* now what in a short time you learn to perceive; and how can you expect to learn from words things which your eyes have roamed over without conveying any ideas about them to the brain? When you read, read Lyell at once. Skip what you don't understand; you will find it out afterwards. I see I have fixed the year 1846 for my return; I am afraid, however, it will be the spring of '47.

April 26. We have lain quietly at anchor here, repairing our own boats and the provision ship. She is now nearly ready for sea, and talks of proceeding. I fear it will be a long while before you get this, as she will probably be forty days to Singapore, and altogether four or five months will elapse. Still, if you get it in September, and write immediately, I may get your answer in January or February, before sailing on our second trip. I believe we shall not see New Guinea this year. I have not done so much yet in the collecting way as I expected and intended. My dredge was eaten by the rats all except the iron, and there never has been time to make me a new net. The few shells I have got have accordingly been from the beach. I have a few very pretty bird-skins; and we have now alive a very pretty little kangaroo, brown and gray, that goes hopping about the poop, and a splendid black cockatoo, spotted with brown about the head and crest, and his tail barred with yellow and red. He was wounded, but seems recovering, and already feeds from the hand. I must now close. Give my best love to my mother, whose health, I hope, continues to improve. . . . Remember me to all friends, and believe me ever your most affectionate brother,

J. B. JUKES.

H.M.S. Fly, Cape Upstart, April 10, 1843.

MY DEAR BROWNE,—Yesterday, after four months absence from the civilised world, our provision ship reached us, bringing food for the mind and body. . . . You will doubtless see the journal I send by this opportunity, and therefore I need tell you nothing of my movements. I am glad to find the Wolverhampton branch of the Dudley and Midland Geological Society flourishes. I sent to our old society from Hobarton an account of our ascent of the Peak; but as it was directed to H., and I see by the papers he was leaving town, I know not whether the Literary and Scientific Society ever got it. I should be happy to give you a paper for the Geological, if I did not think postage too heavy. Would a short sketch of the structure of a coral reef do, do you think, with a map and sections to be enlarged? Talking of coral reefs, should anybody accuse me of telling lies by saying there are coral reefs near Newfoundland, assure them I never meant to say so. I find such an assertion in my book at the end of the natural-history notes, but how it got there I know not.

April 24. The vessel stood in need of some repairs, and is delayed till now; and such a rotten ship is she, that it is doubtful, I think, if she reach her destination. But if you *don't get this, pray answer* it nevertheless. ... Five years' exile appears now a long term to look forward to. But this will be my last lengthy trip, I hope; I shall confine my ambition hereafter to the regions of railroad or steamboats, say Europe and North America. Well, it is now four bells in the first watch, which means that it is ten o'clock P.M. in our part of the world; and as I always, or generally, go ashore at daylight to bathe, I find my eyelids rather drooping. I shall therefore conclude, and with hearty good wishes for your health, wealth, and prosperity, am very truly yours,

J. B. JUKES.

It will be seen in the following journal and letters that the Fly, after commencing her work at Sandy Cape, and surveying Swain's Reef and the Capricorn group of islands, anchored in Port Bowen, the appointed rendezvous. From this place several boat-excursions were made in search of fresh water and to explore the coast, in one of which Captain Blackwood, Dr. Muirhead, and

Mr. Jukes appear to have been almost in sight of the great river of Queensland, the Fitzroy, as they visited the south head of Port Bowen, towards which that river makes a curve. It was probably hidden from them by the range of hills which is mentioned as 'everywhere terminating the view at the distance of ten or fifteen miles from the coast.'

Port Bowen appears to have been the first point of the coast at which the Fly touched after leaving Port Stephens, and from that place their course lay to the north; so that the Fitzroy, which empties itself into Keppel Bay, about sixty miles south of Port Bowen, was not seen by them. In the abstract of the sailing orders issued by the Admiralty to Captain Blackwood, and given by Mr. Jukes in the 'Voyage of the Fly,' the annexed passage relates to this part of the coast; and certainly 'a week or two,' if construed literally, would not have allowed time for much examination of its rivers and bays. 'As you have stated that during a former voyage you remarked a more than ordinary population on the coast of New South Wales about the latitude of 20°, from which you inferred a large tract of fertile soil, perhaps traversed by some considerable river, you are hereby authorised, when engaged on the Barrier Reef in that latitude, to devote a week or two to the examination of that district.'

Since that time the extensive province of Queensland has been separated from that of New South Wales, Brisbane, its capital, being 7° south of the portion here included in New South Wales. Respecting this region, Mr. Jukes says in a note to the Voyage of the Fly: 'After twice circumnavigating Australia, and visiting all its colonies, especially those of the southern coast, I look back upon this tract between 22° and 20° with still higher expectations than before, and certainly have never seen any part of Australia, *near the sea*, of equal fertility, or of nearly so pleasant and agreeable an aspect, or combining so many natural advantages.'

It is needless to remark how fully these expectations have been fulfilled. The river mentioned at page 191 on the west side of Upstart Bay is probably that marked the Burdekin in the recent map of Queensland published by Stanford. The following note on the geography of this part of Queensland has been kindly furnished by C. H. Allen, F.R.G.S., who lately passed some time in that colony:

'It would be very easy to sail past Cape Capricorn and across Keppel Bay to the north without obtaining a sight of the Fitzroy, as it is necessary for vessels intending to enter that river to turn round and steer a course almost S.W., and the whole coast appears shut in by ranges of hill from 1000 to 2000 feet high. This river, the second largest in Australia, is navigable for steamers and vessels of 500 tons register for a distance

of nearly forty miles, and at Rockhampton, the large
town which has sprung into existence at the head of
the navigation, there is a depth of twenty or thirty feet
at the quays, whilst the river itself is about 400 yards
in width at that point. Strange to say, this broad
stream, which drains a district as large as all England,
was discovered from the *interior* by one of the pioneer
squatters of Queensland, who, in striking across country northwards, was pulled up by the river, and eventually he succeeded in following its course to the
ocean.

'I notice also that the Fly does not appear to have
entered or even sighted the waters of the Pioneer River
in latitude nearly 21°. The wonderful development of
the sugar-growing industry in this part of Queensland
fully bears out Mr. Jukes' estimate of the fertile district
of country, which here extends for a great distance
along the sea-board of the continent.'

Journal.

Nov. 24, 1842. Left Sydney.

Nov. 27. Anchored at Port Stephens, where Captain
King (he who was captain of the Adventure with Fitzroy)
lives.

Dec. 17. Left Port Stephens, and on the 28th anchored at Sandy Cape or Breaksea Spit, where our work
was to commence. We had some communication with a

tribe of natives here, but I was unwell, and did not see
them. A trifling and temporary interruption to my usual
robust state of health detained me on board till January
7, 1843, when I landed on one of a small group of coral
islands called Bunker's Group. It was a low sandy
island, about three-quarters of a mile across, with a
circular reef stretching out three or four miles from it
on the south-west side, enclosing a shallow lagoon of
clear sea-water. The island was encircled by a belt of
trees, on which myriads of boobies and noddies built
and roosted. They sat still on their nests within three
feet of you, and even suffered themselves to be taken
by the hand. Tern, gulls, and two kinds of cranes—
one blue, like a heron, one pure white, with most elegant plumage—flew around us. Turtle in the greatest
abundance were taken; and I found one in the wood
sleeping, that suffered me to sit down on him without
being disturbed. Turtle-soup, turtle-steaks, turtle-pie,
and stewed flippers were our regular food for some
time.

Jan. 9 *to* 19. During this time we were engaged in
surveying a new group of small coral islands, which we
had discovered a little to the north of the Bunker's. We
called them the Capricorn Group, as that tropic passes
right through them. We had most lovely weather,
and frequently landed on the islands, being generally
obliged to wade to them over the reefs on which they

stood, sometimes half a mile, up to our middle in water. A coral reef is a very different, and in itself a much less interesting and beautiful object than I had expected ; neither are the colours of the living corals so splendid as is often said. The mass of the reef is a dark-brown, close, hard rock, with here a patch of white sand, and there a lump or a mass of living corals, some of the smaller of which are grass-green, yellow, or red. I found also a calcareous breccia or limestone, made of fragments of shells or corals. When sailing among the islands, however, the reefs and the islands often formed most beautiful objects, as seen from the poops or mizzen-top. The little island, crowned with trees, is surrounded by a smooth space of clear shallow water of a bright grass-green colour, round which is a narrow fringed margin of white foaming surf, sparkling in the sun, dividing the lagoon from the rich clear blue of the deep water outside. To those who have only seen the muddy waters of our own seas, the brightness and depth of colour of the clear ocean water, blue in the deep, green in the shallow parts, is almost inconceivable. If you can form in your mind's eye the picture I have endeavoured to draw, and add to it a nearly vertical sun and a sky without a cloud, with a little brisk refreshing breeze from the south-east, you will have a good idea of our Capricorn Group. You may add the Fly and the Bramble, with their white sails and awnings, at anchor, and six or eight boats of different sizes, likewise fitted with white awnings, the men in white frocks and trousers, sailing in different directions, sounding, or carrying the surveying-officers; and in the stern-sheets of one you may put me, with a white shooting-jacket, panama hat, luxuriant beard and moustache, gun, and collecting-basket.

Jan. 19 *and* 20. A gale of wind, clouds, heavy sea.

Jan. 21. Found a fragment of wreck on one of the most northerly islets, and on the trees were carved 'The America, 1831,' 'Mary Ann Broughton, June 1831,' 'The Nelson, Captain E. David, Nov. 1831,' the soles of a pair of child's shoes, fragments of crockery, &c. We supposed that the America was wrecked in June, and that the crew, &c. had probably been taken off in November by the Nelson; but where they procured water we could not tell. There were no bones nor signs of graves.

Jan. 22 *to* 27. Partly sailing and partly at anchor in an open sea north of the Capricorn Group, with heavy breezes and rough weather. Lost one anchor; but on 27th came within sight of more coral reefs without islands on them, supposed to be the southern extremity of the Great Barrier Reef, which stretches hence to Torres Straits, a distance of 1000 miles. We anchored a little to the west of them, and here passed five very anxious and uncomfortable days, as it came on to blow from the south-west (an unusual quarter), and we had, from 28th to Feb. 2, the heaviest gales any one on board had ever seen within the tropics—with the reefs to leeward, a tremendous sea, and nothing to trust to but the goodness of our anchors and cables. The Bramble parted one cable, but luckily brought up with another before she drove on the reefs. Had she once touched on them, she would have gone just like a glass bottle thrown against a wall, and a few chips or broken planks would have been all we could have hoped to have seen of her.

Feb. 2. What a relief! A lovely morning, smooth, and clear skies. Up, top-masts, top-gallants, and royals! Heave away on the capstan; up with the anchor, and away we go again!

Feb. 3 *to* 9. Running along and surveying outside portion of this reef (called Swain's Reef). It consists of a chain of long oval reefs, with passages between them, into one of which we ran and anchored every night.

Feb. 10 *and* 11. Tied by the nose, the anchor having caught under a coral rock; and we broke three hawse-pipes before we could get it up.

Feb. 12. We now ran into the reefs, and during two days sailed through them by narrow channels towards the west, in order to rejoin the Bramble, who had left us on the 3d to run along the inside edge of the reef.

Feb. 14. Anchored in Port Bowen (our rendezvous), where we found the Bramble. This place was visited by Flinders, and described by him as a good harbour, with fresh water—a thing we were beginning to stand in need of. We found it too shoal to get far in, and his fresh-water gully had only one hole just behind the sand-beach, containing a few gallons of dirty water. The country around was partly hilly and rocky, partly low and sandy, with no abundance of wood and no appearance of verdure. It was, to be sure, mostly covered with wood of some sort, but it was poor and brown. We found on the slopes of the hills many gullies and watercourses, on the sides of which was good grass; but not a drop of water, and not a single living thing except grasshoppers, flies, beetles, and ants. Torrents must sometimes pour down here; but it was now the dry season. The shape of the country about was picturesque. Many lofty peaked ranges or detached hills in different directions, and at high water the bay was a spacious and noble sheet of water; but at low tides it was nearly filled with sand and mud. . . . We stayed at Port Bowen till Feb 28, repairing damages and surveying the harbour. I shall, however, give one excursion in detail.

Feb. 20. At eight o'clock in the morning went away with Captain Blackwood and Dr. Muirhead in captain's gig, with four men, to visit the south head of the Port.

We found it gradually narrowing and winding like a river among muddy flats covered with mangroves, but approaching the foot of a range of hills, from which we were not without hopes of finding a river or some fresh-water stream. We saw a dense smoke rising in one or two places as we drew near the hills, apparently signal-fires. At half-past twelve, having sailed twelve or fifteen miles, we saw a little opening among the man-grove-bushes, where we were enabled to land, and got immediately on a small rocky mount, between which and a low ridge there was a native path. Taking our guns, we ascended this ridge, and saw that our water shortly ended among mangrove-swamps, without any river; but towards the south I saw water some miles off among the trees. As it was worth while determin-ing whether this was fresh water or only the head of Shoalwater Bay, we determined to sleep here, and go towards it in the cool of the morning. Returning to our little mount in the mangroves, we got our things out of the boat; and while Frethy, the captain's cox-swain, prepared dinner, and the captain went to shoot curlew, the doctor and I took our guns to explore the country a little. At the back of the mangroves was a sandy flat or beach, along which a well-beaten native path led both north and south from our encampment. We went towards the south, and shortly came on recent footsteps of a man and a boy, who had evidently been

N

following us along as we rowed, but had jumped aside
into the bush on our landing. We followed this track
back a mile or two. The path crossed in one place a
beautiful grassy flat with a watercourse running through
it, now as dry as a bone. Beyond this we saw other
tracks of four men running; but they were one or two
days old. Returning to our camp, we dined; but before dark found the mosquitoes and sand-flies so annoy-
ing, that we three determined to go some distance up
the ridge to sleep, in hopes they would be fewer. Soon
after dark, however, they attacked us again; and we got
up and made a great fire, and lay down in the smoke. We
got no sleep till nearly twelve o'clock, when, just as we
had dozed off, up started the dogs, barking and running
backwards and forwards, and up started we in a deuce
of a hurry, expecting a spear through our bodies. We
all ran against each other looking for our guns, which
were piled against a tree. It turned out to be Frethy
coming to see if we wanted anything. After a hearty
laugh we lay down again, but once or twice heard
noises in the bush.

Feb. 22. After a hasty breakfast by firelight, we
made preparations for a start. Just as day broke, we
heard a prolonged distant cooey in the woods. It had
a harmonic tone, as if from several voices. We left two
men, with orders to get everything into the boat, and
shove off into the stream; and taking the other two

with their muskets, a little biscuit, and a bottle of
water, we set off before six o'clock. We held a little
to the N., and then struck into the woods and climbed
some rocky hills two or three hundred feet high, having
a deep narrow valley on our right. At the head of this
valley we found another wide and more gradual one,
opening to the W., and just in our line. Here we
started two great kangaroos, but the only kangaroo dog
with us (Blueskin) was not in sufficiently good condition to catch them. We now proceeded down the valley, which was grassy open woodland; and I shot a
very pretty bird, which I think is a new species, if not
a new genus. About five miles from our camp we came
on a watercourse, in which was salt water, and presently after on mangroves, to which we could not see
any termination except blue mountains in the distance
beyond them. We got involved among muddy streams
in crossing a part of these, and nearly stuck in the
mud. As we knew now that it must be all salt water
before us, we returned, and had some pleasant rain,
which refreshed our hot bodies, as well as the parched
earth and vegetation. We kept rather more to the left,
and descended by a valley full of the most luxuriant
grass. On getting to the beach behind our mangroves,
the rain cleared off. About eleven o'clock we came in
sight of our mount, and saw about fifteen black figures
come trooping over it one after another, with spears

and waddies in their hands, and post themselves at the edge of the mangroves and wood, as if to oppose our progress.   Captain B. laid down his gun and went forward to parley with them ; held up a green branch, and made signs of peace.   As, however, they did not come out of the bush, he ordered us to advance, on which they slowly retired up the ridge.   We took off our hats, chattered, imitated their cries, and made grimaces to them, in order to show we were friends.   We counted seventeen of them.   Having a common knife in my pocket, I gave it to one of them next me, and shook hands with him.   They kept up a great talking, and seemed pretty good-humoured, but rather distrustful and afraid of us.   On coming to our camp, we found the two men with all the things safe in the boats, except the boat-stove and coppers, which the blacks had taken away.   It appears that soon after we left, two natives came down and scouted, and went away again ; and that about an hour before we returned twenty suddenly surrounded them with brandished spears.   Having got everything into the boat except the stove, which was both hot and heavy, they shoved off ; but the natives swam off and surrounded them, trying to take the boat sails and awning, but not proceeding to actual violence.   One laid hold of one of the men's muskets, which, though cocked and pointed against his breast, he still pulled at.   The men gave them bread—which, after examining, they threw away—and prudently humoured them, forbearing to fire; for, as they said, they were so many they feared to fire and only kill one. While we were seated on the mount, lunching, and hearing this story, the natives were clustered outside the mangroves at the foot of the ridge, talking and laughing; and one fellow sat on the branch of a tree watching our motions. Suddenly we heard the click of iron, as if they were moving our stove; and picking up our guns we rushed out, and saw it under a tree. Seeing us angry, they retired slowly; and pointing to the stove, we scolded them and shook our guns at them, and waved them away. This latter motion they understood; for taking their spears, and two or three of them picking up fishing-nets (very neatly made like shrimp-nets), they pointed higher up the water, and went away, as if to go fishing; and the doctor, in the benevolence of his heart, said he had no doubt they meant to bring us some fish. Everything was now perfectly tranquil and quiet for two or three hours; when, having dined, and the tide having risen high enough to enable the boat to float, we prepared to go away. Half in joke, I said I should like to see them again; and going to the front of the mount gave a loud cooey. To my surprise, it was instantly answered, and the chattering and jabbering began as thick as ever in the old place, scarcely 100 yards from us. We now took a tin canister and a bottle or

two as presents, and went to have a talk to them. They
retired up the ridge; but on our showing them the
presents, four or five gradually approached and took
them, and we became pretty friendly. We sat down
under the tree where we had slept, and five or six sat
opposite, and we all commenced talking and laughing.
We pointed to the ashes of our fire, and made signs we
had slept there last night; on which they nodded their
heads, as if they were fully aware of that. I then went
through motions of drinking fresh water, imitated a kangaroo and a corroberry dance—all which amused them,
and they seemed to understand, and pointed in several
directions; but we could not learn where the fresh
water was. Presently an old gray-headed man came
down, to whom they all pointed; and calling him towards us, we put a red cap on his head. He bent forward his head with great gravity while we adjusted it,
and as soon as it was on, he sat down by us with a very
dignified air, taking no more notice of his own party,
but talking in a low tone to the doctor, and pointing to
our dogs, of whom they seemed afraid. Six or eight
more fine young fellows, all armed, with white sticks
through their noses, now came hastily up out of the
mangrove bushes near our camp, where they had evidently been lying in ambush; and the others rose and
seemed inclined to press upon us. I accordingly, by
Captain B.'s wish, picked up an oyster-shell, and calling

their attention, fixed it on a tree; and making them all
stand back, fired with ball, and luckily smashed the
shell. At the sound of the gun, some in the rear half
threw themselves down, but those in front stood firm,
while I made signs that with this weapon I could kill
them all. As they were now increased to twenty-four
men, all great naked stalwart fellows, we thought it
prudent to leave them while we were friends. But first
I got a spear from a man who had had a knife given
him, gently laying hold of the spear, and pointing to the
knife repeatedly. He was rather reluctant to part with
his spear, and laughed and shook his head, as much as
to say, ' I'm afraid you're doing me,' but at last let me
have it. On taking leave they imitated our waving
hands, and we reiterated their cries; they did not follow
us down the hill, and after one or two farewell cooeys
we rowed off, and got on board our ship soon after
dark. On the 22d, the captain gave me a boat and a
party to go and visit some hills up the western arm.
This expedition was chiefly remarkable, first, for the
beauty of the view from a hill-top; secondly, for the
whole party being so entirely done up between 6 A.M.
and noon as hardly to be able to reach the boat. I was
obliged to lie down on a rocky beach two hours, and get
the boat sent after me. The rocky gullies, the heat
and closeness of the woods, and the want of all water,
make this country very toilsome to travel in. . . .

March 4. The natives came to us unarmed, and
quite peaceable. In the afternoon the ship came in
sight, and anchored some miles off, it being very shoal;
and we continued here till the 13th, completing the
water of the ship. During this time I made two excursions with Captain Blackwood, and we found a very rich
grassy forest-land everywhere along shore behind the
mangroves, stretching to the base of a lofty and continued chain of hills, which bound the view everywhere
from the coast at the distance of ten or fifteen miles.
There is no lack of fresh water; and the greenness of
the grass was delightful, after the living hay we had
seen to the southward.

March 13. Continued our course to northward surveying; and on the 14th, 15th, and 16th, I had another
interesting cruise in his boat with Captain Blackwood,
examining some deep bights, where we hoped for a
river. They were noble harbours at high water, but at
low tides unfortunately all dry sand or mud, the rise
and fall of the tide being twenty-five feet here. As
ports for anything but small craft accordingly they were
worthless; but the country around was not only very
picturesque, but exceedingly fertile, rank green grass
growing up to our middles, even on the tops of hills
300 feet high; palms and other trees of tropical kind
growing here and there in dense jungles, among woods
matted by huge rope-like creepers and climbers. The

general character of the wood, however, was open light timber. The rocks are all hard quartzose, gray wacke, porphyry, greenstone, &c.

March 19. Anchored under Cape Hillsborough, a lofty promontory 900 feet high, which I scaled and found a singular variety of rocks.

March 20. Away with Captain Blackwood exploring a bay; our pointer Don taken with a fit ashore, and nearly lost in the woods ; obliged to cut his ear open to bleed him. All the dogs done up from want of water and heat of weather and the incessant attacks of mosquitoes. A snake, six feet three inches long, killed.

March 21 *to* 25. Exploring the shores of Repulse Bay, and Whit-Sunday Passage up to Port Molle. A fine harbour between the islands and the main, where we stayed till the 29th. We had here a melancholy occurrence. A man named Dowling, coxswain of the pinnace, and one of the best men on board, had been ill two or three days of a low fever; his hammock was slung under the forecastle, and early on the morning of the 26th the man attending on him, having gone below to get a drink of water, came up again and found his hammock quite still, but presently after, on going to speak to him, it was empty. Search was made everywhere, but neither then nor since was he ever again seen. It is supposed he had seized the opportunity to get up and slip quietly out of the port with the intention of drown-

ing himself. A very strong tide was running, and
though the place was dragged, and search made along
shore, his body was never found.

March 30 *and* 31. Sailed along surveying, and anchored under Cape Upstart. This is our appointed
rendezvous to meet a vessel which was to sail from Sydney on the 20th of March, to bring us a supply of provisions and all our letters and papers up to that time.
We have now been four months without seeing a single
soul but the few black fellows we have met, or hearing
what the rest of the world is about. Cape Upstart is a
bold eminence, 2000 feet high, with flat land inshore
of it. It is rocky and barren; but in a gully is what
we prize above all things on this dry coast—some fresh
water. During these four months rain has not fallen
more than four times, perhaps thirty hours altogether.
A bright blue sky, with a fresh cool breeze from south-
east, continues day after day, so that we think it quite a
treat to get a wetting in a shower of rain. A mean between this climate and that of England would be perfect. Nevertheless, this is very healthy. We sleep
ashore on an open beach without a fire, and if we take a
coverlet, we take off our jackets and roll them up for a
pillow. Even in the thick of the mangrove swamps—
places which in Africa or the West Indies it would be
certain death to stay a night in an open boat—we experience no ill-effect, and our sick-list consists merely

of bruises or cuts, more especially among the men, cuts
from corals or rock oysters on the feet when landing in
a surf. The country too has had some variety in it:
rich grass, breast high and green, dark-green tropical
jungle, with palm-trees and great tangled creepers. The
greens are quite refreshing to the eye after the brown
trees, brown rocks, and brown grass about Sydney. As
to the *morale* of the ship, we are all as comfortable as
can be expected. Little disagreements, either in the
messes or between the authorities, occur occasionally, as
in all communities, more especially where individuals
are jammed together for months with little to divert their
attention from each other's failings. We are united
and pleasant in the gun-room, however, and I, as usual,
am on excellent terms with everybody, like a prudent
and well-behaved person. When we left Sandy Cape a
newspaper was established called the *Circumnavigator*,
in which were some excellent articles, and most capital
and humorous illustrations—really almost equal to
Cruikshank's—from the pen of Melville our artist. These
latter, however, have ceased, I am sorry to say. It appears fortnightly, and I dose the lieges with papers on
natural history, which, being full of hard names, are
generally praised and passed over. Of course you have
seen the comet. Aird and I saw it first, the night we
anchored at West Hill in the pinnace on March 3d. It
was then near the sun, and very magnificent, its tail

having a length of twenty degrees; but now it is gradually fading in lustre. It must be a new one. I have made a map of our course from Sandy Cape to Cape Upstart, as no maps you can get access to will give it you large enough. I shall enclose it if I can.

April 1, 2, 3, 4, 5. The rainy season has suddenly set in, commencing on the 3d; very heavy and continuous rain.

April 6. The pinnace, which had been left behind under Ince and Aird to trace up the coast from Port Molle, came in. At one place the natives had attacked them with stones, and after much forbearance, Ince fired with shot, and wounded one.

April 7. I went away with the seining party, under Mr. Weeks the boatswain, to a bight four or five miles up the bay. The day was fine, but hot. When we landed, leaving them to haul the seine, I walked over a sandy point, and along the beach, shooting sea-birds. I heard a native cooey in some mangroves two or three miles ahead; but catching sight of some very large and strange birds, as big as flamingoes, I went on, taking the precaution to load with ball. In about two miles I came on a creek which looked muddy, though shallow, and as the mangroves were thick, I took a long shot at the birds, and returned. I had scarcely walked twenty yards back, when a shout arose among the mangroves a quarter of a mile beyond me, on the other side of the creek, apparently from a considerable body of natives, who had evidently been lying in wait for me. I returned the shout, waved my hat, and seeing our boat coming round the point two miles off, walked slowly to it, and told them the natives were out. I then returned to our landing place to lunch, understanding they (the men with the boat) were coming too. Without my knowledge, however, they went on, and on coming to head of the bight, the natives came to them. Four unarmed came first, and a man named Hardy went and gave them a couple of fish; more then came out, and seemed inclined to surround him, on which he turned back. The moment he showed his back they threw the fish at him, and were rushing on him, when another man named Grub fired with shot, and dropped the foremost, on which they all ran, the wounded man scuttling through the shallow water on his hands and knees. Fifty or sixty then showed themselves on the beach, and our boat returned. I am very sorry I was not with them, as I could either have saved the effusion of blood, or should have been convinced of its necessity. With proper management, namely, confident and lively looks and actions, laughter and grimaces, I think I could have managed them. I go no more alone, however; indeed Downes, my botanical assistant, would have been with me well armed, but was on the sick-list.

April 8. Rain.

April 9. Rain; but our provision ship is arrived with papers and letters.   Huzza !

H.M.S. Fly, inside Barrier Reef,
Off Cape Weymouth, N.E. Coast of Australia,
July 2, 1843.

My dear A.,—Before you receive this I trust you will have had a large packet from me, containing an abstract of my journal . . . I shall therefore just go on as if I were sure you had had it.   Since we left Cape Upstart, which we did the first week in May, we have been utterly alone, not having seen any vessels barring two wrecks on the reefs here.   This morning, however, a sail was reported, and a large barque is now seen outside the reefs, looking for an entrance; we have sent off our whale-boat to pilot her in and bring her to an anchor near us, and if she comes in safely, I hope to have letters ready to send by her.   I will give you at once a succinct statement of our proceedings.

May 2. After taking leave of our friends on board the William, which was going to sail from Cape Up-start the next day, Captain B., Ince, Harvey, and myself went away in the cutter with eight men and six days' provisions for a cruise.   We had a pretty rough but tolerably interesting one, and returned to the ship on May 6.   The principal incident was a very friendly interview with the natives at Cape Cleveland, who even brought their ladies to have a look at us.

On May 10, the pinnace not being yet ready for sea, Captain B., Ince, and I set off in the whaler to find a river reported on the west side of Upstart Bay. We went up a long mangrove inlet, missed the river, but had another very friendly interview with a tribe of natives : seventeen men and three women sat by our fire all night. Having only one day's provisions with us, we were obliged to return.

On May 13, Captain Blackwood offered me a boat to explore the real river, and I set off, taking Ince, Evans, and Melville with me. We had a narrow squeak crossing the bar at the mouth of the river, on which was a very heavy surf, but found a fine river inside.* After proceeding six or eight miles, however, it became too shoal to go farther except by much labour. A tribe of natives now came down, with whom we went a mile or two on foot; in returning we nearly had a scuffle with them, but found the other friendly tribe at the boat. One of these friendly people had particularly attached himself to me the other night when they slept with us, sitting down at my head and keeping off the mosquitoes; and now when he saw me he came and embraced me, making a cooing purring noise, and showing me to the others as *his* friend. We returned to the

* This must have been the Burdekin.

ship on the 15th, and on the 17th finally left Cape
Upstart.

On the 18th, we anchored at Goold Island, where we
remained till the 1st of June. The natives here were
all in canoes and at first very friendly, coming off to
the ship, where they were very well treated, and had
clothes, biscuit, &c. given them. I was here given the
command of two boat expeditions. In the first, Shadwell and Porcher accompanied me, and we explored and
surveyed an inlet cutting off Mount Hinchinbrook from
the mainland, and were away five days, three of which
were very wet; but by help of the rain awning, although
in a small open boat, we managed pretty comfortably.
In the second, Ince and Pym went with me to explore
an opening like a river (three other rivers having been
found, but of small extent). In going in we were again
nearly embroiled with a party of natives, and they began to throw stones while we were bothered in shallow
water. I was just aiming at one impudent rascal with a
musket when Ince fired with shot and peppered one
fellow, on which they desisted; although, after picking
out the shot and washing his breast and shoulders, the
fellow did not seem to care for it. We found nothing
but jungle and narrow mangrove creeks; so on the third
day we came out and went to a small island near, to
wait for the ship, which was to pick us up next day.
Here in the morning six black fellows came across in

canoes, and after being disappointed in not receiving
presents, they retired into the bush and came forward
again, armed one with a spear and the others with
stones, to attack us. We absolutely stared at their
impudence, and laughed at them, as we were eight to
six; to avoid the necessity of killing one of them, however, we got under weigh and they followed, shouting
at us, along the shore. When a good distance off we
fired alongside of them both with shot and ball, just to
show them we could hurt them if we liked, on which
they scampered off. On the vessel picking us up, we
found that even our friends at Goold Island had attacked a boat the night before with spears and stones,
without any provocation, in the most treacherous manner, and knocked down Mr. Weeks, the boatswain, with
a stone, only desisting when one of their own number
was wounded and another killed by the fire of our men.

On June 4, I went with Captain B. and Bell to Endeavour River, where Cook repaired his ship. I was
glad to have seen the place of which I had so often
read when a boy of ten years old, and which is a classic
spot in the history of marine discovery. On the 6th,
I slept with Evans on the top of Lizard Island Peak,
1200 feet high; nothing remarkable except the lovely
view both of sunset and moonset, and sun rising over
the mirror-like sea.

We now took up the survey of the reefs some

o

twenty miles from land. They stretch in one continuous line of breakers, with only narrow openings here and there, at that distance along shore; inside is smooth water about twenty fathoms deep, full of coral patches and sand-banks, outside of them the sea is unfathomable, and the whole swell of the South Pacific Ocean, urged by the continual force of the trade-winds, bursts on their outer edge. Went ashore with Captain B. at Cape Melville, and saw a party of seven natives, with whom we had a very friendly interview, but in shoving off they threw a spear after us from the top of a rock, which fell in the water near us. I confess I was rather savage, and should have shot one fellow had not Captain B. said, ' Don't hit him ; fire to the left ;' which I did, and cut a twig off just over his right shoulder, on which he scampered. The spear was a very ugly weapon indeed, eight feet long, pointed and barbed with three barbs.

June 24. The Bramble being ordered in to the mainland at Cape Direction, I jumped up from breakfast, and, only stopping for a pair of boots, got into a boat and on board of her, to have a cruise with Yule, her commander. The next day we anchored under Cape Direction. Two natives came alongside in a canoe, to whom we gave presents, and they were very friendly. Went ashore in two boats, taking five men with muskets to protect us, and two in each boat with muskets to

mind the boats. I took Yule's double-barrelled gun—
a nasty cheap Flemish piece, which by some care I got
into as good order as possible. Dr. M'Clatchie had a
pistol besides, while Yule, Pollard, and Sweatman were
unarmed, carrying chronometers, sextants, &c. We
went up a hill about 500 feet high, on which observations were to be taken; and after taking them observed
five more natives, besides the two who had been on board,
below about the boats. We therefore descended. I
cannot give you all the particulars of what followed
without entering into great detail, but in a rocky broken
place I had directed the men in advance to halt for the
stragglers, and had gone back to look for and protect
Yule, who was lame, when I saw a black fellow brand-
ishing his spear through the bushes at a man named
Bailey, the last of our party. Shouting to him, ' Come
on !' I cocked and presented; but, alas, both my barrels missed fire. The native flung his spear and pierced
our man, and disappeared in a gully of jungle before
any muskets could be brought up to bear on him, my
piece again missing fire as he was running away. As
soon as he was out of sight, on my trying it again, *it
went off*. A distant shot at the villain on his emerging
from the other side the gully was all the satisfaction we
got, as we could see no more of them, and the country
was too broken and sheltered for us to attempt scouring
it with any success. The man Bailey lingered for three

days, when he died, the spear having gone through the back-bone and part of the lungs. Bitterly did I regret not having my own gun, when I should either have prevented, or at least have revenged his fall. All forbearance on our part is now blown to the winds, and in any future interview the first suspicious movement will bring down instant punishment upon them. We shall, however, have little to do with the shore; probably nothing at all, as at the end of this month we expect to knock off work and bear away for Port Essington, whence we shall go either to Timor or Java for refreshments, and then to Swan River and Hobarton. We shall probably be in the latter about October. The vessel has now got safely inside, and we are both running to an anchorage under the reefs. All hands on board are pretty well and comfortable, with the exception of some cases of rheumatism and lumbago among the officers, and some slight appearance of scurvy among the men. I have been constituted caterer of our mess, after Shadwell and the doctor had tried in vain to give satisfaction in that important office; and you would laugh to see me giving orders about dinner, taking lists of plate, glass, table-cloths, &c., settling matters with the steward, and bullying the cook, as in duty bound. I have just sent our compliments to the captain of the bark (whose name is Regia, *i.e.* the bark's name) and the gun-room officers of H.M.S. Fly, &c., happy to see

him to dinner, &c. &c. It being Sunday, the captain
and two of the midshipmen dine with us, as is usual on
board a man-of-war; and our dinner will be preserved
soup, do. salmon, roast leg of fresh pork, boiled shoulder
of corned do., preserved jugged hare, peas-pudding, rice
and preserved fruit tart, macaroni, cheese—Madeira,
Tinto, and Cape hock, with bottled porter, being the
fluids; so, you see, we don't starve, though we have
been seven months at sea. Indeed, I can't say that I
feel particularly anxious to go into port at all. With
plenty of books and lovely weather, and occasional
cruises in boats, I pass my days happily, and don't see
any use in the rest of the world, except to bring us
provisions and fresh water now and then and a few
letters from home. I think I ought to have been a
sailor, as I verily believe I should have made a good one
and enjoyed the service much. If I had about 2000*l*.
per annum, I should just buy a good yacht, and live in
her, cruising about the world here and there, according
to the seasons, or as one place or another came into
my fancy. You remember the work-box you fitted up
for me; it is very useful. I have learnt to sew with a
thimble, and have actually darned two rents in a pair
of plaid trousers, with some neatness but more strength.
I intended to have sent you another chart of our course,
but have not had time to construct one yet. And now
farewell again. I need hardly tell you to give my kind-

est love to my mother, who I hope and trust continues
as well as when I last heard from you. . . . Give my
love to my kind friends in the ancient and dirty city of
Brummagem; and with best love to yourself, believe
me, my dear A., your ever affectionate brother,

J. B. JUKES.

H.M.S. Fly, off North Coast of Australia,
Aug. 27, 1843.

MY DEAR A.,—From July 3 to August 11 we were
employed surveying the reefs, without anything particular happening, except our touching at Murray's Island
and having an interview with more than a hundred natives there.   They have cocoa-nut trees there, though
there are none in Australia, and they are altogether a
different race from the Australians, living in cottages and
cultivating yams, &c.   We got a few things from them,
and they were pretty friendly, though very noisy and
clamorous.   It was on that island that the two surviving
boys from the wreck of the Charles Eaton were found.
The Charles Eaton was the vessel in which young Clare
of Wolverhampton and others were lost and eaten by the
natives.   On August 14, we landed at Booby Island,
which is a bare little rock.   We visited the post-office,
but found nothing except a book containing notices of
ships having passed.   There were several such messages
as, 'Mrs. G. begs her kind love to the next lady passing

through, hopes she has had a comfortable passage, as
Mrs. G. has had, and will be exceedingly happy to see
any lady passing through at her father's house, No. 15
——street, ——.' One lady begged her 'kind *remembrance* to the next lady passing through.' From Booby
Island we entered on a calm tranquil sea and light
winds, forming a sensible contrast to the rolling seas
and strong breezes we had had on the reefs. It felt,
however, very hot. We landed at Port Essington on
the 19th, stayed a week, and left yesterday morning.
The land there is low, level, and for the most part dry
and barren, except in the swamps. Captain M'Arthur,
one lieutenant, and thirty-five marines have now been
stationed there alone for five years, living in wooden
houses and reed huts, with only two gardens, a few
buffalo and goats, and an annual supply of hard biscuit, salt pork, and rum from Sydney. The climate is
mostly fine and healthy, but during November, December, January, and February continued rain and thermometer at 90°. Up to this year they had had no sick-
ness; but last rainy season having been very heavy and
protracted, they are now all laid up with intermittent
fever and ague, and look like yellow skeletons. The
natives around them are all friendly and useful, as an
exemplary punishment was inflicted on them formerly
at Raffles Bay, close by, for their unprovoked hostility.
There is some sporting round, but only fancy a five

years' existence in such a hole! And yet our assistant-surgeon, Dr. Sibbald, has exchanged with Dr. Whipple, who was assistant-surgeon there, and now comes with us. As they were not able to give us sufficient fresh beef and vegetables, we are now going to Coepang, and thence, I believe, either to Flores, Sumbawa, or Lombock for refreshments; and as a letter-bag will be sent from one of those places, I am, you see, preparing for it. We have now delicious weather; the sea perfectly smooth, the ship quite steady, and yet slipping along six or seven knots an hour before a light easterly breeze, which tempers a little the heat of the sun, bringing it down to 82° in the shade, while in the sun it is over 90°; and this is one of the coolest months in the year here.

So much for our proceedings. We are all well in health, but longing for a little change, and for some intercourse with civilised society, and for news, from which we have now been absent more than nine months. We expect to arrive at Swan River in October, and at Hobarton in November. Of course I am reduced to mere vague guesses as to your proceedings at home since October last, and can only hope all things are going on well.... I do not know when I shall feel cold again—probably not for three or four years, when passing round Cape Horn on our way home. How I am to manage your winters, I do not know, for I find this

hot weather quite congenial to me now. I have visions
of taking lodgings on my return in some old farmhouse
a mile or two in the country, where perhaps I might
keep a pony, smoke my pipe, read books, and commence
old bachelor, passing a few months now and then on
the Continent geologising, and eking out my income
by occasional literary employment. Some such existence seems likely to be my ultimate destination. It is
the first time, however, I ever formed a plan for the
future, and therefore the reality is likely to be exactly
the opposite of all this, and I may be sent on a geological survey of Kamtschatka, or Guatemala, or the
borders of the Caspian Sea. At all events, I shall have
had quite enough of Australia. I hope you are having
weather now at home somewhat resembling our climate
here. If so, how beautiful England must be! No country equals it, even in this quality. The palm, the
plantain, and the other far-famed trees of the tropics,
are certainly surpassingly lovely, especially by blue clear
waters, in perpetual sunshine, with perhaps noble hills
rising behind them; but their beauty is more like that
of a picture to be looked at from a distance and at rest.
A nearer view, when walking among them, most probably shows a dry, dusty, parched up soil, or a dank
swamp; woods either scanty and devoid of shade, or
matted and tangled with parasitical plants, so as to be
impenetrable. No green lanes, no shady walks, above

all, no meadows, no lovely corners or smiling fields, such as England abounds in, and whose beauty is more easily felt than described. Still, even to relish English beauties, you should have been abroad. Then, how should you like tropical insects ?—to say nothing of myriads of ants in all trees, that either bite, sting, or smell most abominably ; of cockroaches, of all sizes up to three inches long, swarming in all houses, beds, rooms, boxes, cupboards, bread-baskets, running over your dinner-table, or dashing into your plates and tumblers ; only think of the incessant buzz and fever-bringing bites of sand-flies and mosquitoes, obliging you when on shore to sleep in a muslin hive. To make up for all these discomforts, however, and ever consoling us for the heat of the day, how glorious is a tropical night, or a morning before sunrise !—a cool clear sky, with a gentle breeze fanning your temples, and a delicious dew falling around you ; the stars sparkling like gems through the liquid air, and the moonbeams glancing and flickering on the rippling water ; and this not occasionally only, but night after night for months together.

Well, I mean to leave a little space to announce our arrival at Timor, so I will now conclude. Give my mother my kindest love and remembrance ; I sincerely hope she remains as well as when I last heard from you ; but this I need hardly say. Remember me very kindly to —— and ——. I hope you have had many

pleasant botanical rambles, and perhaps that you have
even taken a hammer to Sedgley or Dudley. I wish I
could join you in one to-morrow, and skip back again.
. . . Tell Browne and everybody else to write to me as
fully and as often as they can; never mind how much
chit-chat or small-talk—it is all valuable here.

Sept. 18, 1843. We arrived at Timor on August 31,
and spent four days there. It is a very beautiful island,
and I enjoyed myself much, more especially as the
geology was very interesting. It was a much more favourable specimen of tropical scenery than I had seen
before, and the costumes and manners of the Malays
were striking and picturesque. We got there a sufficient supply of fowls, beef, vegetables, and fruits, so,
without visiting any of the other islands, we sailed at
once for Swan River. We expected to have reached
that place on the 25th, but have unexpectedly had foul
winds, and we are not yet more than a third of the way.
There was no vessel at Coepang, so I could not send
my letter. Indeed, they had had no communication
from Batavia (their headquarters) for seven months,
and the only news they had was from English vessels
from China. Swan River has but little communication
with Europe, none at all with Van Dieman's Land or
New South Wales, but a good deal with India; so I
suppose my letter will reach you from that port, *viâ*
Bombay.

Sept. 24. We are just now in chase of a schooner (when within a week's sail from Swan River) which appears to be going to India. If so, we shall send a letter-bag to her, and I shall enclose this. . . .

Half-way House near Woraloo-road to Toodgoy,
Swan River, Oct. 11, 1843.

MY DEAR BROWNE,—I wrote to A. not long ago by a Dutch schooner that we met at sea, but as there is a vessel leaving this shortly for Bombay, I mean to give you a screed hastily. We arrived at Swan River on September 30, but ever since then I have been engaged with one round of invitations at Freemantle and Perth. Yesterday I set off on horseback for a fortnight's independent cruise, and am now in a small half-built inn with two settlers (one drunk), where we have taken shelter for the night. I overtook them on the road, driving some very fine horses into the interior. I have been very much pleased with what I have seen of this settlement: abundance of excellent land, fine corn-fields, good gardens, and a superabundance of stock. What they complain of is the want of money, having no market to dispose of their surplus in. They are greatly in want of emigrants, especially small or large capitalists. It is not a place to make a fortune, but an admirable one to get an independence, and the

people are all clean and respectable-looking. We have
been revelling in fresh beef and vegetables after our
ten months' cruise. We shall sail from this about the
26th for Hobarton, where I hope to get letters from you
all. The climate here is delicious; fine fresh air, thermometer between 50° and 60°, with refreshing showers,
and everything green and beautiful. The woods are
enamelled with most lovely flowers, and although a
broad belt of rocky hills, forty miles wide, occurs where
I am now, running parallel to the coast about twenty-
five miles from it, the country on each side of this is
in general equally excellent for agricultural purposes
and picturesque in aspect. Of course some privation
of accustomed comforts and luxuries must be submitted
to by the young settler, more especially that of society,
as there are only 5000 people in the whole district of
500 miles long by 300 wide. One great drawback is
the abundance of fleas. A new house, or one uninhabited for only a couple of weeks, absolutely swarms
with them, so that white trousers are entirely mottled
on coming out, and even the greatest care fails to
eradicate them utterly, and they are *so sharp*. I am
on my way to Mr. Drummond's, a great botanist, with
whom is now Gilbert collecting birds for Gould; they
are going an excursion of 200 miles into the bush, and
I shall accompany them part of the way, and then go to
York, where is to be a fair and races on the 20th.

The place does not seem to be very interesting geo-logically, but I shall have a pleasant jaunt, as I have such a nice old horse—a regular old bushranger, nearly thoroughbred, a great runner of races, and comes to me when I call him. The surveyor-general lent me some saddle-bags, and I take a feed of corn, and some bread-and-cheese and brandy, and a head-stall and tether rope, with a cloak, to make us both comfortable should we miss our way in the bush. . . . They have just brought in three kangaroos, one of which seems re-markable, and I am getting a boy to skin it for me. I dined off kangaroo steaks, and very capital they are. How get on all the good folks and my old *pals* in Wol-verhampton—B. and C. and R., and the Geological So-ciety? Pray tell them all I should be delighted to hear from them. I fully intended to write a long letter had I had the room to myself, but under the circumstances I cannot. . . . Give my best love to my mother and sister, and all other friends, and believe me ever yours most truly,

J. BEETE JUKES.

SECOND CRUISE. 1843–46.

Abstract of the Voyage.

THE Bramble remained a fortnight after us at Swan River to complete her refitting and endeavour to re- cover some deserters; and on sailing, her commander, Lieutenant Yule, undertook to carry round some specie for the Colonial Government to King George's Sound. In entering this port and beating through the narrow entrance to the inner harbour, she unfortunately grounded on a rock, where she remained four days. It blew hard during part of the time, and after suffering considerable damage to her bottom, she beat over it. She arrived safely, however, in Hobarton, though very leaky; was hove down at Port Arthur and partially repaired, but was obliged to be taken on to the patent slip on her arrival at Sydney. On January 8, 1844, we

left Hobarton again, and anchored at Sydney on the
13th.   As Captain Blackwood had determined on the
erection of a beacon on Raine's Islet to mark the entrance of a good passage through the reefs, he, in
obedience to his orders from the Admiralty, applied to
the Colonial Government for assistance.   The colony,
however, was in such a very depressed condition at that
period, that the only assistance that could be afforded
was the loan of twenty picked convicts, chiefly masons
and quarry men, and of a small revenue cutter, the
Prince George.   This, together with the repairing of
the Bramble, the purchasing and selecting and stowing
away of the large quantity of materials, tools, implements, wooden houses, &c., and the increased quantity
of provisions we required, detained us in Sydney till
March 27.   We then sailed, and after touching at Port
Stephens we rendezvoused at Sandy Cape, where we
completed our water from an abundant supply immediately behind the beach, about seven miles within the
point of the Cape.   We then passed through the Capricorn Group, and the Percy and Northumberland Islands,
to Cape Upstart, making some additions to our previous
surveys by the way.

At Cape Upstart we again filled our tanks and
water-casks ; and after experiencing a good deal of
blowing weather along the north-east coast, we commenced landing the stores on Raine's Islet on May 27.

As soon as the party was landed and the houses and
tents put up, a quarry was opened at the east end of
the island, near the spot selected for the erection of the
beacon. The stone was a coral-rock—an agglutinated
mass of grains and fragments of corals and shells. It
worked easily into square blocks, and promised to be
sufficiently durable. The lime was procured by burning the large shells of the tridacna and hippopus, which
were to be got in abundance from the reef at low water.
One or two wells were sunk in the island; but no fresh
water was procured, although in one of the wells, at a
depth of sixteen feet, the water was only brackish, and
could be used to slack the lime, though very unpalatable
to the taste. As there was no anchorage near the island,
the Fly had to lie about twelve miles off to the south-
west, behind the reefs of the Barrier; and the Bramble,
the Prince George, the Midge, and the boats were
employed in running backwards and forwards with
provisions, stores, wood, and water. Wood had to be
sought for on some of the islands near the main, as
large quantities were used in burning the lime; and
water was procured from Sir Charles Hardy's Islands,
where small dams had been erected at the end of the
valleys, to catch all that trickled down them. Beams
of wood required in building were procured from the
wreck of the Martha Ridgeway, which was lying on the
reefs some twenty-five miles to the southward of the

P

Fly's anchorage. By the middle of September the party on Raine's Islet, which was under the charge of Lieutenant (now Commander) Ince, had completed the beacon. This was erected after the design of Mr. Moore, the carpenter of the Fly. It is a circular stone tower forty feet high and thirty feet in diameter at the base, where the walls are five feet thick. Internally it was divided into three stories, each of which was partially floored and made accessible by a ladder. It was roofed at top by a dome-shaped frame of wood, covered by painted canvas. Its summit was thus raised seventy feet above low-water mark. A large tank taken from the Martha Ridgeway was placed at the side, into which a series of spouts were led from the roof, so that it would shortly be filled with rain-water. Cocoa-nuts, supplied by Mr. Macleay of Sydney from his hothouse, pumpkins, maize, and other plants were set in a garden, and had begun to grow and flourish when the Fly left. During the latter part of this time —from Aug. 14 to Sept. 25—the Bramble was employed surveying Endeavour Strait; and a good part of the ground between Endeavour Strait and Raine's Islet opening, was likewise surveyed by the Fly, the Prince George, and the boats. An abundant supply of water was discovered at Cape York, which was then and afterwards very useful to us.

On Sept. 21 the Fly left Endeavour Strait, touched

at Port Essington on the 27th, remaining five or six days, and then sailed for Sourabaya in the island of Java. She arrived there on October 19, and was shortly joined by the Bramble and Prince George. The latter was dispatched to Singapore to take there some ship-wrecked people, whom we had picked up among the reefs, and to carry and receive letters and despatches. She returned to Sourabaya on January 4.

On January 14, 1845, the Fly and the Bramble sailed from Sourabaya, to return to Torres Strait with the north-west monsoon, leaving the Prince George to follow as soon as she had completed her refitting. We had very heavy weather at first. After a vain attempt to pass through the Strait of Lombock, owing to the strength of the current setting to the northward, we got through those of Alass, and again reached Port Essington on January 27. On February 4, we sailed for Endeavour Strait, and anchored in the entrance of it on the 10th. From this time to April 19 we were engaged in surveying the central and north-eastern parts of Torres Strait, and succeeded in laying down an excellent track for shipping—round the northern extremity of the Great Barrier Reef, through the inner reefs and islands, to the entrance of Endeavour Strait.

On April 19, the Bramble was sent to try and make her way along the eastern coast of Australia to Sydney. This, being against the trade wind, had only been twice

attempted before — once in the Zenobia, by Captain
Lihon, who succeeded with great difficulty; and once
this very season, by a schooner called the Heroine,
Captain M'Kenzie, the success of whose attempt we
were as yet unaware of. Lieutenant Yule not only
succeeded, but added 120 miles to the survey of the
Barrier Reef, continuing it from Lizard Island to the
southward as far as lat. 16° 40′.

In the mean time the Fly and Prince George went
to explore a part of the coast of New Guinea, to the
north and east of Torres Strait. Having surveyed as
much of this as was possible, owing to the difficult and
dangerous nature of the navigation and other untoward
circumstances, we were obliged on June 2 to close our
labours from want of provisions, and proceed to Port
Essington for a supply. We arrived at Port Essington
on June 12, and found there seventy people, who had
been wrecked in coming up from Sydney to Torres
Strait. These formed the principal part of the crews
and passengers of two large merchant-vessels, the Hyderabad and the Coringa packet. They had reached
Port Essington in their boats. As the small military
post of Port Essington could not support this population
long, and no other vessel was likely to put in for some
months, we were under the necessity of carrying them
up to Singapore. The Prince George, after being partially refitted, was sent round to Sydney, taking a few

of the wrecked people who wished to return there; and on June 18 we sailed with the remainder in the Fly, and arrived at Singapore on July 5.

Sir Thomas Cochrane, the admiral of the station, being now at Malacca, we went up there to communicate with him, and then returned to Singapore.

We left that place on August 3, and beat down against the trade wind through the Strait of Banca to Anjer in the Strait of Sunda, where we remained August 19, and sailed on the 20th. The south-east trade carried us to south lat. 30° 15', east long. 89° 05', and strong south-west winds took us into Bass's Strait, and thence to Sydney, where we anchored on September 25. We here found orders awaiting us, directing the Fly to come home, but Lieutenant Yule in the Bramble to continue the survey with, if possible, a colonial vessel as her consort, under the command of Mr., now Lieutenant, Aird. The Fly, having rubbed her bottom against some of the coral reefs, was hove down and examined, and a small schooner called the Castlereagh was purchased for the Government, and fitted out as the Bramble's consort. These arrangements detained us till December 19, when we left Sydney on our voyage home.

We passed through Bass's Straits, and touched at Port Philip on December 19, and remained till January 11, 1845. On January 15, we anchored in Holdfast

Bay, South Australia, and remained till the 22d. On
February 10, after a stormy passage, with strong contrary winds in rounding Cape Leeuwin, we anchored in
Gage's Roads, Swan River, and remained a week. On
February 19, we sailed, having Mr. Hutt, the late governor of Western Australia, as a passenger, and arrived
in Simon's Bay, Cape of Good Hope, on April 6. On
April 15, we again sailed, touched at St. Helena on April
30, and anchored at Spithead on June 19, after an
absence from England of four years and eleven weeks.

H.M.S. Fly, Hobarton, Nov. 26, 1843.

MY DEAR BROWNE,—I am sorry to hear so many of
our old friends are dispersed. If we return by the Cape,
I shall most certainly look out for Russ, but I think we
are more likely to go round the Horn. I am here full
of geological work, and return five nights a week with
a great bag full of fossils from the adjoining hills. On
Tuesday I go up Mount Wellington, a lofty bluff nearly
4000 feet high, just at the back of the town. You
cannot conceive more beautiful scenery than some of
this country. I went up the Derwent about thirty miles
the other day, with three carriages of ladies and gentlemen, and a young lady and myself on horseback, to see
a fossil tree at a Mr. Barker's, and danced afterwards
in a white shooting-coat till one o'clock in the morning.

Sir E. Wilmot, the new governor here, gives a fancy-dress ball on Friday next, to which I go as a lieutenant of the navy, borrowing a uniform from one of my mess-mates. I expect a capital lark. I am going to Launceston on December 3, and return on the 9th; and on the 15th I am going to escort Miss M. and another young lady over the hills down to Great Swan Port. We all go on horseback—a three days' journey, and part of the way rather rough work than otherwise. I shall not stay with Mr. and Mrs. Meredith more than four days, however, as the captain wishes us all to dine together on Christmas-day, and we sail soon after that for Sydney. I believe we shall make a short trip from Sydney, this time going across to New Caledonia and back in a month, and then setting off for the reefs again. We want to keep a magnetic term in New Caledonia. Captain Blackwood hoped this morning that we should have done our work this time two years; so that by the time we have replied to a few more letters each, you may be looking forward to my reappearance among you. Where I shall pitch my camp, or what I shall do, I have no very distinct foresight. . . . I shall be too old, and have been too much accustomed to wander at my own wild will, to follow your good example or submit to the restraints of a regularly civilised life; for I fear I should begin to sigh for the freedom of the open seas or trackless woods of a wild country. But, as ' Billy' says, ' We know what we are, but we don't know what we shall be.' . . .

Instead of subscribing to any newspaper, since the *Staffordshire Examiner* is knocked up, you can send me now and then one of the old *Chronicle*, or any other local paper which may contain anything interesting, though even the advertisements remind me of home, and thus are interesting. Andrews' roller-blinds, Dakin's tea-warehouse, or Sidney's haberdashery, are all interesting in my eyes at the antipodes, and have a home feeling about them, so that while reading them I forget the thousands of miles between.—With best love to my mother and A., believe me your affectionate brother,

J. B. JUKES.

_____

H.M.S. Fly, Sydney, New South Wales,
Jan. 20, 1844.

MY DEAR A.,—We arrived here this morning, and soon after anchoring I received your long and welcome letter. I am delighted to find you so happy. Long may your pleasure continue uncontaminated by care! . . . Your tour must have been a delightful one, but I wish you could have gone to Minchinhampton, as the view there is magnificent—one of the finest I ever saw. . . . Do not be alarmed at anything Mrs. M. said of my talking of settling out here. It must have been merely in joke, as a settler's life would never suit me;

unless, indeed, I got an appointment as geological surveyor, or something which admitted of frequent change
of place. Now, moreover, when one great cloud of distress and ruin seems to have settled down upon these
colonies, settling would be madness. . . .

I will now give you a *résumé* of our doings, or rather
mine, since my last letter to Alfred. On December 1,
the Tasmanians had a grand day, called Tasman's-day,
that being the date of the discovery of the island by
Tasman in sixteen-hundred-and-something. There was
a great regatta, and in the evening a fancy-dress ball
at Government House. The latter was really a very
splendid affair, the dresses being generally much more
magnificent and correct than I should have imagined.

On the 3d, I set off on horseback for Great Swan
Port, which I reached in two days; pretty hard riding
over such a country—one hundred miles of rocks and
bush. I only stayed five days, and then rode back with
C. Meredith. They had made some improvements, but
had been again visited by floods. . . . A great many
bushrangers were out, and they had attacked houses,
&c. In one instance only were they successfully opposed, when one man was killed and two wounded; but
we left Mrs. M. under some apprehension of a visit
from them in C.'s absence. I travelled without arms,
as the safest method, intending, if I fell in with them,
to make a bow, and say, ' Gentlemen, help yourselves.'

After returning to Hobart, I did not again quit it, except on a boat-excursion to Port Arthur. . . .

On January 8, we left Hobart, and went to Port Arthur to join our consort, the Bramble, who only came to Hobart from Swan River on December 10, a month after us, having struck on a rock on St. George's Sound, and been nearly wrecked, lost her rudder, keel, &c. She had been repaired *pro tem.* at Port Arthur. Here we passed four or five very gay days. As I give you accounts of our rough life when on a cruise, it is but fair I should give you a hint as to the gaieties we have to encounter when in port. I am, indeed, tired of them, and long to be off once more out of reach of the distractions of society. We shall, however, remain here a month, while the Bramble undergoes a thorough repair in dock. We shall then sail for New Caledonia, and remain there six weeks; then explore some detached reefs, and return to Sydney in June or July. Then off again, run along the Barrier Reefs and south coast of New Guinea, and perhaps refit in India or Ceylon; so far is a probable programme of the next twelve or eighteen months.

H.M.S. Fly, Sydney, March 4, 1844.

My DEAR AUNT,—. . . After I wrote to you last, I made a flying visit on horseback down to Great Swan Port, and spent a few days with Mr. and Mrs. Meredith.

I heard from them lately, and they seem (at least Mrs. M.) to be troubled with the fear of bushrangers. Indeed, all persons now in the country in Van Dieman's Land must live as in a state of siege, not knowing the hour of the day or night when they may be attacked by an armed gang of ruffians. On Dec. 10, the Bramble came into Hobarton in a terrible state. She had stayed behind us at Swan River, and before leaving was requested by the Government there to carry round some money to King George's Sound. In going into that place she struck on a rock in a gale of wind, and remained four days on it. She lost her rudder, keel, and great part of her bottom planking ; and if she had not been extra strong, must have gone to pieces. She is now on the patent slip undergoing repair. This great detention has caused an alteration in our plans ; and our former design of spending this month in New Caledonia is given up. Our programme now is to sail hence in a fortnight, taking a small cutter, called the Prince George, and thirty convicts, stone-cutters and masons ; to anchor near Raine's Islet on the Barrier Reef, and land a party to build a great tower for a beacon, as the mark for our best passage through the reefs. While the Fly is anchored there, the Bramble and cutter will proceed to explore the S. coast of New Guinea, during the months of June, July, and August. We then rendezvous at Port Essington, and in all probability shall go to Sin-

gapore to refit. Then sail again through Torres Straits
with the last of the westerly monsoon—a passage never
before attempted—and shall then, I conclude, refit in
Sydney for the last time at the end of the year 1845. . .
I have made acquaintance with George Bennett, the
naturalist, who is a surgeon here. I never met a more
delightful person. His house is quite a treat, so full of
books, pictures, prints, and all that is interesting. I
go with him and Gilbert (Gould's agent and collector)
to look at some gigantic birds' bones from New Zealand
this evening. . . .

March 9. I was much interested with a sight of the
birds' bones, which are in very perfect preservation. I
believe they are going home to Owen. I dispatched a
large case of plants and bird-skins the other day to
Gray and Brown of the British Museum, and have four
or five large cases of rocks and fossils in the commissariat stores here. I anticipate an interesting, though
arduous, cruise on the New Guinea coast. The only
drawback is the difficulty and danger of landing. The
natives are armed with bows and arrows, and are numerous and hostile. We shall, however, go well prepared for them.

Pray tell everybody how I should prize letters; the
merest chit-chat is valuable at the antipodes. . . Your
ever affectionate nephew,

J. B. JUKES.

We have not a particularly pleasant prospect this cruise ; for, in addition to the hard work, we have the probability of having to live on ship's allowance only.

H.M.S. Fly, Sydney, March 20, 1844.

MY DEAR MOTHER,—I will not let another letter go home without assuring you how glad I was to see your handwriting once again, and to hear how well and comfortable you are. When I say *are*, I can hardly help thinking of the lapse of time since I am sure that such was the case, but am unwilling to believe that anything has occurred to render it necessary to say *were*. In the mean time I am as happy as I can be away from you all, constantly in pleasant society, with frequent change of scene and place, and that continued gentle excitement which seems necessary to my disposition. I fear great part of 1846 will have passed away before we see again the shores of Old England, when I hope to pass some quiet resting-space with you, at all events before undertaking any other expedition, even if I do not altogether set up my rest in some country corner, and commence old bachelor for life. I hope the many letters I have sent home will have arrived safely. . . .

If you and A. are happy, I feel that I have little else in the world that can seriously annoy me. And now, my dear mother, for another six months' farewell.

May you retain and increase your health and strength, and have many happy evenings to listen to my adventures on my return, is, I need hardly say, the sincere wish and hope of your ever affectionate son,

J. BEETE JUKES.

At anchor off Double Point, north of Rockingham Bay, N.E. Coast of Australia, May 9, 1844.

MY DEAR A.,—According to promise, I begin a letter for you to be taken on by the first ship we may happen to fallen in with. We left Port Stephens April 7, and had a stormy passage up to Sandy Cape, where we remained several days waiting for our consorts, the Bramble and Prince George. We then passed through the Capricorn Group and Percy Islands with more dirty weather; and after filling up some blanks in our survey about Whit-Sunday Passage, we anchored for four days at Cape Upstart to complete our water. We sailed thence the day before yesterday, and are now lying at anchor, in consequence of the thick stormy weather not allowing us to see very clearly. It is not nearly so fine a season as last year was, but I hope it will improve shortly when we get to our work. Nothing remarkable has occurred since we left Port Stephens, except the death of a man on board the Bramble from typhus fever,

of which we have had one or two cases. As I did not
get any more letters before we left Port Stephens, I am
in statu quo as to intelligence from you, and suppose I
must remain so until our return from the coast of New
Guinea. . . . . A ship of this class, however pleasant a
habitation in fine weather, in heat and wet is very disagreeable. Boxed up in small cabins, with all the skylights down to exclude the rain, which, nevertheless,
will find its way in by some crevice or other, you alter-
nately rush on deck to escape suffocation, and dive down
below to avoid drowning. Then there is no place to
dry clothes; and if you take off wet ones, you are obliged to keep them reeking and steaming alongside of
you. Every place becomes at last damp, dirty, dingy,
steamy, sticky, muggy, and miserable. There! there is
a drop of water will fall on to my paper, so I shall shut
it up till finer weather.

May 26, off Sir Charles Hardy's Islands. Our bad
weather continued for several days, during part of which,
however, we lay at Lizard Island, where we got some
pretty good quail-shooting. We then ran down inside
the reefs, and I spent one night on board the wreck of
the Martha Ridgeway on the edge of reef in the midst of
a foaming surf miles in extent. We were two days on
board, getting out her tanks and other matters that will
be useful to us on Raine's Islet. We then ran down to
these islands, where we have been five days rating chro-

nometers and keeping a magnetic term day. This morning at daylight, as we were going to sail, I went ashore
for the last bathe, and while swimming in the water,
saw the ship hoist her ensign, and signal the Midge to
get under weigh. I immediately concluded some vessel
must be in sight, and on getting on board, found two
ships had been seen from the mast-head. They are
now not more than four miles to leeward of us ; and as
it is highly probable we shall communicate with them,
I am rattling off this letter to be ready for them. The
weather has become finer, and if we succeed quickly in
getting the enormous quantity of lumber out of the ship
on to Raine's Islet, I still hope we may run down to
New Guinea, though we now look upon that as doubtful, for this cruise. Our lumber consists of twenty convicts, and an immense quantity of plank timber, houses,
barrows, jumpers, pickaxes, spades, and all manner of
building, digging, and blasting materials. All these
will have to be carried in boats through a heavy sea six
miles, as the ship cannot anchor nearer Raine's Islet
than that. I expect this will take us two or three
weeks, as twenty of our crew are to be landed, and provisions and water for forty men for three months are
no trifling weight. The bread alone will be 2500 lbs.
Whatever may be our ultimate course, however, we are
pretty certain of being at Singapore in October or November. The ships have bore up, hang them ! and as

we cannot afford to run leeward after them, this opportunity is missed, so I must keep this scrawl for the next.
Farewell for the present.

May 29. Off Raine's Islet. I went in the Bramble
this morning to Raine's Islet, and while wading out on
the reef in search of corals and shells, two ships were
reported hove to near the Fly, eight miles down to leeward. I got on board just now wet through, and so
tired I can hardly keep my eyes open. The vessels are
from Hobarton, but bring us no news. They sail at
daylight. We have terribly hard, heavy, disagreeable
work landing things on this little desolate islet, but
hope finally to accomplish our object in building a
beacon pretty well.

And now, my dear A., with kindest love to my mother, yourself, Alfred, and all friends, you must accept
this hasty scrawl from your very tired but very affectionate brother,

J. BEETE JUKES.

Barrier Reefs, Raine's Islet, long. 144°, *lat.* 11° 45′,
June 21 (*shortest day*), 1844.

MY DEAR AUNT,—. . . . We left Sydney March 27.
Our cruise so far has not been at all interesting. We
have had much bad weather; and as our principal object
was to get here to get rid of our convicts and cargo of
building-materials, we did not stop to visit any new

Q

places or make explorations. On May 27, we came in sight of Raine's Islet, on which we were to erect our beacon, and had several days' very hard and disagreeable work landing our cargo. Raine's Islet is a little low sandy coral island about 800 yards long by 400 broad, surrounded by a coral reef, which on one side stretches off half a mile. It is at the entrance of one of the widest and clearest openings through the Great Barrier Reef, which is, along this part, out of sight of land. No part of Raine's Islet is more than twenty feet above high-water mark; our object, therefore, is to build a tower about fifty feet high and twenty or thirty in diameter, as an object to be seen by ships steering for the passage, and prevent their mistaking any other opening for this, or running beyond it. I am happy to say that my report on the stone has been borne out. Good building stone is procured in abundance, wooden houses and tents are erected, lime is made, the foundation stone laid, and the first course of masonry nearly complete. Lieutenant Ince, with a party of forty men, occupies the island and superintends the work. . . . On May 6, as I was wading about on the coral reef, having visited the island for a few days, a brig called the Archibald Campbell came through and anchored near the Fly. I should have mentioned that there is no anchorage near the island except for very small vessels, as the coral reef plunges down into 160 fathoms at once, and the nearest anchorage ground is eight miles off, under the lee of the Barrier Reef.

A cutter called the Prince George, lent to us by the Colonial Government, is the packet and medium of communication between the ship and the island. On June 13, however, the Prince George was ordered into the mainland to get a cargo of wood, and I went with her. As we approached the coast, we saw a vessel aground upon one of the many reefs that lie scattered about inside the Barrier, and on running down to her, found the unfortunate Archibald Campbell, which had passed us only seven days before, hard and fast. By the culpable negligence of the mate, she had been run upon this reef the afternoon of the day on which she left the Fly. The captain of her was now gone away in the long-boat to try and communicate with the Fly, having been previously driven back when attempting to reach her in a whale-boat. There were on board of her Mr. Smithers, the owner, and his wife—a young lady to whom he had only been married two days before they left Sydney—Captain Richmond of the Bombay army, and Mr. M'Arthur, besides the remainder of the crew. They were, of course, very glad to see us, as they were now sure of being taken from the wreck before their provisions and water were consumed. Their brig was now hard and fast, upright, though bilged, and nearly dry at low water—there was therefore no immediate danger;

so after concerting measures for taking them off on our
return, we went on to execute our instructions. That
afternoon we anchored under a little island called Sunday Island, in Margaret Bay, where, finding wood enough
for our purpose, we determined to give up going to the
mainland, where we should find natives, and be obliged
to shoot them or get speared ourselves. We accordingly in two days laid in our cargo, and on the 17th
returned to the wreck. In the mean while the captain
of her had reached the Fly, and the Bramble and Midge
had come down and taken the passengers away, leaving
the captain and crew to get up the cargo, consisting principally of brandy and champagne. On the 19th, we got
back to the Fly, and found all well. We shall have now
accordingly a considerable addition to our mess until
another ship comes through, by which the castaways can
proceed to India. . . . . Everything having been put in
train for completing the beacon, we expect to be ready
to set out in the Bramble and Prince George on July
2 for the south coast of New Guinea, leaving the ship
to support the island party, and survey the tract between this and Torres Straits. We shall probably be
away two months, returning here in September, and at
the latter end of that month shall run away for Port
Essington, Bally, and Singapore. . . . .

June 22. Two ships are reported in sight, and immediately everything is bustle and confusion. I seize

the opportunity of closing my letter in case a boat should go to them, and they not anchor.—Believe me, my dear aunt, your ever affectionate nephew,

J. B. JUKES.

H.M.S. Fly, off Sir O. Hardy's Islands,
July 27, 1844.

MY DEAR A.,—I received your letter of February 28 yesterday, forwarded from Sydney by a vessel that was likely to fall in with us; but I have not heard or seen anything of the one of December 12, or of Browne's of January 1. The first I am not surprised at, as you say you sent it in a parcel. Now if the parcel came by post, the postage will be some pounds; if it did not go by post, are you aware there is a penalty of 40*l.* for every letter so enclosed? . . . Pray always trust your letters to her Majesty's Post-office. Another thing is, you mention my letter ' off Cape Weymouth.' Now I might have written a dozen letters off Cape Weymouth a month or two apart from each other. Please always to give me the date. It cost me some puzzling and much examination of my letter - book to comprehend that, up to February 28, 1844, you had not received any letter of mine of a later date than that of April 10 and July 3, 1843. Now pray, ma'am, consider yourself scolded. . . . At length I have got to the matter of your letter. I am very glad to hear that my mother still

continued well and in good spirits. . . . I shall be very
glad, I am sure, to come and 'complete your felicity,'
and I think I shall be entitled to a few weeks' holiday
when I return, before setting to work in London, as I
suppose I shall have to do if I bring home a good collection. I am, however, still in absolute uncertainty
as to what is to be done with the results of my labours
—whether I am to do what I like with them, or whether they are to go to the British Museum, or where.
We sadly want a scientific department in the Government to take the management of these things. . . .
How you would envy the corals which we get here! The
most magnificent masses of branched corals are now
drying on the poop; but, alas, they are too bulky and
too brittle to get home, so I content myself with small
pieces. I should like much to have seen your great
fossil tree. I am very glad you are intimate with Mr.
and Mrs. Lister. What I saw of them I liked much;
and I think he is one of the few parsons before whom
you may carry your soul in her ordinary dress, without
huddling her immediately into her Sunday clothes.
So poor R. has arrived at Cape Town. Well, I suppose the Table Mountain is as good a place to die under
as any other, and that, I suppose, is what he intends.
Your missourium of M. Kock I saw in 1841. It is all
true, except that the bones are not properly put together. It really was not larger than an ordinary

elephant, and is nothing more or less than a mastodon. When I wrote to you last, I think I told you we were going to New Guinea. Well, that is all knocked on the head, and we have been doing nothing but cruising between Raine's Islet and Sir C. Hardy's Islands, supplying the building party with wood and water. It is now proposed that the Bramble shall be detached to survey Endeavour Straits. If that take place, I shall go in her; if not, we shall linger on here doing nothing till the middle of September, when we shall be compelled to go to Singapore before the wind changes into the westerly monsoon. This period of inaction is becoming most oppressive to me. The sole event to look forward to is the passing of a ship, while it is only at rare intervals I can do anything in the natural-history way. I would welcome any danger or any hardship even that would break the monotony. I am accordingly as dull, heavy, stupid, and spiritless as it is possible—like an old boat drawn up on a mud-bank instead of careering over the bright waters. By the bye, we got stuck on a coral bank the other day, and received two or three bumps that have probably knocked some copper off; if so, we may go to Bombay to be docked, and I shall thus get within a month's post of you. I do not wish it, however, as it would utterly ruin our chance of seeing New Guinea—of which I have still some hopes—on our return from Singapore.

July 28. There is this moment a vessel declared to be in sight from the mast-head. I shall therefore hastily finish this, in order to be ready for her in case she should not anchor near us. . . .

> *H.M.S. Fly, off south side of Jindana, or*
> *Sandalwood Island, Oct.* 11, 1844.

MY DEAR BROWNE,—. . . . When I closed and sent my last letter, which you will receive viâ St. Helena or Boston, I was in the Bramble, being heartily sick of lying at anchor in the Fly near Raine's Islet. We were occupied from August 20 to September 19 in surveying Endeavour Straits, during which operation I was frequently on shore, away in boats, &c., but nothing remarkable occurred. We had peaceable interviews with one or two parties of natives, and were enabled to make a few notes respecting the natural history of them, and of some birds, &c. On September 19, the Fly and Prince George joined us in the straits, having completed their great tower on Raine's Islet. I then rejoined my ship, and the next day we sailed for Port Essington, leaving our two tenders to finish the survey of Endeavour Straits. Ever since then we have had most lovely weather certainly, but continued light airs and almost calms. We did not reach Port Essington till the 27th, and found them much in the same state as before, but the fever had left them. Three commanding officers of

H.M. brig Royalist had successively died there after we left them; but the mortality had now ceased, and they were daily expecting to be relieved, as the new detachment had arrived in Hobart. . . . Yesterday we passed along forty miles of very picturesque and beautiful coast, the south-east end of Sandalwood Island. Mountain ranges in the interior, five or six thousand feet high, were dimly clothed with clouds, while the coasts were formed of steep jagged hills of one or two thousand feet, furrowed in every direction by gullies and ravines, or penetrated by narrow winding valleys full of groves of cocoa-nuts, with other stately palms nodding their heads from the crags above. It was so different from any Australian scenery, that its natural beauties were heightened in our eyes. And then we saw a waterfall, flinging itself over one of the cliffs, not having seen running water before, heaven knows how long. We are now slowly floating (all hands sighing for a breeze, if it came in a gale) over a tranquil sea, under a roasting sun, which will be absolutely vertical either to-day or to-morrow at noon, hoping to reach the Straits of Lombock, and get through them to Sourabaya in a few days. There we shall probably remain till the Bramble comes up, so that I shall be able to see something of the north-east end of Java, and then we go on through Banca Straits to Singapore. . . . Remember me most kindly to all our mutual friends. I shall rejoice to be amongst you all once more, if even for a short time, should my restless spirit again impel me abroad, to be ' blown with restless violence about this pendent orb.'

October 29, Benaahoude Staack, Sourabaya. Another change. We arrived here October 19, and after spending two or three days, suddenly Captain B. declared his intention of not going on to Singapore, but of sending on the Prince George for our letters. I then determined to live ashore to arrange my shells, &c.; and Shadwell, Evans, and Bell have joined me in hiring a house, where I am now residing, with a Malay boy for a servant, and a woman to cook. I am sitting in the front saloon after my morning bath and coffee, smoking, eating bananas, and writing this. There are a few English residents here, and many Dutch, some of whom speak English; but the common tongue of all is Malay, which I am learning as fast as I can. Should you like to see a specimen of the language?—' This morning I got up before daylight'—

Pagi	ini	sahaya	bangun	lagi	glass.
Morning	this	I	rose	still	dark.

I shall not receive the letters from Singapore before December. Next week, Captain B., Evans, and I start on a tour up the country for a fortnight or three weeks. With best love to my mother and A., and kind regards to your family, believe me, as ever, your affectionate brother,   J. B. JUKES.

Fly Hall,' Benaahoude Staack, Sourabaya,
Island of Java, Nov. 1, 1844.

MY DEAR AUNT,—This address will puzzle you as much as it would have astonished me a fortnight ago. After leaving the reefs and touching at Port Essington, we sailed for Singapore, intending to call here on our way. We arrived here on October 19; and four days after, Captain B. suddenly altered his mind to remain here, and send the Prince George to Singapore for letters, &c. We accordingly shall not get them till December. Three of my messmates have therefore joined me in taking a house ashore; and here I am, with two Malay servants who can't speak a word of English, while I can talk very little Malay at present. There are several English residents here—merchants, sugar-planters, &c.; and of the Dutch officers, naval and military, several speak English, and the rest French, but Malay is the common tongue of all. We have, besides our *salon-à-manger* and bedrooms, a cook-house, bathing-house, &c., a long outhouse in which all my natural-history collections are stored, and I am going to set to work at arranging and describing them. I have already fallen quite into the tropical way of life. Rise before day-light; walk into the bathing-house, where stands a large porous earthenware jar of water, cooled by evaporation; take a calabash, and pour it over the head and dash it about you for half-an-hour; then a cup of coffee, a banana, and a cigar; then to work in sleeping-trousers and dressing-gown till eleven o'clock; then breakfast—coffee, rice, fish, fruit, curry, beer, or wine. At twelve or one go to bed for an hour; at five bathe again, dress for dinner, and ride, drive, walk, or pay visits in the cool of the evening; in bed at eleven. The bed is spacious, hung round with mosquito-curtains, and a mattress covered by a sheet. The houses are all flagged—no carpets, but cane mats. The Javanese are a mild, gentle, and intelligent race of people. I am going to make a trip into the interior with Captain B. We are to have horses supplied by Government, or rather ordered by them but supplied by rajahs, with an escort before and behind, more for honour than use. There is no money but copper doits, of which there are about 2000 to a pound sterling; so that if you pay ready-money, you must have a coolie to carry your purse, which would be a large bag of matting. This alteration of purpose in our arrangements obliges me to write all my letters in great haste in order to go on by the Prince George, besides having to superintend a hundred different things in fitting up the house. I am obliged to give you only hasty sketches therefore, and have no time to describe what we have seen. We finished our great tower in September, but I spent a month before that in the Bramble surveying Endeavour Straits. We then touched at Port Essington for four days, where we left our twenty convicts in charge of Ince, our third lieutenant, and also M'Gillivray, Lord Derby's naturalist. We then sailed slowly, with very light winds, by Timor, along the S. coast of Sandalwood Island, a noble-looking country 100 miles by 60, then through the Straits of Alass, between Sumbawa and Lombock, and along Baly and Java coast here. All these places are covered with magnificent volcanic cones, many 10,000 feet high, or as large as Ætna. I did not see Tomboro, as it was 100 miles distant from the W. side of Sumbawa, and the weather was hazy in that direction. We expect to leave this in January, return to Port Essington, and then along the S. coast of New Guinea, through Torres Straits, and down the coral sea between Australia and New Caledonia, and reach Sydney in May or June. This will be our most adventurous cruise in every respect, trying unknown ground, and unknown winds and weather. . . . It seems to be my fate in all my travels to be confined to islands. I have only set foot on one continent yet—Africa, at the Cape of Good Hope —and now I have no chance of landing on Asia. . . .

H.M.S. Fly, Sourabaya, Dec. 29, 1844.

MY DEAR A.,—Here we are still, and as yet have received no letters. We, however, every day expect the Prince George from Singapore, when I hope to do so.

The first part of our sojourn in Java was very pleasant. On November 9, Captain B., Evans, and I, with a young man of Sourabaya, set off for the interior of the country. We travelled first for about seventy miles through Passaronan and Probolingo along the coast to the eastward, driving in a carriage with four horses, or ponies rather. From Probolingo we struck off to the southward on pony-back, with thirteen coolies carrying our baggage on foot, and about ten attendants under a native chief on horseback. We now got into the most beautiful, rich, picturesque, and magnificent country I ever saw or could have imagined—rich plains covered with all kinds of tropical productions, watered in every direction by clear rocky brooks, surrounded by mountains, either in single cones or serrated ranges, from 4,000 to 11,000 feet in height; abundance of game whenever we chose to stop and shoot—jungle-fowl, peacocks, deer, wild pigs, tigers; all the native chiefs apprised of our approach wherever we came; a house and most luxurious entertainments ready for us as we dismounted; escorts, coolies, the whole country apparently, at our command; every trouble taken out of our hands, and nothing for us to do but to ride, shoot, eat, drink, and admire the glorious scenery that everywhere surrounded us. We crossed one great range of mountains by a path that led us through the extinct crater of a volcano, five miles across and 7000 feet above the sea,

and in the centre of which was a small cone and crater
still in action, though when we looked down into it, it
was only blowing out steam, with a roar as of a thousand blast-furnaces. Take a scene on the slope of these
mountains, as they dip into the plain of Malang.

Scene. An open mountain valley, full of coffee-plan-
tations, with small scattered villages, into which opens
a deep mountain glen crowded with the rankest luxu-
riance of tropical vegetation, groups of tree-ferns and
great broad-leaved plants, so as to arch over and frequently hide altogether the full brook that comes flashing and roaring down over the rocks in a succession of
rapids, varied by waterfalls; the road, narrow, steep,
and slippery, as it winds down the sides of the glen,
expands into a broad green lane, with an exquisite carpet
of turf as it opens on the more level lands.   On this
road, mounted on a spirited and excellent little horse,
rides an English gentleman (myself) dressed in shooting-coat, straw hat, and leather gaiters, preceded by
two native spearmen in blue and red on good little
horses. The gentleman, being better mounted, has
outridden his companions, and is now endeavouring to
pass the spearmen, whose duty it is to be fifty yards
ahead, to encounter tigers and clear the road. The
spearmen, enjoying the fun, are trying to keep their
place, which the goodness of the gentleman's horse renders difficult, when, turning a corner of the road, a

gateway is seen with unsaddled horses and native servants grouped about it. The spearmen shout to pull
up, a native gamelang, or band of music, from a building inside the gates strikes up the tune of ' Rajah da-
tang' ('The prince is coming'), the gentleman dashes
through the gate into the midst of the servitors, who
all, headed by their chief, bow and hold his stirrup, and
having dismounted, stalks into the building to the clang
of the gamelang. One takes his coat, another brings
water and a towel, another arranges a chair at a table
covered with delicious fruit and all kinds of sweetmeats
and confectionery, while a young chief comes forward
with a tray on which are tea and coffee. Presently his
companions and the rest of the escort drop in ; and
when all have washed, had tea, fruit, and sweetmeats,
the table is cleared, and on comes a substantial meal of
numberless dishes of poultry, beef, curries, stews, &c.,
with rice and potatoes. Two of the escort produce bottles of beer from their game-bags, of which a stock is
carried by coolies from place to place; and after an
hour's rest and refreshment, fresh horses are saddled, a
fresh escort provided, and the party proceeds, the gamelang striking up with a great clangour ' Rajah jalan'
(' The prince is going').

This is at one of the little mountain villages; but
that same afternoon we rode into Pakis, on the plain of
Malang, with an escort of thirty mountain horsemen

and spearmen in three lines, headed by a widono and two bakkels, whose silver trappings and gold-covered krisses and gay blue-and-red garments quite put to shame our soiled and wayworn English shooting-dresses. Here we could not stir without a guard of honour of six spearmen, and every one uncovered and sat down (their position of respect) as we walked along. We all agreed we never were such great men before and never should be again. Here, too, we found some beautiful ruins of ancient Hindoo temples, dating from A.D. 700 or thereabouts, and a most delicious country. Being 1200 feet above the sea and rain every afternoon, the temperature was delightful; and, as far as mere climate and situation goes, I never saw a country I should more like to live in. No tropical country is so cool and pleasant and healthy as the interior of Java. We saw several English and Europeans of all ages, from children to old gentlemen of seventy, with ruddy cheeks and every sign of health. Indeed, in the interior fever seems to be scarcely known. The people, too, are all so gentle, docile, and good-tempered, and appear so comfortable and well off, that, were it English instead of Dutch, it would, I am sure, be a favourite residence of our gentry. At present, however, there is no society, and it was only by favour of the Resident of Sourabaya that we were allowed to penetrate so far, the country being entirely locked up even to Dutchmen, except the few civil

R

and military officers that are posted at different places. We returned to Sourabaya on 27th of November, after a delightful excursion of eighteen days; but on coming down into the swamps and flats of the coast we began to feel the climate. I made another trip of a day across to Madura to see the Sultan of Bankālang, where we stayed a night, but it was not very interesting. By December 14, I had packed up and shipped off all my this year's collections; and after catching cold in a boat one night, and being up all the next at a ball at Grissek, I had a slight attack of fever, which confined me on board for a week, but I am now quite stout again. This is the west monsoon, and now, after a fine morning, we invariably have heavy rain from noon till night, when it clears off again. Sourabaya itself is a shocking stupid place, and we are all anxious to be off again, and I believe we shall sail to-morrow week. If you get Raffles' *History of Java,* you will see drawings of some of the temples we have seen. His account of the country is quite correct to the present day. . . .

H.M.S. *Fly, Sourabaya, Java, Jan.* 10, 1845.

MY DEAR AUNT,—. . . . Our stay in Java has been very interesting, more especially the first part. The coast is flat and swampy, and the towns dull and stupid, but the interior of the country is the most beautiful and magnificent I ever beheld—every variety of surface and outline in the ground, from level plains to mountain ranges and volcanic cones, from 4,000 to 11,000 feet in height, enclosing here and there rich plains or valleys, thirty or forty miles wide, and 1,200 feet above the sea; the wildest scenery, knife-edge ridges, and deep narrow winding glens and ravines, all clothed in the rankest profusion with every variety of tropical vegetation, from the cocoa-nut and banana of the plains to the groves of bamboos rising in great clusters to 150 feet in height, their stems arching and interlacing overhead on the mountain sides; interminable forests of teak, and trees like the banian, and numerous others up to the bare mountain tops, with grassy ridges, on which grow here and there tufts or belts of lofty pine-like casuarinas. In the glens are tree-ferns and innumerable great broad-leaved plants, generally quite concealing the deep rocky bed of the brook, the roar of whose waters is heard on the heights as it rushes down into the plains. Then the people: from chiefs to coolies, gentle, docile, good-humoured, humble, but ready to enter into and enjoy a joke; the higher ranks polished, courteous, well-bred gentlemen, and all watching to anticipate your wishes and supply your wants. Wherever you stop, a table laid under a large bamboo shed, generally in some picturesque spot, with sweet cakes and all kinds of confectionery, mangos, pine-apples, tea and coffee, with handsome services of glass, china, and plate, purple finger-glasses, fine napkins, and *all for nothing;* with troops of servitors, and the chief of the district doing the honours; fresh horses every six or seven miles; a mounted escort to carry our guns, &c.; spearmen ahead to clear the way, gold and silver trappings to the horses (the chiefs sometimes sport gold stirrups), and gay clothing to the escort. It is moreover a sealed place to all the world. Nothing but our being a man-of-war, and the Resident of Sourabaya being a liberal-minded man, procured us permission to stir a yard, as it is positively against orders for any foreigner, or even a Dutchman without a pass, to go into the interior.

We looked down into an active crater, that of the Bromo, about 7000 feet high; but it was only a small hole in the middle of the old and now extinct crater, which is five miles across, and is called the Sandy Sea. We spent a week at various heights from five to seven thousand feet, and got strawberries, raspberries, and green peas. Then down to the plain of Malang, where, as far as extreme loveliness of country and freshness, pleasantness, and healthiness of air and climate go, I could be perfectly content to pass my life. Fresh green lanes, with short springy turf, brawling brooks, patches of forest, undulating ground, with pretty hills among rice-fields and banana-trees; and every now and then,

in some forest glade, you come upon an old Hindoo
temple in ruins, with elegant architecture, elaborately
carved tracery and ornaments, and graceful, grotesque,
or really beautiful sculpture and statuary. Around
these temples piles of brick, buried in the depths of
the forest, fragments of walls, or the road running
along some ancient brick causeway, all attest the former
grandeur of the ancient Hindoo empires and kingdoms,
all but the tradition of which has passed away. Some
of the most beautiful ruins date from A.D. 700, when
Java was far more populous, wealthy, and powerful
than England. On our return to Sourabaya, I one day
caught cold and had a slight fever, but soon recovered·
We also visited the Sultan of Bankālang in Madura;
but as I am now arrived at the end of my sheet, I must
defer farther detail. . . .

H.M.S. Fly, at anchor off Darnley Island,
Torres Straits, April 16, 1845.

MY DEAR A.,—As the Bramble is now going to separate from us, and beat back to Sydney along the east
coast of Australia, while we, after a short trip to New
Guinea, shall return by the west coast and Swan River,
I am going to send you a brief epistle by her, as it will
probably reach you sooner than by any other opportunity. I left a letter in Sourabaya on January 10

last to go by a Dutch ship, since when we have seen no
homeward-bound vessels.   We left Sourabaya on January 12 in a gale of wind, which, after one or two narrow escapes from getting ashore, took us into Alass
Straits, and I landed on the Island of Lombock. Thence
we went to Port Essington, where we found the last detachment had gone home, and a new one arrived. There
we embarked our convicts again and Lieutenant Ince
and M'Gillivray, who had stayed behind, and proceeded
to Endeavour Straits, which we reached on February
11.   Since then we have been surveying all the
south-east portion of Torres Straits, and the reefs north
of Murray Island.   We have landed on most of the
islands; and on most of the larger ones have had very
friendly dealings with the natives, although they are of
that savage race who murdered the crew of the Charles
Eaton, including Mr. Clare of Wolverhampton, and only
spared a boy and a little child.   No longer ago than
yesterday morning I was seated—with one man only—
among a group of these people at the hut of 'Duppa'
(the man who took care of the child, young D'Oyley),
quite out of sight of the ship, and on the most friendly
terms with them all. I bought, for an empty bottle and
two cigars, from one of the black young ladies a new
dress just finished.   It consists of a plaited band that
goes round the waist, and a deep thick fringe of strips
of plantain leaves, scraped very thin and smooth. When

put on, it forms a petticoat or kilt from the waist to the middle of the thigh ; this, with shell necklace and tassels of red berries, and some seeds strung through the lobe of the ear, with perhaps a streak of white or red ochre on the cheeks or forehead, forms the gala dress of these sable beauties. The only thing the men wear that can be called clothing is oddly enough a wig, formed of a kind of matting with the hair twisted in and curled, and very well and artificially made it is too ; they generally have also an armlet on left arm to prevent the string of their bows from cutting them, and sometimes a headdress of feathers, or a large piece of mother-of-pearl hanging on the breast like a gorget. I bought a wig for a knife and a red-worsted night-cap, and a splendid bow, upwards of seven feet long, and a sheaf of arrows, for two knives. Altogether we have been greatly interested and amused with these people, and like them much ; and were it not that they have an unfortunate predilection for collecting skulls, no fault could be found with them. They have excellent huts, large canoes, and cocoa-nuts, plantains, and yams in abundance, and in all the arts of life are far above the Australians. The fondness for skulls is common over all the islands from Borneo to New Guinea. It proceeds from an idea that of all the skulls they can collect during life-time the owners shall be their servants and followers in the next world. Of course to have a white man as a slave would be a great honou*. Their language is clear, harmonious, and expressive : their very words for ' yes' and 'no' show this. ' Yes' is ' wŏw'—how positive and decisive!—only pronounce it with a nod of the head, wŏw ! Lōlă, on the contrary, is ' no'—one of the most gentle negatives I know, nothing short or snappish about it, and yet sufficiently clear and firm. They don't shake the head with it, but holding up the right hand, shake it by turning it half round and back again two or three times, ' lōlă, lōlă.' We have several hundred words of their language, and they were quite delighted to teach us, and to hear us use them. When I accosted a stranger once with the words, ' Narrow dallee ?' (' What's the name of this ?') pointing to a shell, I thought he would have gone wild, such a shout of joy and surprise he set up, and immediately embraced me, and danced round me. When at Darnley Island last, I went by myself—having a pistol in my pocket—across the island, and came to a village where were several people I had not seen before. They were delighted at seeing me, offered me yams, sat down round me—men, women, and children—brought me cocoa-nut, water to drink; and when I got up to go away, used all their endeavours to get me to stay with them, pointed to a hut, to the cocoa-nut- and plantain-trees, made signs of sleeping there, and lastly brought a fine tall young lady for a wife for me; but when they found even the last argument fail, and I was determined on going back, as the sun was going down, one or two escorted me back to within sight of the ship, and then took an affectionate leave of me, not asking even for a bit of tobacco. By the bye, they grow tobacco, and smoke it in great bamboo pipes in a peculiar fashion of their own ; and whence they derived the custom and the plant puzzles my philosophy. For a few knives I could get beautiful tortoiseshell enough to set you up in combs all your life, but have not much room for it, and fear it may cost more making up than you could buy them for. However, I will bring a little. We have all been in good health with the exception of some cases of dysentery and fever brought from Sourabaya, the former being fatal in two cases, one in each vessel. We are victualled till the end of June, by which time we must get to some port for supplies. I expect that in a day or two we shall stretch over to New Guinea, near Cape Rodney, and then run along the southern coast of that great country for a week or two; but by this day month, at the latest, we must take our final leave of Torres Straits, and shall probably run either to Swan River or to Adelaide, and thence to Sydney. . . . . On our arrival at Sydney, it is highly probable we shall receive orders for home, in which case we shall, after two or three months, refit and turn our head once more towards old England. . . . . I have lost all interest in public news, and England will be a

foreign country when I return; and were it not for my
mother and yourself and my other relations, I should
feel little wish to see it again, so strange and remote
does it now seem to me. You however will, I hope,
find me, if not altogether unchanged, at least, as ever,
your loving brother,

J. B. JUKES.

H.M.S. Fly, off south coast of Timor,
June 21, 1845.

MY DEAR AUNT,—Since I wrote to you last from
Java, we have had a very eventful and interesting cruise,
but it has involved the delaying of all news from you
or any one else to an indefinite period. . . . We left
Java in a gale of wind on January 12, and had one or
two narrow escapes from getting ashore about Lombock.
At length we got to an anchor in the Straits of Alass,
and went ashore, having a pleasant walk to a village of
Bhugis, with whom I found I could talk Malay with
tolerable fluency. We then proceeded to Port Essington, and after taking our officers and convict-party on
board again, ran to Torres Straits and commenced surveying. We were engaged in this during the months
of February, March, and April, laying down reefs and
islands and the passages between them. . . . We had
considerable intercourse with the natives of Murray and
Darnley Islands, and found them by no means to de-

serve the bad character they have acquired. Both bodily and mentally they are far above the Australian natives, and totally different from them, having sprung either from New Guinea, or, as I think more likely, from the islands to the eastward, as Solomon Islands and New Caledonia. They build very good houses, in the shape of a huge beehive, fifteen feet high and fifteen or eighteen in diameter. They have well-fenced gardens, in which they grow yams, plantains, sugar-cane, ginger, and tobacco, and the islands abound in cocoa-nuts and other fruits. They are very quick, intelligent, and bold, have a good easily-learnt language, are well armed with bows and arrows of great size and power, and are altogether a superior people. They are eagerly desirous of barter, exchanging tortoiseshell, &c., for axes and knives. Their women are well made, but not handsome, are well treated by them, and though perfectly frank and fond of our society, would by no means admit of any undue familiarity on the part of our people. Moreover, they wear clothes— an unprecedented thing in this part of the world: a thick petticoat or kilt of plantain-leaves scraped smooth, one of which garments, as well as arms, utensils, &c., I have brought away with me. Having picked up a little of their language, I became a great favourite with them, used to go among them unarmed, and on approaching a village was always received by men, women, and children with shouts of ' Dookés, Dookés !' or ' Dookessa !'

to which they transmogrified my name. I taught the
young ladies to walk arm-in-arm with me, which puzzled them very much indeed at first; and one day when
a number of strangers had landed from Tood and Damood (islands to the westward), on approaching their
encampment, two old friends, Miss Moggore and Miss
Bood, ran to take me by the arm. I shall never forget
an old lady stranger's astonishment at seeing me walk
up thus accompanied. Continued exclamations of 'Wai,
wan, wan, wan!' &c. flowed from her lips as long as
we continued within hearing. On one of our visits to
Darnley Island the two sides of the island were at war
with each other, and I was lucky enough to be spectator
of the battle which ensued. It was a fine sight while
it lasted, but the only execution was a few scratches
from arrow-glances; in short, it was more noise than
work. At last, on April 25, we saw the low land of
New Guinea about Bristow Island, and during May we
explored about one hundred miles of its coast to the
north and east. This was by far the most difficult and
dangerous part of our cruise. The land is very low,
and the sea off it is so shoal that we frequently could
not approach within sight of it. A great mud-bank,
from five to twelve miles in width, runs along the whole
coast. We were accordingly obliged to work in boats;
but even for these there was no shelter, as the trade-
wind blows right on shore, and causes a huge rolling

sea over all the shoals. We found the land to be in-
tersected in every direction with innumerable creeks
and arms of *fresh water*, varying in width from a few
yards to two, three, and five miles, so that a large river
or assemblage of rivers must empty themselves here.
No high land was seen, except at our extreme N.E.
point, where a hill was visible in N., and beyond it
once or twice, glimpses of very lofty mountains at a
great distance were obtained. We had much bad weather and heavy rain, and to add to our anxiety, we
missed two of our boats, containing three junior officers
and twelve men. They had been sent inshore to trace
the coast, with a week's provision, but we could not
find them again. They left us on May 2, and from
the 10th of that month to June 2 we were constantly
employed looking for them. I accompanied two of
these expeditions, in one of which Captain B. and I
penetrated fifteen miles up a fresh-water creek in a single
boat, till we came in sight of one of the huge Papuan
houses, when, after a slight skirmish with the inhabitants, we returned. The whole of these creeks swarm
with natives living in large villages, going in bodies of
several hundred, well armed with bows and arrows,
large wooden swords, stone-headed clubs ; and having
never seen white men before, were daring, fierce, and
treacherous. [A description is here given of the weapons
and houses of the natives.] Every two or three miles

hereabouts was a village containing from one to four or five of these great houses, besides a number of others much smaller and of a somewhat different construction. The land about here was a wet clay, very swampy, covered with a dense jungle, in which were numerous palms, both cocoa-nut, sago, and many others. Pigs were abundant, and we killed and carried off two from this village, as we had no fresh provisions, and were now on two-thirds' allowance of everything. I cannot give you more details of what we saw in this country, or I should have to copy a whole notebook; and having given up nearly all hopes of ever seeing our lost companions again, we left the coast before worse things happened to us. Touching at Darnley Island, however, an excellent old fellow there named Seewhy, to whom I explained the circumstances, assured us that our boats were not at New Guinea, but had been seen by the people of Tood and Damood some time ago going to the southward. I had the greatest confidence in his word; and as Booby Island lay in that direction, and had a store of provisions left upon it, no longer doubted that the boats, having missed the ship, had bore up for that place, which eventually turned out to be the case. When we arrived at Booby Island, we found they had gone on for Port Essington, and that the crews and passengers of two other wrecked ships, the Coringa packet and the Hyderabad, had likewise been on the island, and gone on for the same place. . . . I confess I am getting more and more enamoured of a sailor's life, and regret I did not know the navy early enough to enter it. I see it would have suited me exactly.

June 29, off N. coast of Java. Since writing the above, we have had a pleasant and prosperous voyage thus far. In passing through Alass Straits, we anchored and landed on Lombock for refreshments. The natives were a particularly kind civil people, many of them speaking Malay, although they call themselves Sassacks. The Rajah is under the guidance greatly of a Mr. King (an Englishman, whose agent, a Mr. Herder, entertained us at Taryong Luar, where we landed), and is quite independent of the Dutch. I should like much to travel through these islands, which, at the cost of a few presents to the chiefs, might be done with great facility. . . .

July 2, Banca Straits. On our left hand now stretches a long, low, flat shore, fronted by mud-banks, and clothed with a dense jungle of mangroves—this is part of Sumatra; while on our right half the horizon is broken by detached low hills with round backs, and here and there the tops of trees stretching between them—this is Banca. The distance across is from ten to twenty miles. Notwithstanding our descent on the Lombock shore the other day, and our bringing off two dozen fowls, ten dozen ducks, with fruits and vegetables, we are again running short of fresh provisions; so you may conceive the havoc made among poultry by a set of hungry fellows who had been six months on salt grub, and short allowance of that at last. But then they are cheap enough—beautiful fowls for 4s. a dozen; ducks, 8s. a dozen; bananas, 8s. a horse-load; rice, 4s. a picul (133 pounds); plantains, eggs, &c., for anything or nothing. Mr. Herder told us a common man's expenses in Lombock were about eighteenpence a month; for half-a-dollar a month he would live an idle well-fed existence. If you want to know anything about Singapore, you should read Earle's *Voyages in the Moluccan Archipelago* or *Eastern Seas*—I forget which is the title. He is a very good fellow, and is now in England. His old servant Shadrack has been with us, but is now at Port Essington as interpreter. . . .

H.M.S. Fly, Java Sea, June 29, 1845.

My dear A.,—You will probably receive this letter before two others I have written previously. . . . We have had a very eventful and interesting cruise, from which we have escaped all well, although at one time we thought we had lost three officers and twelve men. These, having been detached in two boats on the coast of New Guinea, managed to miss the ship in bad weather; and, after a vain attempt to find each other, they were re-

duced to make for Booby Island, where a store of provisions had been left; and we were detained three weeks,
in a very dangerous position, on a lee-shore, in shoal
water, searching all the creeks and river-mouths, that
abound there, in our remaining boats, exposed to a huge
rolling sea, and to the continual treacherous attacks of
the natives, who are very numerous, fierce, and implac-
ably hostile. I was in two boat-expeditions, in one of
which Captain Blackwood and I, in his gig, with six
men, penetrated fifteen miles up a river, and luckily
escaped with only a slight skirmish. Another time we
took in the Prince George, having previously lightened
her; and they had the impudence to come off in canoes
and attack us, but suffered severely for their temerity.
We stormed one village, and searched it for any fragment of our boats or crews, but found none. Their
houses, however, were well worth seeing. They have
two kinds, of which the most singular is in shape of a
huge hall or tunnel, 300 feet long by 30 in width. It
is all raised on posts, to a height of about six feet above
the swampy ground; has a perfectly level floor at this
height, neatly formed of bamboos and the thick bark of
a palm; and the roof is 18 feet above the floor in the
centre, sloping down to it at each side. At the front
and back there is an open platform or balcony, over
which the roof projects, with a broad stair to the ground
and three arched doorways to the interior. Inside, a

s

row of cabins runs along each side, containing a fire-
place, bed-place, shelves, &c., with a neat door of split
bamboo; and between every two of these is a small
door or port-hole in the side of the roof, level with the
floor, with a ladder from it to the ground, and closed
likewise by a door or shutter. The centre forms a
perfect promenade in wet weather, being quite clear
from front door to back; the roof is quite weather-proof,
very thick and strong, made of the leaves of the sago-
palm; and everything was quite neat and clean and
comfortable, though rather dark. The natives resemble
our friends the Torres Straits' Islanders; and if as well
acquainted with our power, our wealth, and our good
intentions, would probably be as friendly and man-
ageable as they are. At all events, the discovery of
the many fresh-water creeks we met with, will open
a road into the interior of New Guinea; and a well-
managed boat-expedition would, I daresay, have no
difficulty in penetrating, either by force or favour, or
a mixture of both. I would gladly volunteer for one.
It appears, however, a very wet climate, as we had
continual rains. But I must defer all details, to go
on with our subsequent movements.

On June 3, we returned to Darnley Island, where I
heard from an old friend there, named Seewhy, that our
boats had been seen by the people of Damood going to

the southward; and on June 9, on our arrival at Booby
Island, we found that not only our boats, but the crews
of two ships wrecked on the reef had been on the island,
and successively followed each other to Port Essington.
We sailed for that place accordingly; and found our own
people pretty well, and sixty other people, among whom
were some ladies we had met frequently in Hobarton,
three officers going to their regiments in India, and
some children with their nurse. There was also a part
of one of the crews, to the number of twenty-seven, left
on a sand-bank near the wreck, with provisions and
water for six months. We were accordingly now under
the necessity of making entirely new arrangements;
and Captain B. determined to carry the crews and passengers in the Fly up to Singapore, and send the Prince
George round to Sydney with all dispatch, in order that
a vessel might be sent to fetch the remainder off the
sand-bank. We found, to our great annoyance, that all
our letters and papers had been sent up to Port Essington from Sydney, with a high probability of their being
lost by the way; so that when I am to get any intelligence from you or any one at home, again, I do not know.
Any letters sent to Singapore will have been forwarded
to Sydney long ago; so that we are altogether at cross-
purposes. . . We are in an absolute state of uncertainty
as to our future movements. Our orders are now awaiting us in Sydney; but whether they will be to proceed

with the survey, to take in guns and go to fight the
New Zealanders, or to go and beat the French out of
Otaheite, or to proceed to China, Kamschatka, Cali-
fornia, Cape Horn, or England, we have no means of
judging, nor, as far as I am concerned, of preferring
one to another.   We are sure to be paid off before
November 1847, and the more places we see between
this and then, the better I shall be pleased.   In the
mean time we are having a delightful voyage, with
smooth water, pleasant breezes, fine weather, and the
addition of the society of ladies to our usual comforts.
These are, Mrs. and three Miss B.s—very pleasant
girls, and very fond of waltzing; so that now every
evening we light up the quarter-deck with lanterns,
and waltz to the music of the ship's fiddler and the
captain's barrel-organ, or sit in groups, gipsy-fashion,
on the poop, and enjoy the cool breezes of the lovely
tropical nights, with the stars gleaming overhead, and
the sea sparkling and flashing with phosphorescent light
on the surges that flow along our sides, or the eddies in
our wake.   Mrs. B. is the widow of a captain in the
army, and returning with her daughters to her friends
in India.   She was originally from the north of Italy,
and has mingled much in the best society of Italy and
France in former days.   They have borne their shipwreck and subsequent boat-voyages very well, although
with the loss of almost all their baggage and outfit for

India. We took them ashore the other day on the Island of Lombock, where the inhabitants were delighted to see them, invited us into their houses, and brought presents of rice and confectionery, and birds in cages. We had afterwards a magnificent view of Lombock Peak, a volcano that rises directly from the sea to a height of 11,400 feet, by our measurement. It fully equals the Peak of Teneriffe in majesty of appearance.

July 2, Banca Strait. We are now running between the low swampy mangrove shores of Sumatra on the left hand, and the Island of Banca on our right. We have just passed three vessels, English and American, being almost the first square-rigged vessels we have seen for six months. As we are within $2°$ of the equator, with so much land about us and little wind, you may easily conceive we are rather warm. A temperature of $80°$ is that which now feels most agreeable to me ; but when it comes up to $85°$ it begins to feel oppressive, and at $90°$ we call it hot. We have been so long in warm weather, that I look forward to a voyage along the south coast of Australia with a kind of dread of the cold, as we shall probably get the thermometer down to $50°$. Singapore is in about $1°$ north, so that to-morrow or next day we cross the line, and I shall again be in the same hemisphere with you, and shall remain there probably a fortnight. . . . . Give my kindest and most affectionate

love to my mother, and believe me, as ever, your most
affectionate brother,

J. B. JUKES.

July 7, Singapore. We arrived at Singapore the
night of the 5th, but have scarcely had time to land and
look about us before we are ordered off to Malacca to
meet the admiral, whom circumstances render it necessary for us to see. Where we shall go to next, I can't
say, but will write again soon.

Sydney, Oct. 2, 1845.

My DEAR A.,—We arrived here on September 25,
and long before we cast anchor, I was deep in the perusal of my letters, brought off by the Bramble and Prince
George. I read from before noon till dark night, when
I could no longer see ; and since then I have skimmed
over them again, but still feel my ideas concerning all
things at home in a whirl, and am unable to classify or
arrange the facts in my memory.

The general result is melancholy. Death and other
misfortunes in several branches of our family threw a
weight over my spirits, that the subsequent excitement
of much bustle and business and kind reception ashore,
has not yet been able utterly to heave off. Hardship,
danger, or death in action, or in fight, would be luxury

to my feelings, compared with this slow consumption of
life or its means and enjoyments in quiet domestic
existence. Heaven send I may never have to die in my
bed by decay or disease, but may meet the ' grim fea-
ture' openly face to face in the free air, and with my
blood warm! All my old boyish recollections and associations broken up, or brushed out of sight into the
lumber-room of memory, no more to be renewed or mentioned ! What changes, by the time I return, will have
taken place in every part of my little world ! But I do
not know why I should trouble you with gloomy fancies,
which I suppose visit every head that has seen thirty-
four years pass over it. They will come perhaps too
soon. Be assured I am delighted to hear of your happiness ; it is my greatest comfort and consolation.
Long, long may you all remain so ! I could almost find
in my heart to envy you ; but my craving for change of
scene and freedom of action is not yet satiated. . . . .

I have now re-read all yours, and proceed to answer
them in order. You talk of coming to meet me on our
arrival. I should be delighted at your doing so, but we
shall not know where we shall be till we arrive. . . . .
As to Mitchell's travels : no travels in Australia can be
amusing, although they may be valuable. In unknown
countries and coasts the journal form is essential, as
you would find if you came to tread in a traveller's footsteps. . . . . I do not think I am at all altered in per-

sonal appearance, except by the addition of beard and
moustache, not having touched a razor for two years.
Age tells, of course, but climate, I think, has not affected
me. . . .

Now, then, to bring up our history, and guess at
future movements. After I wrote to you at Singapore,
we joined the admiral at Malacca, where we spent a
week. I do not know whether Malacca has the same
dreamy kind of antique associations about it for you
that it had to me, from the mention of it by the old
navigators. It was once the Queen of the East under
the Portuguese. All that is gone; but yet I was much
interested in it. There are the remains of a fort and a
cathedral with old Latin tombstones of 1550. There is
a half-caste who assumes the name of Alfonso Albuquer-
que, and claims descent from that stern old warrior. It
is a pleasant place even now, and to my feelings prefer-
able to the new, bustling, thriving, garish Singapore,
which reminds me of a piece of an English colonial
town pitched into the tropics, where it has settled among
a few Chinese and Malay habitations. Even it, however, has a history. Sir S. Raffles found it a desert
swamp in 1815; but it once was Singhapura, the capital of a Malay empire, that fell as Malacca rose, and the
very memory of which had almost passed away, when the
genius of Raffles saw the advantages of its situation,
and now it has 50,000 inhabitants, and fleets from all

parts of the world at anchor in its waters. An Englishman knows nothing of the power and grandeur of his
country who has remained at home all his life. England is but the heart of the empire. He only knows
the stature of the whole body who has seen the limbs
embracing the whole world, and people, kingdoms, and
nations reposing under the shadow of its arm.

At Malacca the admiral at one time was going to
send us to Hong-Kong; at another to Bombay to be
docked and repair; but at last let us come to Sydney,
where we are going to be hove down. He is a bit of a
Tartar. Are not they kept in order, captains and all?
He was, however, very civil to us, only rating one of the
midshipmen soundly for having a spot of tar on his
white trousers. We left Singapore on August 3, and
the Straits of Sunda on August 20, taking our last
look of Java the magnificent. We had then been a year
and a half in a constant temperature of between 80° and
90°, and in rather more than a fortnight we were in one
of 42°. Thick gloves, woollen stockings, blankets—
all failed to keep us warm. However, we escaped with
only a little dysentery and catarrh. Off South Australia we had one very heavy gale; but on September 25,
we anchored all well in Sydney. Last August, however, there was a dreadful wreck in Bass's Strait, when
414 lives were lost, chiefly emigrants for Port Phillip.
We found the Bramble and Prince George all well, and

we found *our orders for home.* We are really not going another cruise; but the Bramble is to remain out
and complete the survey, or to go on with it, at all
events. We are not sorry, though I should like to see
more of New Guinea, while Yule is, although by staying out he is sure to get his promotion. We shall be
ready for sea again about the middle of December probably, and shall sail for England in January. Perhaps,
however, we shall go down to Hobarton, and also to New
Zealand, whence we shall very likely bring home Captain Fitzroy. You will see in the papers the account of
the fighting there. . . . . . At all events, we hope to be
home in the summer of next year, and unless they offer
to make me governor of a settlement at Cape York
(which we mean to propose the formation of) I do not
think I shall be tempted to revisit Australia. I am rejoiced to hear my mother still continues so well. . . .

Oct. 4. I have just received by a ship that came in
this morning your letter of May 30. The reason I did
not get your letters was the unexpected delay of the
mails from Calcutta to Singapore, as the steam arrangements were not then complete so far. . . . I am sorry I
cannot tell you where to direct on our passage home, as
we do not yet know which way we shall go, and cannot,
till we hear from New Zealand. I cannot hope, therefore, to receive an answer to this till I get one from your
own mouth in June next or thereabouts. Give my very

best love to my dear mother, and believe me, as ever, your most affectionate brother,

J. B. JUKES.

I saw a letter from Sturt the other day, written in the interior, north of Adelaide. What a country he is in! Sandy deserts interminable, with the thermometer 130° in the shade four feet from the ground, and 157° in the sun. It only confirms my preconceived notion of what the interior would turn out. Mitchell is going again in a few weeks. I am going to call on him to-day.

H.M.S. Fly, Bass's Straits, Dec. 27, 1845.

MY DEAR A.,—We are actually on our road home, and expect to arrive in June or thereabouts. I was so much engaged the last week or two in Sydney, that I had no time to write; but having now brought up my journals and business to the present time, I am clearing off my letter-scores. I passed a very pleasant time in Sydney, and had a capital geological excursion with Clarke down to Illawarra, from which I returned about ten days before we sailed. I had then some collecting work to do in the harbour, and all my kind friends to visit for the last time, and take leave of. I really felt quite sorry to leave Sydney, at which I was rather astonished, as I did not like it at all at first. We sailed on December 19, bound for Port Phillip, from which place we are 100 miles distant, having been bothered with light winds and calms.   We are now trying to beat round Wilson's Promontory.   We shall not revisit Van Dieman's Land, for which I am sorry. . . . At Port Phillip we shall probably stop a week, and I shall dispatch this letter thence by one of the wool-ships. We shall proceed to South Australia, and stay a week at Adelaide; then to Swan River for another.   Then running up into the trades, we shall stretch across the Indian Ocean, and perhaps touch at the Mauritius.   At all events, we shall visit Cape of Good Hope and St. Helena on our way home.   You will probably get this in April or May; and if nothing happens to detain us anywhere, you may expect in June or July to receive a note from me, dated ' English Channel,' announcing our being off the shores of Old England. . . .

GEOLOGICAL SURVEY OF THE UNITED KINGDOM.

ENGLISH SURVEY.

1846–1849.

GEOLOGICAL SURVEY OF THE UNITED KINGDOM.

ENGLISH SURVEY. 1846–1849.

Return to England—Appointment to the English survey—Letter
of Sir H. de la Beche—Congenial companions—Edward
Forbes—Professor Ramsay—Letters to Professor Ramsay
from London, Bala, Llan-y-Mowddwy, Cerrig-y-Druidion,
Yspytty Evan, Penmachno, Bettws-y-Coed, and Wolverhampton on geological work and problems.

THE Fly returned to England in June 1846, and
Captain Blackwood having resigned to Mr. Jukes the
office of recording the history of the expedition, he
shortly afterwards published the *Voyage of H.M.S. Fly*,
and subsequently *A Sketch of the Physical Structure of
Australia*. In preparing his journals for publication,
he would seem to have followed the advice of Professor Sedgwick, given in these terms : ' Don't ever try to
write fine, but tell us plainly what you have seen.
Honest idiomatic English is the stuff to write.'

Before these works were completed, however, Mr.
Jukes entered a new epoch of his life. In August of

this year he joined the geological survey under the directorship of Sir Henry de la Beche, the local director for England being then, as now, Professor Ramsay, to whom Sir Henry addressed the following letter introducing the new member of the staff:

London, Aug. 18, 1846.

My DEAR RAMSAY,—I have had a very satisfactory interview with Jukes, who appears a very fine fellow, and to love knowledge for its own sake. All is arranged; and as our corps is getting a thing *desirable* to join, I have found it *desirable* that he should join on the 1st of October. The pressure for unfit persons is a matter to be avoided. I have explained his duties, and that his instructions would come from you as director for Great Britain. I expect it will be the day after to-morrow before I leave London for *Dol Jelly*, *viâ* Shrewsbury and Dinas Mowddwy, I suppose.—Ever yours,

H. T. DE LA BECHE.

Jukes comes at 9s. per diem. He seems to care very little on what terms, so that he comes. *Good, this.*

Arduous and frequently exhausting as was the work of the survey, it was in itself a labour of love to Mr. Jukes at this period of his life, and possessed the great attraction of bringing him into intimate and hearty fel-

lowship with men of a kindred spirit, who were studying
the same comprehensive chapter in the great book of
Nature. In October he joined Professor Ramsay and
Mr. Aveline at Bala, where Edward Forbes, who held
the office of palæontologist to the survey, soon afterwards made one of the party. Here the outdoor work,
the invigorating air, the congenial companionship, and
the freedom from care and conventionalities, combined
to form a life peculiarly suited to Mr. Jukes' taste.
The friends then made were friends for life, and amongst
the most intimate and valued were Professor Ramsay
and Edward Forbes. Of the correspondence in illustration of this triple brotherhood, the materials are as
ample in the one case as they are wanting in the other,
for unfortunately Mr. Jukes' letters to Forbes have not
been preserved; but the high estimation in which he
held this beloved friend may be judged from the terms
in which he speaks of him when death deprived the
world of that bright spirit. The following is taken from
the 'Address to the Geological Society of Dublin,' 1855:
' The bitter pang of personal regret for the loss of a
dearly-loved and highly-valued friend mingles with, and
renders more deep and acute, our sorrow for the extinc-
tion of one of the great lights of our science. The
news of his death fell upon my heart and that of others
here present as a great public and private calamity. . . .
His heart and his intellect were alike large and catholic

T

in their instincts and capacities. He seemed not to perceive, or perceiving not to regard, the failings and imperfections of men; but to look at their capabilities, and seek for and call forth what powers and good qualities they possessed. It mattered not to him of what party, of what sect, or of what nation a man was; if he had in him any love for natural science, any desire or any power of aiding its extension, he might be sure of hearty appreciation and effectual aid and encouragement from Edward Forbes. It was this spirit of fellowship, this instinct of union and of solidarity (to use a newly-coined word) among all scientific men, which seemed the natural impulse of Forbes' mind, the abstraction of which may be a yet greater loss than even that of his own power of scientific investigation and discovery. . . . He was one of those who, while he would not hesitate to oppose openly and frankly the opinions of his best friends when they conflicted with his own, never in so doing lost a friend, never, I believe, made an enemy of a stranger. Even in cases where much ignorance or much error was displayed, if he thought it was exhibited in an earnest and honest endeavour after truth, he would not only treat the shortcomings leniently in public, but give up much of his valuable time in private, to assisting and instructing the inquirer. I insist all the more strongly on this side of Forbes' character because I think it is incumbent upon all his friends to do their best to cherish and extend this spirit, which he by no means monopolised, but for which he was remarkably distinguished. The rancour and jealousy and bitterness which for so long have been the reproach of scientific men have of late years greatly diminished. It is high time that they should altogether cease to give cause for scandal to the rest of the world, and that those who claim to breathe a higher and calmer atmosphere, to have their powers and faculties busied with purer and loftier subjects than the generality of men, should show the effect of this noble employment and this higher station, by exhibiting their minds and tempers free from the little personal strifes and enmities, the jealousy and uncharitableness of common life, and should be characterised by that generosity and largeness of heart which always puts the best, instead of the worst, construction on the actions of others, always accepts and appreciates the good, and passes by and ignores the faulty and the bad. Men of science should look upon their common pursuit as giving them a claim on each other's consideration greater than what any difference of creed, or party, or country, or wealth, or station, can injure or efface.'

How truly, in thus describing the 'natural impulse of Forbes' mind,' did his friend delineate his own!

Although it was known to the Editor that the correspondence between Professor Ramsay and Mr. Jukes

was very extensive and the confidence between them
complete, she was still scarcely prepared to find in the
letters kindly placed at her disposal such an entire unfolding of her brother's thoughts and feelings as amounts
to an unconscious autobiography. Frank and outspoken
as he was on most subjects, he was yet somewhat re-
served, or at least reticent, of his own feelings, and
only revealed them where, as in this case, he was sure
of the utmost sympathy, and could repose perfect trust.
The office of selecting from materials so ample is one
requiring both *care* and *judgment.* The first has not
been spared, but the other is a faculty of ' unknown
quantity,' and may be viewed in many lights. The
non-scientific reader will perhaps consider that too many
geological details have been retained. These, however,
faithfully portray the operation of the survey and the
kind of life led by the field-geologists, and will doubtless possess much interest for those who know the
valuable results of such minute and careful work.

To Professor Ramsay.

London, Sept. 29, 1846.

My dear Sir,—I should have replied to your note
earlier, but a severe cold has so clogged my brain as to
hinder it from thinking. Thank you for your kind consideration ; I have, in fact, some business which will de-

tain me in town till the 1st. Having been so long used
to a sedentary life, especially in so smoky an atmosphere
as this, I am just at present shockingly out of condition, and incapable of much active exertion. I feel
the cold, too, of the approaching winter most sensibly,
after four years' luxuriating in a perpetual summer. I
have little doubt, however, that the free air of the Welsh
hills, and the use of my legs once more, will soon put
me into a condition for work.—Believe me yours very
truly,

J. BEETE JUKES.

Wolverhampton, March 14, 1847.

DEAR RAMSAY,—. . . . I went yesterday to Stourbridge on the top of the coach, and acquired a brutal
cold and sore-throat by the operation. The railway cuttings there have greatly disappointed me. The wretches
have taken so high a level, that, instead of cutting
through the junction of the C.M. and the Y.R.,* they
actually *bank it over.* Only conceive such a piece of
brutality! Nevertheless, they will have to cut farther
on, pretty close to the junction, and perhaps into it, as
the ground rises, thank goodness! pretty steeply. There
are now three railways coming into this district this
summer, the whole of which should be watched from

* Initials of Coal Measures and Young Red (sandstone).

time to time. These are the Oxford, Worcester, and
Wolverhampton, the Stour Valley, and the South Staffordshire Junction. Facts of great importance, both
practically and scientifically, will turn up on all of
them. Near Stourbridge at the present moment the
outcrops of two or three beds of coal are exposed, dipping everywhere from the Y.R. at an angle of 10° to
30°. There are also some beds of red and variegated
marl, interstratified with the true coal-measures; a fact
not of common occurrence in this country, I believe.
I have always been highly interested in the, as yet, un-
solved problem of the condition of the coal-measures
beneath the Y.R. plain of the centre of England.
There are facts to be picked up here and there, round
the edges of the present coal-field which tend to solve
it, viz. on the southern edge of the Derby and Notting-
hamshire, and of the North Staffordshire, and round all
the edges of the Leicestershire, Warwickshire, South
Staffordshire, and Shropshire fields, as also along the
thin skirt of poor coal-measures running from Cheshire
to the Forest of Dean. It is certain that great denudation took place in the C.M. before the deposition of the
Y.R., and that even the present system of valleys in
the carboniferous formation is of that early date. In
many instances, what is called by the miners the boundary fault of the present coal-fields is no true fault at
all, but merely the abutting of the Y.R. against an old

cliff of C.M. In other cases, however, the boundary of the two seems to coincide with a fault, as in the cutting of the Grand Junction near Wolverhampton something of this kind was to be seen [Two pen-and-ink sketches are here inserted in illustration of the subject]. But I cannot be sure that even that was a true fault, as it might be débris of the old cliff, though, as far as I recollect, there were no pieces of coal in the beds of the Y.R., nor any pebbles—all fine sand. There are in other places true Y.R. beaches round the coal-field. Now they have already in two or three places sunk boldly into the Y.R. hereabouts at a distance of some miles from the present fields, but hitherto without piercing it. It would be a great feather in our caps if we should be able to give a decided opinion on the state of the coal beneath the Y.R., and where the best place was for making the attempt at reaching it. The materials for forming that opinion only want collecting. The period of denudation is still farther narrowed by the lowest beds of the Y.R. passing down conformably and transitionally into the coal-field of Warwickshire, with the middle beds of the Y.R. resting unconformably on both. Excuse my telling you what I daresay you already know, but I just want to recall it to your mind, to show the great importance of seizing the opportunity of collecting facts in the Midland counties at the present moment, when they can be got. The opportunity may never occur again, as they are about the last batch of railways now. I set out for Bala to-morrow, and should have done so to-day after receiving your letter, but thought I had better give my cold one day's nursing here than have, perhaps, to give it a week at Bala.—Ever yours truly,

J. BEETE JUKES.

Kind regards to the staff.

Bala, March 16, 1847.

DEAR RAMSAY,—Just arrived, and got your letter; *obstupui, steteruntque comæ, vox faucibus hæsit.* Nevertheless you are right in the main; but as I am here, would it not perhaps be better for me to finish this job, which I could do in a month, and then set to work at the Midland coal-fields? I did not like to press the matter before, because I thought you might think I was working a dodge to try and get into my own country. As you have opened it, however, I can say freely, I think no part so *urgent* as the South Staffordshire field, one so rich in so small a space, and so rapidly being worked out. Still, I think if I begin in April, I can entirely spifflicate it in the course of a summer. I will think it over to-night, as I fear to hear the mail coming up the street, and will write you more fully to-morrow.

March 17 (St. Patrick's day in the morning, half-past six A.M.). I have cogitated it over, slept on it, and cogitated again. The result of my deliberations is the following: It is highly important that sections should be kept of all the railway cuttings in and about the Midland coal-fields, especially near their boundaries. These sections, when *complete*, will be most useful documents in the survey of these coal-fields. There are now in progress the Trent Valley Railway, from Stafford to Rugby, to be opened in May; the Oxford, Worcester, and Wolverhampton, now cutting; the South Staffordshire Junction; the Stour Valley line; the Chester and Shrewsbury, and the Shrewsbury and Wolverhampton lines. Now these *all* want looking after, as well as some other lines and branches now going on. It would require somebody with a good horse to make the entire round of them, first of all rapidly, and then to take accurate sections of the most important spots. If, therefore, I were immediately to commence the survey of the South Staffordshire field, this work would have to be done by some one else. Again, the South Staffordshire and the other Midland coal-fields will be good winter's work, because much of it will consist in poking down pits, and going over mining plans with the 'agents,' copying their faults, and tracing the out-crop of coals, &c.

These premises being granted, it appears to me

that if Captain Ibbetson, or any one else who is up to the dodge, can look after the cuttings on this group of railways during this summer, the survey of the South Staffordshire field might without injury, if not even advantageously, be deferred to the autumn. . . . Believe me, I take your offer as a very kind one. The Midland counties have always been a great object of geological ambition with me, and rather than that they should be consigned to any one else, I will be off immediately ; but I think I should feel more equal to them after a summer's work and sectionising here than I do now. Still, I am of course entirely at your disposal. Say the word, and I am off. If Captain Ibbetson cannot go to these railways, or if from any other reason you think I had better return to Staffordshire, ' Ay, ay, sir.' In that case, could you not contrive to give me a meeting either before or after your going to Builth ?

A southerly breeze sprang up yesterday, and brought an infernal heap of clouds and deluge of rain here. It looks a little better this morning ; and if there is any chance, Gibbs and I set off at half-past eight for the country beyond Moel Migynau.—Waiting your decision, believe me yours very truly,

J. B. JUKES.

Bala, March 20, 1847.

DEAR RAMSAY,—I just write to ask you to ask Salter whether he knows any actual limestone south of Moel Migynau. If so, let him say as near the spot as he can on the map. It will save me a deal of bother. Thursday being a model day, G. and I had a long ramble all over Moel Migynau and round about it, and down the valley we took our way,

> Where Afonfechan's waters play,
> But devil a limestone therein lay.

The consequence was, I got so blistered about the heels, that when I pulled off my stockings, I pulled off my skin, and so got laid by the heels. Not being able to walk, I took a pony on Friday, and rode along the road up to Bwlch-y-groes, and found fossils by the roadside above Tyn-y-frou, but no limestone. To-day being wet and windy, we rested; and I shall not be able to do much till my heels get skinned over again. The fact is, I am rather out of training, and must get gradually into my work. I shall work Aran Benllyn country from Llanuchlyn, where is a public, when the equinoctial gales cease. It is now all in the clouds. In the mean time I shall work the Hirnant district. G. complains of pain in his side, and says this winter has nearly knocked him up. My dog Governor is too fond of mountain lamb. He killed one yesterday by the lake. . . . The assizes were held here yesterday, but there was nothing to try. Two men blew two trumpets vilely out of tune, and as nobody did anything to them for it, the judge walked off. Fancy two asses braying, in a high key and a low one, and trying to make their voices harmonise without success, and you have just the noise they made. There is a big-little feud here about a fishing society who want to preserve the river. There is to be a public meeting at the Town Hall, 'to stand up for the rights of the people of Bala.' Several Joneses have promised to address the meeting. . . .

Just got yours of yesterday. I think *Captain Ibbetson* is right; but there is a railway, which I believe to be the Oxford, Worcester, and Wolverhampton, now cutting both about Stourbridge and Dudley. At the latter place it passes through Wenlock shale and *trap*, I believe in *junction in one cutting*. The old line from Birmingham to Chester can't be done now. It is all covered with earth and grass. There are only two important cuttings in it. Would that I had sketched them at the time! but it was want of money, not want of will. The only part of the Trent Valley important will be, where it cuts the two coal-fields of South Staffordshire about Rugeley, and of Warwick near Tamworth—that is if it does cut them, which I don't know.

March 22. We yesterday tried to carry on the Hirnant to the northward, but could get no fossils north of

Aber Hirnant. I completed rocks thence up to the river. Salter's information will be useful. The country south of Moel Migynan is so covered by drift, that there is little or nothing to be seen till we come to the Dinas Mowddwy road.

March 23. Are you keeping the fast? It is a regular Sunday here, and I have forborne to go out from respect to the prejudices of the people. Gibbs found the continuation of the limestone of Y-Gelligrin below the road towards the house, showing a clear nice little fault of about fifty feet perpendicular in the proper direction. There is also another clear shift of twenty yards horizontally in same field. He also got it in the Aber Hirnant road across the brook, so that we can trace it now, almost without a break, from the road near Pont Cennant to the south side of Y-Gelligrin. Then comes the big jump on to the north side of Moel Fryn, where it is again almost continuous by Maesmeillion to the valleys east of Moel Migynau.

Bala, March 30, 1847.

DEAR RAMSAY,—I have received the books, &c., and thank you much. When will this winter end? I walked home this afternoon in a heavy snowstorm. I have, however, discovered the clue to the momblement on the other side of the lake; and by aid of the limestone and

an ash-bed, shall be able to lay down some faults and
guess at others. I have satisfied myself of the unity
and persistency of the ash-bed. My notion of writing
a paper was only a temporary whim; but I shall be
very happy to contribute my mite towards any more
general affair you may contemplate. My fancy was to
give only a page or two of explanation on the queer
twists and twirls there will be in the map, just saying
how they come there, whether there was one bed or
two, and whether the one was persistent, &c. If, however, I can work out the faults with sufficient certainty
to lay them down, the map will explain itself very
nearly. I have written to William, giving him my
notion of things about Walsall. By the bye, was my
paper read at the Geological last time?

March 31. Just got yours of to-day, as I have been
staying in this morning, laying down my work. Alas,
it is all confusion more than ever confounded. The
fault that throws *down* the limestone throws *up the ash.*
De'il bake them together! . . .

It appears to me that Winter, not satisfied with the
way in which he has done his work this year, is going
to do it all over again; or else it is next Winter that
has somehow got a start before next Summer, and come
tumbling in here to the aid of his expiring brother.

Bala, April 20, 1847.

DEAR RAMSAY,—AS I had half an hour to spare to-day after coming home, before dinner was ready, I have amused myself with just tracing off against the window the enclosed copy of the present state of the Bala limestone and ash-beds. You recollect the pretty red lines we drew for the strike of the rocks? There's rather a difference between them and the truth. I should like to see some one draw-in the faults. I think in a geological examination, this would make a good problem; only it would puzzle the examiner as much as his pupil.

I am strongly of opinion that the two ash-beds are the same, or at least, if there are two on the west of the lake, the upper one is the same with that on the east of the lake. I am inclined to suspect also, that the Rhiwlas limestone is just the Bala, as I am sure that the limestone you and I traced last year is nothing else; and Gibbs is of that opinion too, spontaneously.

I think that, after the rocks were bent into that remarkable twist that smashes them towards the north, a fault ranged down Bala lake, and dropped the so bent and broken beds to the west. If we had the survey on the six-inch scale, I have no doubt we could make it out; but it is impossible to unravel it completely with the one-inch. I believe now there is not a bit of ash anywhere I have not got in, except some bothering beds just under the limestone of Y-Gelligrin, which I am going at to-morrow. I shall then charter a pony, and run down to the northward towards Dinas Mowddwy, where, if it please Heaven the beds run pretty regularly without these infernal smashes, I shall soon polish off the remainder. Gibbs has been all over the Caradoc sandstones, and reports no fossils. Thank goodness, I found some to-day, beyond Moel-y-gamedd, only just above the black slates, and shall send him there to-morrow. Then I shall send him to the southward; so I have got work for him for the present. My book is progressing. The first volume is nearly printed off. Boone says it will be finished to-morrow.*

April 23. I did not half believe the story of the Bala limestone coming down this way; but it's a fact, and no mistake. By Blaen-y-pennant and Cwm Dinewyd, there is a regular band of Bala limestone, six feet thick, striking off in this direction. When I got down here, I also heard of limestone on the Dolgelly road from this. At Blaen-y-pennant there are lots of fossils, but they appear to get scarcer down here. Do you know any limestones between Dinas and Dolgelly? If they turn round that a ways, I shall want t'other sheet.

April 25. I have decided, unless you order to the contrary, that Gibbs should go down to Llan-y-Mowddwy or Dinas Mowddwy—whichever he likes best—some time this week, and remain there for about a fortnight, at all

* This refers to the *Voyage of the Fly.*

events. My inferences respecting Rhiwlas, Llwyn-y-ci, and Eglws Anne are not inferences at all. I have not been in that country yet. I only presume to guess from the smashed condition of the beds to the northward, of which I sent you a copy, that such would turn out to be the case. We shall see. I mean now to go to Llan-y-Mowddwy to-morrow, and finish the tail end of the Bala limestone and accompanying ash-beds, if any; then to try and join that piece with the one about Moel Migynau, although, from the difficult nature of the country, I expect it will be almost impossible to do so. I shall then commence at Rhiwlas, and drive the whole country right before me to Cerrig-y-Druidion, as I shall then have the necessary data for so doing. I shall be chargeable in horseflesh, or else I shall never get the work done, it lies now so wide. I have given up walking carefully over beds. You can never see the faults that way. I have had so many things like this (section), the fault being afterwards proved by other facts.

I shall trust entirely to the limestones and ash-beds when laid down on map.—Yours, in haste,

J. Beete Jukes.

Bala, April 28, 1847.

My dear Ramsay,—I have made up my mind now. I had never been to see the limestone in Rhiwlas

U

grounds before Monday. At the very first glimpse of
it, I could have sworn it was the Bala. I have now
been over the ground between here and Eglwys Anne
carefully, and the whole thing is clear. There is but
one bed of limestone in the whole country, and some
three to five hundred feet below it, is a bed of ash more
or less calcareous; the limestone also, in places, contains streaks and patches of ash. There may also be
another bed of ash somewhere below the limestone,
but that is at present doubtful. There is one considerable but not enormous fault ranging down Bala Lake,
and dropping the country to the west; many minor
faults, more or less complicated, but mostly parallel to
each other, cross this main fault at right-angles, and
some also run parallel to it; the pieces included between these faults have settled at various angles, producing dips in different directions, and there have been
considerable contortions or bendings beside. There is
nothing improbable in this statement. The Bala limestone being thrown from Y-Garnedd or Gelligrin to
Rhiwlas does not involve so great a throw as would its
being pitched to Llwyn Tollyn or Bwlch-y-Tyno. It
must have been thrown to one of those places as it goes
nowhere else. Suppose, before denudation took place,
the limestone of Y-Garnedd rose at a slight angle (for
its inclination there is very moderate) into the space
over the lake and the Dee, a fault of five or six hundred

feet would easily drop it to Rhiwlas. On the supposition that the limestone and ash are not the same beds,
you have to account for a multitude of pieces of limestone, of precisely similar mineral and fossil character
(and that mineral character a very peculiar one, so as
to produce a distinct physiognomy in the weathered
surface that can be recognised half a mile off), each
piece of limestone having the same thickness, but not
one of the pieces being half a mile long, and some of
them only twenty yards; and that ash-beds set on
below the limestone, always at about the same distance
below, always as peculiar and similar in mineral composition, and always varying in length with the bed of
limestone above. But even then you would not solve
the difficulty; for sometimes the ash and the limestone
approach each other and lie cheek by jowl, sometimes
with apparent conformability, sometimes at right-angles.
You would have farther to suppose that in adjacent
tracts of country, as, for instance, to the northward, the
various pieces of.limestone and ash, instead of setting
on at various levels throughout a thickness of many
thousand feet, suddenly arranged themselves each into
one line and went on regularly, and that the thickness
of the whole mass was suddenly compressed into a few
hundred feet. In this last remark I am again running
a little ahead·of my work; but I am almost certain it
will be verified by the run of the rocks to the south-

ward towards Bwlch-y-Groes and Dinas Mowddwy.
Gibbs yesterday found the limestone in three places
on Bwlch-y-Groes, all in a straight line. The state of
the case will be this. The rocks come up from the
southward to the eastward of the range of the Aran,
with a pretty regular and high dip to the east under
the Caradoc sandstone. As they range to the northward, the lowest or western portion of the rocks are
brought within the influence of the disturbances, and
become more and more affected by them as they approach Bala Lake. The highest parts and the Caradoc
still run on unaffected, till they get into this line of disturbance north of Llanderfyd, where you found the great
fault in the Caradoc. The disturbance there seems to
have been concentrated into one blow. If you run from
this point a line (N.E. and S.W.) down Bala Lake,
does there not also seem to be a shift of the traps?
Does it not look as if the traps of Aran were cut off by
it and dropped to the westward? Moreover, it coincides
with that remarkable valley running that way. Father
Adam says, the traps of Aran come in at a slightly
higher level than those of Arenig. He is quite right
in his conclusion, it appears to me, provided there is no
fault, but if there is, the fault would produce the appearance. A great fault, which acted as one split on
thick hard compact things, like the Caradocs and the
traps, might easily spread into a number of small faults

when passing through a comparatively soft and incoherent country such as the Bala rocks.

I believe it will be impossible on our maps to show all the faults, or even to say to ourselves precisely where they are. We must trust to the general effect. How capitally, however, such an example as this shows the necessity of our survey! Sharpe, or even sharp-eyed old Sedgwick himself, than whom no man had a quicker eye for a fault, or a tighter grasp for the bowels of a country, might have gone over here with their hasty traverses a dozen times, and fired scores of papers at the Geological Society, and only succeeded in puzzling themselves, perplexing their hearers, and utterly flabbergasting their palæontological brethren (confound that long word, it tires one to write it; and after the effort of spelling it, I can't spell even a little one!) I wrote to Forbes yesterday to know if he had any objection to all the limestones being the same, or fragments of the same bed, telling him at the same time I did not care a rush whether he had or not. Do you recollect you and Aveline and me, first seeing the limestone at Creigiau Isaf one day, when Gibbs and William were hammering at it, and my swearing it was the Bala from the look of it? From their report of the fossils, I gave it up afterwards as the Rhiwlas, but I shall have now a greater respect for lithological character than I ever had before; at all events whenever that character is peculiar, and the result of the combination of a great number of modifying circumstances, as I believe was the case with the Bala limestone. You see, I treat this as a certainty, because it would be useless to deny that I am sure of it. I will give you a week under my guidance to come to the same conclusion. Well, you see I have got my inferences to go at last; but it was not by attempting to drive them through every gap, but spurring them slap across the country, and leaping over all the stumbling blocks that came in the way. It is blowing a gale of wind with infernal showers, making it impossible to go to Bwlch-y-Groes, but I have still some small minute work about here with patches and fragments of ash, &c. One thing that convinces me I am right in this country is, that I can prophesy respecting it; *i. e.* when I find one thing, I know where to look for another, and generally find it. I have one or two little tests of this kind still unworked. I spare you the rest. —Remember me to Aveline, and believe me thine ever,

J. Beete Jukes.

Please to put this scrawl up and keep it a bit, as it is partly prophetic, and I shall be glad to see if it comes true.

Llan-y-Mowddwy, May 5, 1847.

My dear Ramsay,—I have finished my work here *pro tempore,* and to fill up the vacuum after dinner I am going to scribble to you. I left Bala on Monday morning. During the last three days I have done five miles and a half of limestone, running in a straight line north 33° east. The last piece seen was within a mile and a quarter of Dinas Mowddwy, about Bryn Sion, in the Cowarch Valley. That, however, I am sure is not the end of it, but it gets thin and concealed, and probably runs down in the flats of the Dyfi. I have accordingly left it for Gibbs to stumble across in his fossil explorations. It lies exactly as I expected, unbroken, but at a mean angle of 45°. A large part of its outcrop is concealed by the black bogs of these barren moors; and it so happened that it invariably chose to make its appearance from under them, at the edge of the steepest and most precipitous places, and then plunged sheer down into valleys that turned me sick to look at them. In fact, I worked yesterday for some hours in an almost perpetual state of funk, and had no sooner got out of one stew than I was obliged to get into another. Gibbs went up a precipice to-day that I did not dare to watch him on; and yet I believe it was nothing to any one, but such a brain-spinning, topsy-turvy, toe-and-heel-staggering rascal as myself. I hope it will please Heaven that the Bala limestone will never get into such places again, or I will throw my hammer at it, turn round, and have nothing more to say to it. By the bye, between the Bala limestone and the Aran there appears to be some re-

verse or westerly dips, which are not down in the map.
I got one S.W., and the other W.N.W., on a hill called
Drysgol yesterday, and Gibbs reported another to-day
on the south side of the hill called Pen-r'allt-isaf. The
valley that goes up from Blaen-y-Pennant to the foot of
Aran Mowddwy affords a very fair section in the brook,
although the hill-sides are covered up. There are no
ash-beds either above or below the limestone hereabouts,
although the limestone is ashy in places. It alters its
character very much at Blaen-y-Pennant, where is a
blue calcareous slate, muddy, and not at all calcareous-
looking, but quarried and burnt for lime. It recovers the
Bala character, however, to the southward, and for a mile
south of the place, where —— said it died out. The
fossils accompany the limestone in full force as far as
the Cowarch Valley at least, and for all I can see, they
are likely to go on to Machynlleth. The first night I
slept in this little public, after amusing a kitchenful
of people all the evening by trying to talk Welsh, I was
awoke in about half an hour by a tingling which I found
to proceed from a myriad of fleas. I feared there would
be hardly enough of me left to get up in the morning.
As soon as it was daylight, I jumped up ripe for slaughter, and killed a dozen. The rest escaped, but I put
myself into bed again as a bait, and when I felt them
busy, made another onslaught; and so on till I had made
a visible diminution in the numbers of the enemy. Last

night, being knocked up with hard work, I slept in spite of thunder or fleas, and only found four or five this morning, so I hope to have some little peace to-night. How brutally cold it was on the hill-tops yesterday, in spite of the sun! My big dog Governor always goes with me, and is the admiration of all the peoples. I knocked the old fellow up yesterday.* Tell us what you mean by fossils on Dinas Mowddwy.

May 8. Very shortly after we parted the other day, Gibbs again struck on the limestone in the bed of the

---

* This was the Newfoundland dog previously alluded to, his unfailing companion on the Welsh hills during many a long day's work. Great affection for animals had always been a trait in Mr. Jukes' character, and met with the usual response from those creatures—so dumb in words, so eloquent in action and expression. Dr. Ingleby, who visited him in Dublin in 1861, and sympathised in his cousin's love for animals, thus writes:

'It would do your heart good to see Beete with the beasts in the Zoological Gardens. Not only is he fearless, but he has a strange power over them—by affection, not by fear. The leopard plays with him like a great cat; he scratches the two tigers with his bare hand, and takes all sorts of liberties with the old lion, the lioness, and her cubs. There is also a hyena, who fears every one but Beete, and is quite attached to him; but of all the creatures (and Beete knows them all personally), the leopardess is the most friendly. When she sees him coming, or hears his voice afar off, she positively dances for joy; and when he comes up welcomes him with every sign of doating affection.'

Mrs. Jukes fully sharing her husband's fondness for animals, their house in Dublin was like a menagerie on a small scale, and the breakfast-table was attended by dogs of various sizes, a cat or two, and a very lively and affectionate cockatoo.

Dyfi, just where we had urged it was likely to go, and traced it for half a mile from near Aber Cowarch to the 'w' of Tyn-y-Pwll, within half a mile of Dinas Mowddwy. He says it is as thick, strong, and characteristic there as anywhere near Bala, with some six or eight feet of 'banded' beds, and some hard gray crystalline limestone underneath them, the thickness of which cannot be seen for the river. Not a soul in the neighbourhood knows of it, and they want lime shockingly. It lies precisely where I had drawn a pencil-line for its probable position, if it went on unbroken by faults or curvatures. On returning from Llan-y-Mowddwy, I brought it on a little farther northward in the regular straight line north 30° east, as far as half a mile north of 'bwlch' in Tan-y-Bwlch. We have it now, therefore, running in a directly straight line for seven miles parallel to the Aran. To the south it seems to be either broken or bent suddenly, as there are some due south dips near Dinas Mowddwy. To the northward a strong ash-bed suddenly comes in at the top of Craig-yr-ogof, and the limestone is for the present gone to Jericho, and I can't go after it till it please Heaven to raise the thick curtain of clouds and mists and storms in which the hills are enveloped. About the top of Craig-yr-ogof there are the biggest, brutalest, brownest, barrenest, boggiest moors in the country; and just when I was in the dreariest centre of one the other day, it began to

blow and rain like fury. Yesterday it rained, and I
rested; to-day I could rest no longer, so marched out
into the mist, and got a soaking for my pains; but I
think made out the stumptailedness of one piece of ash
that had puzzled me. But why am I bothering you
with all this now? Will it not have to be talked of in
the day when we meet each other face to face? What I
want to ask you now is a question concerning the rules
of the service. Can we ask, or can you grant, leave of
absence in case of the death of a relative? My uncle,
whom I left in London, is, I hear, very ill, and not
likely to recover. Should he not do so, it is probable
that it will be necessary for me, as the only one acquainted with his affairs, and the one in whom he has
always most confided, to go up and superintend his
funeral, &c. How am I to manage in such a case? I
was asked this week to come to the wedding of a cousin.
This, however, I decidedly negatived, on the plea that
'the service did not admit of it.' The other affair, however, should it occur, will be no party of pleasure, but
one which, if it devolved upon me, I should think it my
duty to undertake. This question I want answered by
return if possible. It appears our letters go by Aberystwith. Yours had been three days on the road, so that
I could get an answer from Paris just as soon as from
Builth.

On May 4, I stood over shoe-tops in snow. I am now

just going to correct the last sheet but one of the *Voyage of the Fly*, of which I am heartily tired.   I begin, now I see it in print, to fear it is a very trashy affair. However, it will perhaps sell all the better for that.

If William is with you, ask him what he has done with my maps. He was to send them to ' A. H. Browne, Wolverhampton ;' but Browne has not received them. Hooker wants them, as he goes down there directly.— Believe me yours ever,

J. BEETE JUKES.

Bala, May 23, 1847.

MY DEAR RAMSAY,—Last Thursday the weather became fine, it having rained for several days previously as hard as it could pour. I accordingly set to work, and have had three days' regular labour. I found more ash-beds N. of the Llaithgwm one, and traced them in as far as Llwyn Gwegian. The next day I got a pony, and went N. over Cader Benllyn and Moel Eglwys, all slate dipping S.; thence down to Cerrig-y-Druidion, slate with a westerly dip. Thence to Cader Dinmael, where the Bala limestone is in greater force than near Bala, but is much broken, and goes sprawling over the hill in all manner of directions. Then returned to Cerrig-y-Druidion to sleep. . . . While on Cader Dinmael, I seemed to feel as if some great change was about to

take place in the character of the country, and I see
Sedgwick draws the upper Silurians over all the country
round by Pentre Voelas and N. of Cerrig-y-Druidion,
so that what becomes of our rocks Heaven only knows.
I have just read over Father Adam's paper on North
Wales, and I see he has looked on the physical structure
of this country with his accustomed accuracy and saga-
city. His 'enormous dislocations' are positive facts,
only, being more enormous than he imagined, or at
least being more frequent, he has taken for different
beds what are in reality repetitions of the same. Sharpe
insists on the Hirnant limestone in two or three places;
so though I don't believe in it, I must have another
look at it. But I mean first to get out the Bala limestone thoroughly as a base. When I have done that, I
think I had better just ride over to Llanwddyn on the
one side, and to Yspytty Evan on the other; as Father
Adam says the Bala limestone is at the former, Sharpe
at the latter place. While I have its characters in my
eye, I think I had better determine these points. What
say you? I shall have two or three pounds horse-hire,
but as it is only five shillings a day, that is not much.

I heard from Selwyn yesterday, but his letter had
been six days on the road. Gibbs has not found any
more limestone yet, but describes very large trilobites
on Moel Dinas, much larger than any at Bala, but
other fossils growing much scarcer, as if dying out. I

am not at all satisfied about that country, and think
that when the boundaries are traced in, you had better
take a month to traversing it about and about. As soon
as I hear that Smythe is there, I shall go over for a day
or two and have a squint at it myself.

May 28. Being too utterly sleepy to answer you last
night, I have arisen at five this morning, partly for
that purpose, and partly to lay down yesterday's work.
Selwyn came here the night before last. He is now
gone to bathe in the lake, taking my big dog Governor,
and his own spaniel Carlo, to pull him out again.
Perhaps you will hardly believe it, but I am almost
knocked up. I have had a week of downright toil, never
returning till 8 P.M., and never sitting down during the
day; consequently I am stiff in the hips, shaky in the
ankles, and altogether groggy. I have succeeded in
carrying the ash across the big Cwm Chwilfod fault,
although in a most spifflicated and broken condition;
but the limestone won't show. Selwyn and I hammered
every piece of rock yesterday, from Cryniath and Erw-
feirig to the top of Cwm Cywen, and from Pentre-cwm-
da to Llawr-cwm, and the limestone ain't nowhere no-
hows.

I contemplate to-day resting myself on a pony, and
riding between that and Pont-y-glyn-diffwys, sleeping
at the Goat Inn or Cerrig-y-Druidion; and to-morrow
making a dash right out north, and seeing what those

rocks really are. If the Caradoc sandstone curves round
there, it is very shortly succeeded by something else,
as they bring flags here, from a place a little N. of the
Corwen sheet, which are covered with beautiful Creseis
and Cardiola, and other fossils quite different from any
in these rocks. While this fine weather lasts, I mean
to make vigorous play; and if my own legs won't serve
me, why I must borrow four more. If I can only get
the limestone up to Pont-y-gd, I shall sleep happy to-
night.

June 6. I got yours of the 2d yesterday. I have
now devoted a solid week to the country between Cwm
Chwilfod and Pont-y-glyn-diffwys; and as far as the
position of the limestone goes, I am as wise as ever.
There is a public at a place called Tyn-y-nant, where I
shared a bedroom with the coachman of the Holyhead
mail, and there is the Goat Inn. I have slept every
other night at one of these places, vibrating between
them, and spending the whole day in the field, either
hunting or cogitating. It's a queer spot. I have got
the ash-beds on, although much broken; and I have
looked at the Caradoc in your boundary, which is quiet
enough. . . .

I am sorry you find so much perplexing work where
you are; it only shows how necessary it will be to publish a second edition of some of the earlier sheets,
worked by the light of the more modern ones. . . . On

Selwyn's return, we'll have a go at the maps, polish off
the west side of the Bala sheet, and then proceed to the
desolate moors of Cerrig-y-Druidion.—Ever yours,

J. B. JUKES.

Cerrig-y-Druidion, Aug. 22, 1847.

MY DEAR RAMSAY,—I am very sorry to hear your
foot is still so bad. I hope you will shortly come to a
better understanding.* 'Ash, sir, ash,' is still the go,
a little piece here and a little piece there, here a strip,
and there a patch, and then a sprawl. I have, however,
now driven the beggars into a corner, and fairly succeeded in jamming them up into a narrow passage between the traps on the one side and the Caradocs on
the other, only a mile and a half wide. What has
become of the limestone from the top of Cader Dinmael
to Yspytty, Heaven only knows. You would get an
exceeding curt note from me regarding the maps. I
was in haste when I wrote it, having fallen asleep the
night before, when I intended to write it. And how is
the Forbesian hero? He of the long bones, ' the starve-
ling, the elf-skin, the tailor's yard, sheath, bowcase ;
O, for breath to utter what is like him !' I slept at

* This refers to a severe sprain which Mr. Ramsay got while
with Sir H. de la Beche on Glyder-Fawr, Caernarvonshire, and
of which he was trying to get cured while on a visit to Mr. and
Mrs. Johnes of Dolaucothy, Caermarthenshire.

Bala on Friday night, and Mrs. Jones inquired very particularly after both of you. She hoped Mr. Forbes was better; he did look so very ill when he was there. I should recommend him to drink whey while in Wales; my dog Governor is growing fat upon it, and it might suit the Forbesian anatomy. ' Good counsel—marry take it, take it.' I verily think I could write an entire letter this morning of Shakespearian quotations. Did you never feel your brain in a complete ferment and hurly-burly, with all kind of three-cornered scraps and brecciated angular fragments of poetry flying about it in all directions, and jostling each other at every angle, without any kind of purpose or arrangement? ' Wit, whither wilt?' I heard from Darwin yesterday. He wants some pedunculated cirripedia; but I have told him all my spiritual exercises, procured on the Australian coast, are now buried in the deep, deep sea of the vaults of the British Museum. By the bye, in the second review of the *Athenæum* of my book, they have a go at me about shooting the black fellows; and say, ' Mr. Jukes should recollect it is the triumph of the moral and cultivated man to subdue such resentments,' &c. What a lark! Fancy their addressing me gravely as a moral and cultivated man! How I should like to get the chap that wrote that, in a boat-cruise on the coast of New Guinea; keep him out for three days in a heavy sea; feed him on salt-beef, rum, and tobacco; make him sleep

x

on a board in a flannel-shirt and no pillow; and then take him into a scrimmage with a lot of black fellows. I'd then ask him how he felt in his morality and cultivation, and whether they sat easy on his stomach, or not.

Well, here comes the post with the newspapers, which saves you from any farther infliction of my nonsense; so with best love to Forbes, believe me ever yours truly,

J. BEETE JUKES.

After another careful look at the ash-dyke, I think it possible it may be the result of a complicated kind of fault, but if so, it's a very rum one. There is a clear fault in a quarry hard by, that produces a distant approximation to that kind of thing.

The following note, taken from Mr. Jukes' address to the Geological Society of Dublin, in 1855, already referred to, may serve to illustrate the difficulties of surveying the country, from which most of these letters were written:

'I had once been working hard for about five weeks, trying to understand and delineate on the one-inch map a complicated bit of mountain ground a few miles south of Conway, in North Wales. It was made up of inter-

stratified slates, sandstones, and felstones, with large
and irregular masses of intrusive greenstone, the exposed parts of each being frequent, but not continuous.
Many a weary day had I climbed the sides and clambered
along the crags of a hill, some five or six miles in
length, by two or three in breadth, and the highest
peak of which was not more than 1800 feet above the
sea, trying in' vain to reduce to order the seemingly
endless complexity of its structure, and having at length
on the map as curiously complex a patchwork of incongruous colours and unnatural forms as Punch, had he
turned geologist, could have devised; when one evening,
as, after a hard day's work, I was descending a steep bit
of ground, almost in despair at all my labour seeming
to be thrown away, I hit upon the clue to a great fault
or dislocation. I had only time then to verify the observation, but it gave me at once the solution of all the
puzzle; and in two or three days I was enabled to map
the whole district, with as near an approach to accuracy
as the scale of the map admitted of. The country was
chopped up by a series of large parallel faults, that
were quite easy to be seen when once the clue to one
of them and its bearings were obtained, but which there
was nothing to render à *priori* probable, and which
could not have been discovered without that thoroughly
exhaustive process of examination which I was enabled
to apply to the district. I have ever since regretted

that, in my haste and joy at acquiring a right notion, I
obliterated all my former work from the map which contained it; for I should have been glad to preserve it now
as a curious instance of the contrast between laborious
hypothesis and the simplicity of natural truth.'

Yspytty Evan, near Pentre Voelas,
Denbighshire, Aug. 28, 1847.

MY DEAR RAMSAY,—As usual, I send you a report of
proceedings.   I told you I had traced the ash-beds into
the narrow passage between the traps and the Caradocs;
I have now got them through it. In the first place, they
confirm the unconformability of the Caradoc, inasmuch
as I get the upper ash running in a straight line for a
mile, and approaching in one place within a quarter of
a mile of the base of the Caradoc, the dip of the ash being 15°, that of the Caradoc 25° nearly in same direction. This proves that the cross of the limestone is
under the Caradoc, which is confirmed a little farther
on by its appearing from under it very near its boundary.   In the next place, I get two lines of ash, one
broken, the other continuous, but each pointing directly
for a mass of stratified *trap*, rather thicker and more
massive than the ash, but not greatly so, and occupying
the same relative position in the slate-rocks as the ash-
beds do. These traps, however, do not appear to ex-

tend very far to the west, but the country is much covered with drift and bog. A little farther west (about a mile) there is, however, below the limestone, about the same place as the upper ash, a line of detached patches of a most curious rock. This is sometimes a regular flaky ash, at others a hard siliceous trap, full of nodular concretions about the size of apples, and being either agates or chalcedony. Each concretion is formed of concentric coats of white compact milky quartz, and they are in fact agates. They are frequently so thick as to compose the whole mass of the rock. . . . If inclined to speculate a bit, we might say, on the supposition that the ash-beds are really the product of sub-aerial ashes, that the wind was westerly during the eruptions, and blew the ashes from where the traps are, towards the east. On, to my mind, the more probable supposition—that these eruptions were submarine—the currents must have set to the eastward, and drifted the débris and lighter materials in that direction, forming what we call the ash-beds. Anyhow, it appears that this is a most curious and interesting country when it comes to be worked out in detail. It requires, however, very close work and much walking. How the engraver will ever get in all the bits of ash scattered over the country, and the colourer avoid missing them, I don't know. I hope by this time that you are at last on the mending tack. Don't make yourself uneasy; the summer

is done as far as you are concerned, and therefore you
have only to make yourself comfortable, and take care
you don't contract a permanent lameness by working
too soon. I have supped full of praise; in fact, it is
very well I am in a little quiet place here, where nobody
regards literary reputation, and where accordingly the
inflation consequent on so much flattery can quietly
work itself off without producing any evil consequences.
*Daily News, Morning Post, Atlas, Spectator, Morning
Herald,* &c. have all been praising my book; the latter
is quite sugary as regards myself personally. I really
wish some one would set to work to attack me, if it was
only to prick the flatulency of self-conceit that is ' blowing me up like a bladder.' The *Atlas* does have a dig
at me for shooting, or wishing to shoot, a black fellow;
and the *Morning Post* spits on my notions of protection
and monopoly in Java; but both these things I expected in much worse fashion than they have given me,
so that they fail of their effect as flabbergasters. If
they'll only contrive to sell the book, so that I may
ballast with a little gold against all these puffs and
breezes, I shall be content. By the bye, we get the
Bala fossils here for Forbes, but not very plentifully.
Gibbs, however, reported something fresh yesterday,
along with many univalves, turbinolopsis, and heads of
illænus. Some orthoceras and encrinites; no cystidea
yet. Love to his lengthiness and yourself.

Yspytty Evan, Sept. 6, 1847.

I have just received yours of the 2d, per Aberystwith, Shrewsbury, and Bangor—a pretty little round. Last week's *Athenæum* has never arrived at all; it is probably still on its travels round about the principality. I shall be compelled to write to the *Caernarvon Herald* and touch up the Post-office. I am sorry that your understanding is still in such a crippled condition, and that you are compelled to ride the high-horse in consequence. Mind he does not come down with you and damage your other extremity; in our work a man might very nearly as well be without his head, as laid by the heels. I hope when you arrived at Kington you received a pretty long letter from me anent these parts. Since then I have traced two or three miles of limestone, which in one place rests directly upon a great boss of trap, in others has a large ash-bed for a supporter. The curious knobby trap made of agate nodules still runs on in detached lumps about same distance below limestone as did the ash-beds in the Bala country. So far all well; but on approaching the Caradocs, I am in a shocking state of confusion. . . . I had a very hard day on Saturday, and drew at last two boundaries half a mile or more apart from each other, and can only say that on the farther side one is certainly Caradoc, the other Bala; but what is the intermediate, I do not know. Then there are such dips as N. 20°, W.

at 80°, N.E. at 10°, N. at 70°, S.W., and all higgledy-
piggledy. There is a great lot of massive hard sandstone south of this place, and whether above or below
the limestone, be hanged if I know—(stop till I fill
my pipe; so!) Neither can I connect the two pieces
of limestone that lie one south and one west of this
place. Altogether, I feel rather in a fix. However, as
usual, if he won't run, he must jump; and if I can only
get a week or two more fine weather, I'll ram the spurs
into him, and make him go somehow.

On the morning of the 2d, having heard something
of partridges being seen while they were cutting the
corn at the back of the house, Governor and I arose up
early in the morning, I carrying my gun, and walked
quite promiscuous-like into the stubble. Old Gov. begins sniffing and wagging his tail, makes a bolt, up
springs the covey, and down comes one. While marking them down, however, the old rascal (it being his
first essay in the business) had very nearly swallowed
the bird, thinking it was knocked down solely for his
gratification. Walked on, Gov. busy. Down comes a
brace; secured one whole, and whipped him for spoiling the other, and showed him how to handle them
gently. Ground belongs to Lord Mostyn. I met the
old gentleman one morning. He was very civil, and
said when we came into Flintshire he should be happy
to show us any civility that lay in his power. Mean to

have a regular day before I leave this. Partridges are
a very good alterative when on a course of mutton. . . .

Winter has set in here, and has put me considering
on the future. It is very evident that I sha'n't be able
to finish this tract of country, including Llanrwst, before the snow comes. Now, is it expedient that I should
winter at Llanrwst? There is a wide tract of country
to the west there that can only be done in fine weather,
and, I think, with a tent; as there is not a house near
it, and the hills nearly as big as Snowdon (to be sure,
that may not be Bala beds), but Selwyn thinks they go
up to Dolwyddelan, another wild tract. Now, if you
think it better, I shall of course be happy to stick at it;
but really it appears a kind of country in which winter
work will not be very productive of results to the survey.
Would it not, then, be better that I should take the
winter in the South Staffordshire coal-field, the out-
lines of which I have already mapped, and where I could
be occupied in preparing pit-sections, &c., from the
different coal-owners' documents? Rain or fine won't
make much difference there, as the coal-field is all covered with rubbish, and I shall have to trust to the
mining plans and old butty colliers for the outcrop of
beds and course of faults, &c. &c. Just turn it over
in your mind. Don't think I am working a dodge.
Never mind which would be pleasanter to me, but just
decide which is most expedient for the survey. I thought

it better to mention it now, in order to be prepared as to lodgings at Llanrwst and other arrangements. Of course I must carry these beds up to some definite point before leaving, so that I, or any one else, could pick them up again. Selwyn went to Capel Curig yesterday. We shall meet shortly, and hold a consultation. I am happy to hear Forbes is looking so blooming. I suppose there is no chance of your or his reaching these parts this autumn. Very quiet this place. All gone to bed, and nothing to be heard but the roar of the Conway, swollen by the rains. Good-night; and believe me ever yours most truly,

J. BEETE JUKES.

Sept. 14.

DEARLY-BELOVED BRETHREN,—When the fossil man picketh up a fossil, and bringeth it unto me, and behold the form thereof is strange unto my eyes, then shall I make a sketch thereof, and send it unto the elders, and they shall pass judgment thereupon (c. x. v. 6 and 7). In the words just read you have, my brethren, the direction of the writer as to what is to be done in a case that unfortunately does not often arise, and so forth.

Gibbs found the other day in a wall what I take to be a pseudocrinas. He found out where the stones came from. Its place is below the Bala limestone—pro-

bably four or five hundred feet below it. I send a rough
sketch of the thing, but we have also the lump of stone
out of which it came, and which shows some curious
features; more especially the mark as if a soft bag had
proceeded from the centre of ' *b*,' like the dome-shaped
bag in a comatula. There are also the arms, of which
those on ' *b*' are casts. A piece remains in one; my
sketches are too rough, however, to describe it. The
plates seem hexagonal, but they have all suffered in the
squeeze it has had. I send the sketch, as Forbes may
wish to have the specimen directly. Gibbs has also
found the limestone between Penmachno and Dolwyddelan, where I sent him the other day. Penmachno is
too far from the Caradoc boundary to be a good place
for me. Indeed, I could get no rooms, unless I applied
for Lord Mostyn's shooting-box, which a gentleman told
Selwyn he could get for us. Meanwhile, it is getting
expensive in this confounded tourist country. They
wanted twenty-five shillings a week for two little poking
rooms at Bettws-y-Coed in a small cottage that was a
public—merely for the rooms, and find yourself. By
the bye, both Gibbs' and my compass are spiflicated.
I am going to try and remagnetise them, but fear I
sha'n't succeed, in which case I must send them both
up to London. Would it not be a good dodge to have
a few spare needles and cards at the office in boxes
made on purpose, which could be sent at once in such

a case? They have spare ones on board ship always. I don't think I can go on without the compass to fix myself.—Your worship's most obedient,

J. BEETE JUKES.

Yspytty Evan, Sept. 22, 1847.

MY DEAR RAMSAY,—In your last note you talk of the ' massive genuine erupted lavas ;' and then you say the parts we now see *never were at the surface* till exposed by denudation. Now, according to my notions, that is a direct contradiction of terms, since lava cannot be *lava* unless poured out not only *at*, but *on to* the surface. Perhaps you merely meant by ' lava' igneous rock generally. I think, however, it is bad to get into the habit of talking loosely on these points, since it is apt to mislead the uninitiated. It is very evident that our *geological* terms for igneous rocks have very rarely any reference to their chemical constitution (even the mineralogist considers two things as totally different minerals which are almost chemically the same, as mica and horneblend), but have reference principally to differences caused solely by the different conditions under which they were placed. The same identical mass of matter may form granite or obsidian, just according to circumstances. . . .

I doubt whether even the term submarine lava can be admissible except in those instances in which a stream of lava, after flowing on the land, reaches the sea, and falls into it. Igneous rock poured out under a *deep* sea can never be lava certainly. I enter into this question because I am sure the time is fast approaching when the whole nomenclature of igneous rocks will have to be revised, and we ought to be up to the mark and ready to enter on the question. I see some foreign chap has been saying that quartz veins are often injected while fluid from heat: what I have seen near here leads me to agree with him, and also that they have altered the rocks through which they pass. Wherever we find large quartz veins here, the rock is either genuine trap —compact, smooth, splintery, siliceous stuff—or else a sandy or siliceous rock, that is hardly distinguishable from trap. On the other hand, some of this compact stuff contains fossil shells; so that I am sometimes tempted to consider the traps (of the Arenig, for instance) as merely a mass of rock altered by heat, consisting, we will say, originally of such materials as a low degree of heat would easily affect, and containing perhaps here and there small eruptive points, near which the rocks have really been fused. Such a degree of heat might not produce any greater change on the other rocks, such as the sandy and muddy slates, than we know has been produced, viz. their intense consolidation and their slaty cleavage. Under this view, all porphyry which merely contained crystals dispersed through a

compact base would not be genuine igneous, but merely
a highly metaphoric rock, same of hornstone and all
similar compact rocks. This would make crystallisation the test of igneous fluidity, except in the case of
lavas, obsidians, &c. poured out under the air. Barometer rising, so I'm off.

(8 o'clock, P.M.) Lost the thread of the above discourse, so we'll call another subject. I've had a hard,
perplexing, and most unsatisfactory day's work. To
describe it, I must quote *Punch :*

Report.

The commissioner required to report on the Caradocian boundary, in the neighbourhood of Pentre Voelas,
begs to report accordingly :

There is no Caradocian boundary in the neighbourhood of Pentre Voelas.

(*Signed*)   THE COMMISSIONER.

Not that I mean to assert there is no Caradoc. There
is, on the contrary, abundance. Still less do I affirm it
to be boundless, seeing that a mile to the westward
there is none. Where the boundary is, however, and
what it is—whether formed by a lot of jagged faults,
with long needle-shaped pieces of the two formations
dove-tailed into each other—whether pieces of the Caradoc sandstone are pitched in below the Bala limestone,
or pieces of the Bala limestone thrown atop of the Cara-

doc, Heaven only knows. I give it up. All I shall say is, ' it is somewhere hereabouts,' as the merchant-captain said of the ship's place, when he spread his hand over the chart. It must be worked from above, by tracing the beds of the upper part of the Caradoc, which I leave the illustrious Aveline to perform. By the bye, I see in some of the papers in the *Geological Journal* they talk of the Terebratula navicula as confined to the Ludlows. Will you tell Fubbs we found it near Cerrig-y-Druidion in a bed of semi-calcareous slate, associated with Rhiwlas fossils, and which may be in the parallel of the Bala limestone. It's all nonsense those Palæonts trying to confine bits of fossil shells within two mathematical planes. I believe now that these Bala beds, after sweeping round the north end of the Arenig trap, begin to be affected by several big contortions. They run in a short broken trough up this valley. The next valley to the west, that of Penmachno, is on the axis of an anticlinal; the one beyond that, Dolwyddelan, on one of a synclinal; and up that these Bala beds will run—I believe to Beddgelert from Selwyn's account, and that they rest with an easterly dip against the east flank of Snowdon, according to yours. The most perplexing point is to get them to turn the corners nicely here towards the Caradoc. However, we made a good find the other day of limestone and trap north of the Conway, just where they were wanted. I think your ideas about Gibbs good. One thing is plain : it is mere waste of time to remain amongst these high hills in the winter—they are often enveloped in clouds and storms when it is quite bright and fine even at Cerrig-y-Druidion. Well, I should say you'd had enough ; so good-night to you and his lengthiness.

Yspytty Evan, Sept. 25, 1847.

Yours of the 21st arrived yesterday. I think it will be just as well to take the South Stafford coal-field first, and clear it out entirely. It is a separate district, an island in the Y.R., extending from Rugeley to Stourbridge, and from Wolverhampton to Barr Beacon. For this purpose, therefore, sheet 62 will be sufficient. From Coalbrook Dale to the Clee Hills is another natural district, which, especially about Bridgenorth and Bewdley, will be more important from the deductions to be drawn from it than from the value of the beds themselves. I look upon the Bridgenorth and Bewdley coal-field, and that between Coventry and Tamworth, as *the districts* to enlighten us on the true relations between the coal-measures of the Y.R. In each of these coal-fields there is a perfect passage from the C.M. into the Y.R. itself, and an unconformability consequent thereon. Now it is clear the lower Y.R., conformable to the C.M., represents the rothe todte liegende beds ; but whereabouts

is the place of the Mag. L.? Murchison says this latter
is represented by some bands of cornstone which occur
well up in the horizontal beds of the Y.R. But I think
there is a perfect passage and conformability from those
beds up to the lias, and that consequently these red
horizontal beds are the Banter sandstone, the lower member of the Trias. I suspect also that the Mag. L. ought
to come in, in the place of the break in the Y.R., and
that the absence of the Mag. L. was caused by the movements of disturbance going on in the Midland districts
while that bed was being deposited on the north and
south. It is another case above the coal similar to the
absence, or almost entire absence, of the Mountain L.
below the coal, and probably due to similar cause, viz.
that either the bed of the sea was in process of disturbance, or that having been disturbed, it formed dry land,
&c. A corollary of the thorough working out of the
Midland districts, and comparison between N. and S., tracing the marine and fresh-water beds, &c., may perhaps
be, some light thrown on the origin of coal as connected
with movements of elevation and depression. Another
deduction of great practical importance will be the state
of the base of the Y.R., and of the carboniferous rocks below it. One might almost say the latter question was one
of vital importance to the future greatness of the empire.
Bravo! hurrah! strike up, band, ' When Britain,' &c. &c.

As to driving these beds here into a corner, it's no

Y

go. They won't stick, but slip out again on one side,
and are off for Bedgellert.   One thing, however, I did
yesterday—being a fine day, for a wonder—was to dove-
tail my work into Selwyn's, and solder them together
into a compact mass.   What do you think of the ash-
beds of Bala being identified with the ash-beds to the
north of the Ffestiniog slate quarries.   I have not yet
quite all the links of the chain, but enough to satisfy
myself of the fact.   I told you my upper ash passed
into trap, and then into the hyperbolic rock.   This
rock goes on capitally at just the proper distance below
the limestone, and they both cross the anticlinal of the
Penmachno Valley, and dip under the synclinal of that
of Dolwyddelan.   I have got a very calcareous gray
ashy trap, or trappy ash, in the place of the lower ash-
bed of Bala and Cerrig-y-Druidion, just at the top of
the black slates; and this, in like manner, goes from
this side of the anticlinal to the other, and then strikes
off for Selwyn's long narrow lines of ash, of which Gibbs
says the lower is just the same material on Moel Far-
llwyd. . . . I think if I work these beds all north of the
Conway, and east of it, and all between the Conway and
the Lledr to the south and west, I shall have got them
to a very convenient boundary.   A fine fortnight at
Bettws would do this. They are from five to seven miles
away from here now, across the most up and downiest
kind of country into the bargain.

A strange and marvellous History of a Temptation,
and what befell thereon.

On the day of the week called Wednesday, and on the
22d day of the month which men call September, there
came unto me early in the morning a great devil, in the
likeness of the keeper of the Lord Mostyn, a mighty
shooter of birds, and he said unto me, 'Sir, if it seemeth
good unto thee, let us now rise up early on the morrow,
and go into the land of Gwydir, to the place called
Trefriw, the which country belongeth unto the Lord
Willoughby d'Eresby, and the keeping thereof is in my
hand. And it aboundeth in grouse, and in partridges,
and in woodcocks, and in hares, and in all manner of
game; and I will bring with me my shooting-iron, and
do thou also bring thy shooting-iron, and we will shoot
exceedingly over all the land.' Now this he said tempting me, and the thought thereof was pleasant unto my
mind. Then rose up the good spirit called Conscience,
that resideth in the chamber of the breast, and said
unto me, 'Nay, but thou canst not do this thing, and
go away from the work on which thou art employed.'
But to this answered the little devil Inclination, that
lieth along with Conscience in the chambers of the
breast, and ruleth with him over the heart of man; and
he also began to deal subtlely with me, and said, 'Canst
thou not also do thy work there, and see rocks which

it is necessary for thee to see, and acquire the knowledge thereof?' But Conscience answered and said,
'How many rocks, now, wilt thou see when the dogs
and game are before thee, and thy shooting-iron in thy
hand; and how great will be the knowledge of them that
thou wilt gain?' And the good spirit Conscience gained
the mastery in this matter, and prevailed over the great
devil and over the little devil, and set their counsels at
naught. Then saith the great devil, 'If this seemeth
not good in thy sight, lo, on Monday next I shall come
here, even unto this land which is round about thee,
and shall shoot therein. Wilt thou not take thy pleasure
here for one day only, and show forth the goodness of
thy shooting-iron and the skill of thy right hand?'
And the little devil Inclination, being grieved that he
had been worsted in the former contention, here hastened to bestir himself; and he rose up quickly, and
cried with a loud voice, 'Yea, this is good, and pleasant, and proper, and moreover it is right; and it shall
be done.' And he withstood Conscience boldly and
with much clamour, and defied him to say anything
against it, and refused to hear what he had to say; and
he overcame him, and flabbergasted him, and put him
altogether to silence, so that he was both afraid and
unable to speak. Then did I agree with the great
devil; that it should be even as he had said; and he
went on his way, having comforted himself with a vessel

full of gin. And the little devil Inclination and the good spirit Conscience sat down together as of old, and reposed themselves in the chamber of the breast, and became friends even as before.

During the writing the above true and marvellous history—more marvellous perhaps to you on account of the appearance of any Conscience at all—the clouds and the rain have cleared away, and I am off to a little bit of work about two miles from here.

Nine o'clock P.M. Having dined, smoked two pipes, drank three cups of tea, and a jug of water, I feel myself equal to the task of finishing this terrific letter. The fact is, I've used up all my note-paper, one advantage of which form is, that it compels you to study brevity, whereas in one of these large sheets, not only do the sentences get most unmercifully spun out, but the pen is apt to get visited with a kind of delirium tremens, or morbid excitement to writing, and running altogether away, refuses to stop till it comes to the end of the paper.

Concerning a certain 'phossilipherous indiwiddle,' as Forbes would call him, there is no great necessity for one in the Wolverhampton district—at all events, not immediately. I can set quarrymen and miners to work, who in that district are fully up to it. If, however, I find it necessary, I can apply for one. My principal work, I take it, during the winter will be the collection of pit sections and tracing the faults. There will be but few out-crops of beds, as the thick coal is the only one worked over the largest part of the space. Boundaries I already know very nearly. There will be, however, some close work of that sort required in obscure places on the edges of the field, and in its northern portion, of which absolutely nothing is known to any one, beyond the mere fact of there being coal-measures there.—Believe me, as ever, yours,

J. BEETE JUKES.

Yspytty Efan, Pentre Voelas, Denbighshire,
Oct. 1, 1847.

DEAR RAMSAY,—You justly complain of my long letters; but during the rain, being all alone, I was glad to do anything for an amusement. Wet weather is the cause of more writing, than fine of hard work. As Shakespeare says, 'Peace is a getter,' &c. I forget the greater part of what they were about, but will just notice one or two points in yours. I look on the term 'trap' as a generic one, loosely applicable to all such rocks as basalt, greenstone, porphyry, or lava, the whole of which, as well as syenite, granite, and all other igneous rocks, are apparently the same *identical stuff*, differing only from the different conditions under which they have been placed. I should restrict the term 'lava'

to those masses which could *be proved* to have flowed
under the air. Many thick lavas, if only the bottom
part remained, would no doubt be called basalt or greenstone, and such I should continue to call and consider
them, until *they could be proved to have flowed in the
open air.* Thus in Van Dieman's Land I always in my
notes spoke of a mass of cellular and scoriaceous igneous rock as 'cellular trap,' until I found fossil trees in
it in position of growth, and evidently enveloped in it
while living. I then held that this cellular trap was
proved to be lava. The surrounding greenstones I
simply styled 'greenstones,' leaving the question open
to farther proof, whether they were part of the real
lavas, or an older exhibition of igneous rock, formed
beneath the sea or beneath other rock, and therefore
not lava. The same rock may be both basalt and lava,
or porphyry and lava; but if we apply the latter term
to it, we, according to my notions, say that the rock on
which it rests was dry land at the period—an assertion
involving so wide and important a condition of things,
as to require the most clear, strong, and undoubted
evidence before we can admit it. For my part, I have
yet seen nothing in North Wales which might not have
been formed in the depths of the Atlantic, according to
my own notions of these matters. You talk of ashes
lapilli, big stones, dust, and volcanic mud, to which I
answer, 'Proof, proof, proof!' Except the big stones,

which prove nothing, I do not know that I have seen
one of these things. Mind, I say I do not know, neither
do I know that I have not. They may have once been
such things ; they certainly are not so now, therefore I
want proof that they were. Where is the proof that
there was dry land in the space now the British Islands
during the Silurian and anterior periods ? If proof
can be brought, then I would at once admit (what I
do not now deny) the possibility of the subaerial nature of these igneous products, though even then they
must be most singularly metamorphosed from their
former condition, if they resembled once the volcanic
products of the present day, or, at least, any I have seen.
Fossil terrestrial plants would go a good way. Take
our traps—they rest on thick marine beds, they are
covered by thick marine beds. Where is the proof that
they were dry land in the interval, or where was the
dry land from which they proceeded ? The traps of
the Clee Hills were, I believe, lateral injections into a
mass of coal-measures, the upper part of which is denuded. In speaking of the traps of the Arenig as
metamorphic, I only spoke of an *extreme* case, which I
was sometimes tempted to believe, not of what I really
thought. You speak of layers of ash occurring between
' unaltered slate,' as a proof that the ash is not metamorphosed from its original state. There could be no
better proof that it was. Surely if the ash were actual

porphyry, it could not be more changed from its original condition of a trappean detritus (as I suppose) than is the slate changed from its original condition of loose silt. I should use the metamorphic condition of the thin beds of slate to prove the necessarily metamorphosed state of the other beds. Whether that metamorphosis was caused by heat, wholly or in part, is quite another question, though I believe it most probable that heat set the other agencies at work. Just beyond, you exactly express what I mean, when you say, ' good crystals may well occur in ash. Render this compact, and it becomes porphyry.' For 'ash' read ' igneous detritus ;' for ' render compact' read ' metamorphose ;' and the two propositions are identical. Apply heat and pressure to the loose Silurian sands and clays in Russia, you change them into these kind of things, as in tracing them to the Urals they are found to be changed, and as in tracing the clay and shale Silurians of Shropshire about Wenlock to the metamorphosed region of North Wales they are found to be changed. There is, I believe, no passage from C.M. to N.R‡. in the Coalbrookdale coal-field proper, but in the extension of it to Bridgenorth and Bewdley, also in the Manchester coal-field. In the north, the mag. lime. is as well marked as the great oolite in the south. You may walk along its escarpment overlooking the coal-fields from Nottingham to Sunderland, its beds being perfectly re-

gular and most tranquilly deposited, often a fine-grained
crystalline stuff like loaf-sugar to look at.

Here's another long letter! I've only the old excuse, that I had not time to make it shorter. And now
what do you mean, you villain, you and that sallow
spectre that now haunts you, by setting the whole
survey at me? 'I'll be revenged on the whole pack of
ye.' Five attacks have I had to parry and lunge back·
as well as I could. But what do you mean by keeping
my letters? All letters containing this sort of nonsense
pass direct from my hands into the fire; pray you do
likewise, otherwise I shall have to send for a 'Complete
Letter-writer' to model my correspondence upon. I'll
look out for some lecture-room specimens for you. I
can give you also some good rocks, granite, &c., from
Newfoundland when I go home, if you would like them,
or rather you must come and choose them. Looking
over your letter again. Your diagram would at once
almost convince me that none of the igneous rock was
lava. I should say the horizontal part was just as
much an injected mass as the perpendicular, and that
there might have been thousands of feet of stuff on it
when it was formed and now denuded. In short, I
don't believe there is any lava in Britain, but that all
our igneous rocks were formed either beneath the sea
or beneath other rock. There is no occasion for a term
descriptive of submarine flows of igneous rock: they

would be greenstones or basalts or porphyries. The detritus from such masses or beds, and the powdering and splintering and splitting into thin small flakes at the surface, which they would suffer when first ejected, would just form the ash-beds; and as great currents must necessarily be caused under any depth of water, and even waves of translation of considerable magnitude, we should have not only fine detritus, but large blocks spread over the surrounding sea, perhaps for many miles. In short, I think mere mineral character just as favourable to the *submarine* as to the *subaerial* origin of these things. I don't care which is proved, but I don't see why one should be assumed more than the other.

Yspytty, Oct. 5, 1847.

DEAR RAMSAY,—I can now recur to note-paper. As to the trees in lava in Van Dieman's Land, it was the fact of their having been evidently *charred through in situ*, and reduced to white ashes before silicification, that first led me to believe they were enveloped in lava while living. All this was fully expressed in my paper as it stood at first, but I was forced to cut all out except the bald facts, and a great many even of these. I should never have thought of publishing such a meagre thing as it now is, had I not got savage and determined to try if *I could get anything* printed about Australia. I am very sorry I did not put all my notes into my book instead of trying to make papers for the Geological. Not only were the trees charred, but the lava formed concentric coats round them upright; Strelecki traced the roots of one resting on one bed of lava, enveloped in another, &c. I feel that our trap controversy has had the proper effect of smashing the barriers with which a man's own ideas are apt to get surrounded, and allowing our notions to have a fair view and get acquainted with each other. As you say, the difference between us now is small. I thought you stuck out for all the ash-beds being the result of volcanic matters blown from a *subaerial* vent clean into the air, drifted by the winds, and falling in the sea. While not denying that some of them *might* be so derived, I stuck out against their *all* being so, or even the major part of those which I have been working. When you admit the probability of some of the ashes, or most of them, being of entirely subaqueous origin, you admit all I am arguing for. My notions of the origin of the ashes I have seen are these:

1. They are all of igneous origin, *i.e.* they are what Sedgwick calls 'recomposed igneous products'—of this no one has any doubt (except perhaps Sharpe, who has notions of his own).

2. Beds of 'ash' can be traced to have an intimate connection with beds of trap, to come up to and join

into them. Now I think volcanic forces were at work
on the old sea-bottom, burrowing here and there about
it, and every now and then squeezing out masses of
melted rock, which would flow slowly along the bottom
in various ways. Now we know very little of what
would be the *exact* result, but I think we may say these
masses would be irregular in shape, rather uniform in
lithological constitution, which would differ both from
similar subaerial masses and from the same masses
injected into rocks. They would hold the middle place
between the two; they would have more pressure than
the subaerial, but cool more rapidly than the injected
(*ceteris paribus*). Suppose a mass of liquid feldspathic
rock issuing from a vent in a sea-bottom. When it came
in contact with the water, it must surely be cooled
rapidly and the water heated rapidly. Its surface, being
so rapidly cooled (instantaneously we may say), would
be split and shivered, not only into small fragments,
but into actual powder. It would be, as it were, weathered at once, and to a considerable depth too. Then
the water being rapidly heated, almost formed into
steam, in spite of the pressure would ascend (and actually expand into steam as it neared the surface), and
currents would be formed of a force far exceeding any
we have in tide-ways or surface-currents. Great force,
however, would not be wanted, since they would have
only shattered and powdered materials to act upon,

though much of these materials might be carried up by
the ascending currents, and thus get exposed to wider
acting influences, and spread over much of the sea-
bottom. In this way I should conceive that any small
submarine vent must have its ejected trap almost immediately converted into what we call ash; except just
its nucleus, which, having been better protected, would
be still trap, but with every gradation between the two
things.

In my notes on the Bala beds, I have described the
ash-beds as having three varieties : 1, the flaky; 2,
the brecciated ; 3, the crystalline. Now the occurrence and the mingling of these things would all be
caused by the process supposed above. In some of the
brecciated ashes, some of the angular fragments have
crystals of feldspar in them, and are true splinters of
trap. The flaky ashes were not originally formed so—
the flakes *lie at right-angles or oblique* to the beds, like
cleavage ; it is a cleavage, or, as Darwin would call it, a
foliation, like that of mica slate, affecting a peculiarly
fine-grained originally powdery material, but, together
with the foliation, thin plates, strings, and finely-crys-
taline bits of minerals have formed, in the same direction as the flakes. Now it was this kind of detritus
that I alluded to as the origin of the ashes, not ordinary
wasting of a sea-cliff, or of any land, but a detritus
formed wholly under the sea, and at any depth (perhaps

the deeper the sea the more detrital matter, the fiercer the currents, and the greater the bobbery altogether; but that's no matter). Whatever temperature we suppose the deep sea to have, there must be a vast difference between it and melted rock; and if the two are brought suddenly into contact, there must be a mutual action. Shove a bar of hot iron into cold water, you *scale* it; do the same with a trap-bed, and you'll *scale* it, and the scales are the ash-beds. Them's my notions, sir. I never said the North-Welsh slate rocks were ordinary detrital matter metamorphosed into trap; I was always thinking of this kind of action. I am not up to the Clee Hills, and was only thinking of Staffordshire, and supposed it to be the same. As to 'lava,' the term always used to be restricted to melted rock flowing in the open air. I was protesting against an innovation, and thinking of other friends of mine, more than you while attacking your use of it. One can't explain everything in letters, and I am afraid I often pitch into one man, while I am fancying him another that I can't get at. Those confounded artists won't go away from Bettws-y-Coed, and I'm almost reduced to a standstill, as I could do in one day from there, what will take me a week from here. Gibbs can't find a place at all there; every house is crammed full of amateurs of scenery and salmon-fishing. I sha'n't shoot any more; the keeper dodged me yesterday. Since he has found I *can* shoot,

he won't go into the best ground with me, but on some
pretence or other walks away over hills and banks
' where the partridges *don't* grow.' The Holyhead mail
stops to-day, and Telford's great road is abandoned to
desolation, save when at times some solitary car, with
stone-collecting cove or tourist fraught, jogs slowly o'er
its now grass-covered space, and wakes the pikeman to
a brief delight, &c. I don't very clearly know how I
am to get out of this country. I'd send my baggage
by carrier, and walk by Denbigh and Mold, only it
would be expensive, and all out of my own pocket. I
must car it either to Conway or Corwen. I heard from
Selwyn to-day. It appears that fossils descend and
traps come up, in the series about Capel Curig; moreover Barmouth sandstones come up, as I have sandstones not very unlike those of Barmouth immediately
beneath the Bala limestone. With this sort of riggish-
ness Snowdonia will puzzle us.—Yours ever,

J. BEETE JUKES.

Yspytty, Oct. 8, 1847.

DEAR RAMSAY,—Just got yours of the 4th, and as it
rains virulently, and I'm tired of copying maps, I'll
answer it. My ideas of metamorphism are certainly
vague, but I believe they are more appropriate on that
account, as it is a vague thing. In the widest sense of

the word, not only all rocks, but all matter is metamorphic, since 'nothing is but has been something else' (Old Play). In its technical geological sense, however, I'll take your definition of it—'a metamorphic rock is a rock greatly altered from what it was when first consolidated.' Now that certainly includes all slates, because when first consolidated they must have been shales like coal-shales or lias. Into my notion of metamorphism, however, there always enters the idea of a wide-spread and general action. I should hardly call a mere baked rock by the side of a trap-dyke metamorphic. For instance, coal baked by trap into coke could hardly be called metamorphic, except by the straining of the term. A truly metamorphic rock can only occur in a region where the action, whatever it was, that produced the metamorphism, has acted widely and probably for a long time. Supposing heat to be at the bottom of it, which it probably is, no truly igneous rock may appear in that region at all. I suppose it to be a deep-seated cause, which has determined a certain amount of heat towards a particular region at or near the surface for a long time. That heat may not have literally melted any part of it, or if it has so affected some portions that were very readily fusible, other and intermediate portions, nearer perhaps to the source of heat, may not have been melted on account of their refractory character. The heat, or the cause of it, will

z

have set other agencies at work, chemical, or mechanical, or magnetic, or what not. In some of the beds a molecular action may have taken place, allowing a change in their particles, such as crystallisation, the formation of mineral veins, or the production of cleavage. Each of these may be produced in parts suited to them, and not in other parts traversed by the same agencies; just as in a great mass of rocks all the fine-grained beds are affected by cleavage, and not the coarse ones. I have no doubt these Silurian slates were once just the same stuff as the soft clayey shales of the S.E. of Shropshire. I believe, then, that these in North Wales are all metamorphic. Had the oolites of Somerset and Gloucester, or any other formation, been here at the time the metamorphism took place, they would now be lithologically just what we see here, except the true traps and igneous rocks, of course. I mean they would have been slates, and flagstones, and gritstones, and crystalline limestones, all more or less affected by cleavage. I think this general metamorphic action took place in parts of Britain at a time anterior to the mountain limestone. It has taken place subsequently in other parts of the world, as since the oolitic period, in a part of the Alps; since the cretaceous, in the northern Andes. It is probable, perhaps, that the same amount of metamorphic action on any particular small piece of rock can be produced, either by an in-

tense heat in a short time, or by a very small elevation
of temperature when very long continued and widely
spread. There will of course be every gradation from
an intense and long-continued action, that would reduce
any rock into granite, to one so slight that its results
are inappreciable. It will therefore often be difficult to
decide whether any rock has been subjected to this
action or not; but we may almost assume *à priori*, that
in a metamorphosed region every fragment of rock of
the requisite age has been subjected to this action. I
should therefore say that every true slate-rock, every
formation any part of which is affected by ' cleavage,'
must have been more or less affected by this metamorphism; and that, as we find its effects in the cleavage
of one kind of rocks, we may look for its effects in some
other change that has taken place in other kinds of rock,
forming part of the formation. In these ash-beds, for
instance, the *transverse* flaky structure, the segregation
of plates of semi-crystalline minerals, parallel to the
flakes and transverse to the beds, are probably the
result of metamorphic action. It is possible that even
the parts that are now compact trap with disseminated
crystals may be metamorphic, but that is an extreme
supposition. If we admitted that, we might suppose
the stratified traps of the Arenig also metamorphic;
but that would by no means invalidate their *previous
igneous origin* (*i.e.* a submarine detritus of igneous

rock), and the metamorphic agent may have traversed the subjacent and superjacent slate-beds without doing more than sealing up their 'lamination,' and giving them their cleavage. These, recollect, are speculations to give you my notions of what metamorphism could do. Pray don't suppose, however, I lay claim to these notions for my own; I believe I have got them from Sedgwick, and Lyell, and Darwin, and Phillips, &c., and I have not yet seen anything published to supersede them. I do not see either that your ideas of metamorphism can be very different. I should never dream of giving a different colour to a rock on account of its metamorphic character, though in extreme cases some mark might be advantageous—I mean those extremes when it is difficult to say whether the rock was not originally an igneous one. Well, I'm tired of writing; what shall I do next?—Yours ever,

J. BEETE JUKES.

Penmachno, Oct. 15, 1847.

MY DEAR RAMSAY,—This fine weather induced me to come suddenly over here, where I have taken up quarters in a small room (bedroom and parlour and all) leading out of a small kitchen of a small public. I have to depend on passing strangers to explain to the people what I want, though I get on pretty well with signs and odd words. I return to Yspytty to-night, and on Tuesday next go to Bettws for a fortnight, by which time I shall have completed all my present lines. It so happens that the last time I saw Selwyn, we were mutually wishing that you would send us both into Staffordshire together. I have not the least wish to play solo, and should gladly see the whole strength of the survey concentrated on that district. The more the merrier. I have written to Selwyn to ask him to join me at Bettws. It would be expedient we should have a little work together here, as my Bala beds apparently lose the limestone, the ashes pass into traps, more traps come in, and in fact our lines coalesce just about Bettws.

Since I wrote to you, I have been quartered in the *Quarterly*, which snubs and pooh-poohs me a good deal, and wishes Captain B. had written the book, and wants to hear more of the *fortemque Gyan fortemque Cloanthum* that trod the deck; to which I might answer, that we had no such gentleman on board at all. The faults they find are no doubt the true ones—such as want of the history of the voyage and of anecdote, and all that; but I thought I had explained the reason. . . .

Let us come back to our mutton (ah, I wish I could!). I think the notion of taking coal-fields for winter work, for those who are engaged among mountains in the summer, an excellent one generally. We ought to have a good body on the survey, if not the whole of it, well up to coal-measure work in all its bearing and details. It is our most legitimate practical operation. Besides, the winter is almost thrown away among mountains, and a man works all the better for having had a little change; and he also works all the better before the change comes, in order to finish up to a certain mark before he starts.

Penn Fields, Wolverhampton, Nov. 4, 1847.

We arrived here yesterday, after a series of adventures, and mishaps, and rows. We started in a car and a cart (the latter for luggage) to Conway. At Conway they told us the mail, being the only coach, was always full, and absolutely declined to take luggage beyond a carpet-bag or so. We hesitated what to do: whether to go back to Bangor, and then by sea next day to Liverpool, or to push on as well as we could. The former mode would be much the most expensive for ourselves, and the latter for the survey, and we decided the survey ought to stand it. We accordingly started for Chester in an open break sort of a thing with two horses, and got to Abergele. There we fell in with a surly fellow of an innkeeper, who refused to forward us and baggage with fewer than four horses.

We had no end of a row with him, and at last he agreed
to send us, in an old barouche with a pair, seven miles
to St. Asaph. In the scrimmage we left a lot of luggage (my desk, &c.) in the boot of the break; at all
events, it was *non est* when we got to St. Asaph. Here
the man was civil, and helped to pack our baggage in a
chaise; but he also insisted on four horses, and at last
we were compelled to take four, for the first six miles
up the escarpment of the mountain limestone. Fancy
two poor geological surveyors with four posters! Arrived at Holywell, we had an awful bother with the
receipts, which they at first refused to sign, and then
insisted on filling up, and spoilt them. It was a pitch-
dark night, and we had hardly gone a mile when we
found the driver so drunk he could hardly sit on the
box. Ya! hip! He flogged away full gallop, passing
houses, donkey-carts, coal-pits, fires, and dark hollows,
down to a ferry over the Dee. We did not know a step
of the road, or I'd have had him off his box and driven
myself; but I knew that among lots of roads only one
took to the ferry. However, at last we reposed ourselves in the White Lion, about ten at night, and came
on here the next day. We have reduced our receipts
to about one shilling a mile apiece, and hope they will
be allowed. We arrived here with only a few shillings
between us. North Wales is now just in the transition
between coaches and railways, and the innkeepers all

grasping at anything they can get. I hope to recover our baggage, especially my desk and carpet-bag, in the latter of which was seven or eight pounds' worth of plate. How I hate quill-pens! I can neither write nor spell with them. I hope you will contrive to come and see us here.

Penn Fields, Wolverhampton, Nov. 11, 1847.

. . . Three fine days have enabled us to spiflicate this end of the Silurian ridge, and get boundaries of Y.R. and C.M. for two or three miles. At last heard from Hooker, just before he sailed. *Blackwood* and *Examiner* are very civil to me.*

Dec. 27. Percy's coal-pit, that Selwyn and I went down, is a most interesting place—lots of trap veins traversing coal. The coal is often but little altered, even close to the trap. I have got a lump for the Museum, including some of both. Then there was a fault most clearly shown, like a diagram, and that fault was coincident with the extension of the trap, *i. e.* beyond it the coal was undisturbed by trap. On the trap side of the fault, the coal changed from ten yards thick to four yards, with another wild coal twenty yards above it. I wish you had gone down with us. But I must go again, and get more minute details. . . . I find one of the railways I wrote about in the spring, *is finished*

* In reviewing the *Voyage of the Fly.*

without Ibbetson's seeing it, viz. the South Staffordshire Junction. I'll go and look at it. I must ride
over that country to pick up stray facts of which I have
heard.—Yours ever,

J. BEETE JUKES.

Penn Fields, Wolverhampton, Feb. 15, 1848.

MY DEAR RAMSAY,—. . . I made a grand haul yesterday in a lithographed mining plan of a part of the
coal-field, and a capital lot of sections to be depended
on. I am overwhelmed with affairs. I have survey en-
gagements in all directions : in the coal-field to go and
look at points, or coax bits of information out of different people. I have two lectures to deliver in Birmingham on the two next Mondays. Here's a letter come
from Herr Otto Roettig, and I suppose he's following
it; and down comes Captain Ibbetson last night, and
wants to whirl me off into North Staffordshire. You
must have had a pleasant party at Darwin's. I should
have liked much to have made one. Has Selwyn sent
down his paper and maps ? That is another breaker
ahead, right in my track, which every now and then
comes into my head, and carries me away from coral-
reefs, Australians, pit-sections, Polynesians, outcrops,
pretty girls, faults, money-matters, coal-measures, German geologists, dinner-parties, and the new red sand-

stone boundary.—Believe me, when I have time to think
of you, ever your true friend,

J. BEETE JUKES.

Feb. 18, 1848.

I have just returned from North Staffordshire, where
I have been with Ibbetson. It has cost me a horrid
lot of tin, and I forgot to take any receipts with me.
However, I have made out the two points I wanted.
First, there's a fresh-water limestone, with microcon-
chus, and only one little bed of coal over it, and a known
section of 1000 yards of C.M. below it. There are other
measures and coals lower down in the north of the
district, passing down into the millstone grit. I hold
this fresh-water limestone to be the actual top of the
carboniferous series, and the North Staffordshire coal-
field to be the *only one* in Britain where the whole
series of C.M. can be proved to exist. I'll draw out
the case more at length in future. Secondly, the boundary fault is, I think, an actual big downthrow, subsequent to the deposition of the new red, because there
are a series of faults on both sides of it, both in C.M.
and in Y.R., all throwing down in the same direction.
What do you mean by my having any objection to your
reading my share of our paper ? I should consider it
necessary that you read and approve the whole before it
is sent in. I shall have no time to tackle it till Tues-

day next, when I will squeeze it into about three sheets
of letter-paper; so tell Selwyn to look out for his being
mortally transmogrified.

March 3, 1848.

Writing to Sedgwick the other day, I told him that
Selwyn and I were going to read a paper on North
Wales, and that I was very glad of it, as relieving me
from all restraint in writing to him. In his reply, he
says that I may well differ in details from him, as he
only made one or two traverses, &c., and had no time
to follow the beds longitudinally and connect the sections. 'But,' he says, 'I did put North Wales into
good approximate order, and ought long since to have
published the details. That ugly abstract of papers
(published in the *Journal* about three years ago) was
done by Warburton, who would not even let me see the
proof-sheets, and, strange to tell, I have not read it over
since it was printed.' 'The last short paper was by
myself; but rheumatism, gout, &c. &c.; so that I have
done nothing hardly in the writing way for some years.'
Now to this letter I have made the reply enclosed; but
so fearful am I of overstepping discretion in this matter (how I do *hate all concealment!*), that I would be
obliged if you would look it over; and if you think it
too open, put it in the fire, and tell me so. The first
sentence refers to matters not geological; all the rest

might be advertised in the *Times* as far as I am concerned. What fun this is in France! Is it thought there will be any fighting, in which we shall have a part? If so, will a geological corps be of any service, think you? I'll make one with all my heart.

Dudley, March 8, 1848.

. . . . There does not appear to be any chance of war by the papers; indeed, I suppose nobody but the French would dare to attack us, and I should be sorry to fight with them now. *Vive la République!* . . . I hope with all my soul the Italians of the north will kick those beggars of Austrians over the wall of the Alps, into the depths of the Black Forest. Stirring times; but we Englishmen seem to have done all our fighting, and can't afford to pay for any more diversion that way.

Dudley, March 17.

I have had some very good days' work here over a most complicated bit of ground, which, by the help of a brick of an ironmaster, I have got pretty well disentangled. It is chokeful of faults, varying from ten yards to one hundred. By the way, the lower coals here are almost gone, the beds thin out, and although the place of the coal can sometimes be found by a 'slum'

or a ' smut,' there are none of them workable. Now
should I mark the outcrop of them ? If they never got
thicker, I should take no notice of them, as a matter of
course. There are also here some trappean sandstones,
rounded grains of basaltic stuff, but these contain nodules of a calcareous ash in some places. There is also
a kind of trappean breccia. They are thin, irregular,
and bothering; must I mark 'em in, think you ? I
certainly think Selwyn can't do better than come down
here.

Dudley, March 19.

I shall look to to-morrow fortnight, April 3, as my
time for starting hence. I should like a day at home
to pack up and arrange, and then, hey for the Berwyns!
It is, however, a *terra incognita* to me, so please to
send maps and information. Perhaps, however, Aveline
will be there by that time, so that I can join him. I'll
work hard to knock off the Bala limestone and ash-beds
that make their appearance therein ; but I take it, if
they are complete up to the valley of the Dee this summer, it will be as much as they will, for I anticipate
difficulties as we go north, big faults, &c. However,
nous verrons. . . . In the afternoon of yesterday I
knocked off three miles of trap boundary along the
Rowley Hills. I have been reading Arnold's *History of*

Rome during meal and smoking times; an admirable book, and a great pity he died before he finished it.

———————————

Llanwddyn, April 26, 1848.

DEAR RAMSAY,—We've made out this corner by the help of the Bala ash-bed, which I caught by the tail one day. It is rather queerious, and I'll send it to you on a bit of tracing-paper. There's an anticlinal line, or rather what I expect Darwin means by an uniclinal line. Next we want to know whether it would be fair to charge the survey with our postman. There is no post-office within eight miles, nor will there be at Llangynnog. We send a boy three times a week, and he'll charge a shilling or eighteenpence a day, say four or five shillings a week. Without this, even official letters might not reach us for a week or two. . . . You will see in the tracing that the Bala ash-bed coming down from the north, broken by a few small faults, gets suddenly affected by a very sharp anticlinal, that makes it strike about S.W. In that direction it cannot be traced far, and probably is altogether concealed by upper beds. But towards the N.E. it runs on, gets more and more perpendicular, till at last it is absolutely so, and then bent over so as to dip the wrong way, and all the superior beds, as also the Caradoc and Wenlock shale, dip at it.

Llanrhaiadr, June 13, 1848.

Where are you? I conclude the rain or the pretty girls detained you at Tyn-y-Groes. I have had one day here, and two miles N. of this place have found the Bala limestone!

> ' Not that the thing is either rich or rare;
> One wonders how the devil it got there.'

Moreover it is two or three hundred feet thick; fossils the same as Bala, but with large trilobites. According to the dip of the traps and ashes a little W. of this spot, the lower slates (base of the trappean series) should have come out here, and, lo, the Bala limestone! I expect Father Adam must be right in his big N. and S. fault. I had two hard days before leaving Llangynnog at Aveline's sprawl of trap to N.E. of that place. I could make nothing of it at all; it is full of beds of slate, which seem sometimes to cross it, but at others, end abruptly against trap both ways, and some run altogether across the strike of the others. I could form no hypothesis to fit the things at all. It seemed as if the trap lay horizontally on some tilted slate, but sent many beds for short distances conformably into the slates, while some slates were caught in among traps, but I could never see any bedding either in slate or trap. I gave it up at last, and mean to find a tract where they lie undisturbedly, and I can study the beggars. Evidently this is not a district to be done in a

hurry, so you must e'en have patience. . . . I have pretty
tolerable quarters here, so far as house and people go,
but the customers are the noisiest, drunkenest set of
rascals I ever met.   After keeping quiet for two days, I
could stand it no longer, but took command of the
house, went into the kitchen, and threatened 'em all
with the constable, the lock-up, and the stocks, and
astonished them not a little.   To-day I caused a farmer
of the neighbourhood, and a cooper who boasted of being
' a rich man,' to be ejected *vi et armis* into the street.
There is neither constable nor magistrate, and the whole
set seem to me a drunken lot of blackguards.   If the
rain keep me in the house much, I'll swear myself in
' special,' and, by heaven! two or three of the biggest
of them shall have a night in the lock-up.   I see I could
do it; for the big cooper's look of surprise, and alarm,
and deprecation, when he found my hand on his shoulder, and himself advancing to the door this afternoon,
almost vanquished my wrath, which had boiled over.
I mean to speak to parson Williams about it, however,
and if necessary, have a police-officer.

Llangedwyn, July 28, 1848.

DEAR RAMSAY, — Fate is conspiring against me.
After so much wet weather and enforced idleness, we
have got a fine gleam, and I am laid up with a bad

cold, &c., and can't take advantage of it. Had we a
summer like last year's, we should long ago have finished
all this country, and been working from Llangollen.
As it is, I hope to be moving there in about a fortnight.
Since I wrote, we have rather had reason to alter our
notions as to the ending of the Caradoc in the locality
you mention, and to think it may end at a fault; but
still it does not greatly alter my conclusions. The
country is obscure. We thought (both Aveline and myself independently) that the Caradocian, coming up from
the south, passed into the Meifod beds gradually, by
the thinning out and disappearance of the coarse sandstones, and their change into fine grained sandstone,
shales, and flagstones. The Meifod-bed fossils, however, are decided not to be Caradoc; therefore, 1st, the
Caradocian do not pass into the Meifod beds, or 2dly,
the Caradocian, if passing into the Meifod beds, are
not Caradoc.

We want, therefore, some direct and independent
evidence either, 1st, that the Caradocian are Caradoc,
or 2dly, that they pass into the Meifod beds. Is there
any evidence of the first to be got anywhere? If not,
had we not better get some evidence, if possible, of the
latter? Now for that purpose I was thinking, that
when Forbes comes down, Aveline and I had better join
him at Meifod and Llanfyllen for a few days, show him
the fossils, &c., and look together with him at the lime-

AA

stones of Mathyrafal, &c., which Sedgwick puts into his sections, but which I have not seen, nor is any marked in the map. . . . I suppose, after all, the only way fairly to work it out is, to disentangle the Meifod group of beds, and endeavour to trace their calcareous bands. Not a very attractive job, and it opens the Welchpool sheet again. I'll keep you informed of our movements, of which I am rather uncertain. We must walk over to Llangollen to engage rooms there, as, being a tourist place, there may be some difficulty; and how gladly I shall get into lodgings again, out of these horrid publics, Heaven only knows! We frightened these people here yesterday morning, and to-day they got us breakfast by a quarter-past eight, which they consider a triumph of early rising.

Llangollen, Aug. 26.

Yours of the 24th just received. Rains like blazes. Forbes is to be married next Thursday, and comes here the week after. I give you two sections. I expect the ash will be the Upper Bala ash-bed; but whether the sandy calcareous beds below the flagstone are the base of the flagstones, the top of the Bala, or the representation of the Caradoc, we can't yet say from the survey, and it is one of the questions we want Forbes to solve. . . . We have had altogether two days' work this week.

What are we to do? Our work now lies very far, and over some great hills. I have already got a cold from Thursday's work, so that I fear, if I were to keep at it, I should merely lay myself up when the fine weather does come; and yet it is downright intolerable to be imprisoned here. If one had books or society, one would not perhaps complain; but in this social and intellectual desert, it seems a sheer waste of existence. Heigho! Forbes is a lucky dog! Did you hear anything of Father Adam at Swansea? I have not heard from him for a long time, and can't conceive where he can have got to. I have a great mind to take a day-ticket to Bangor and spend to-morrow with you; but then cash is running low. Old Gov. seems the happiest of all the three, as he lies curled up like a nautilus on the hearth-rug. Well, there's half an hour spent, so farewell.

Llangollen, Sept. 3.

Your of the 31st followed me to Llansaintfraid-Glyn-Ceiriog, where Talbot and I stayed three days to work the ash-beds. Why, you are flying into the uttermost parts of the earth! However, I conclude the post will follow you. We are happy to hear that no sections are to be run this year, as now we can finish the Oswestry, Bala, and Llangollen, and perhaps the Corwen, sheets comfortably before going into winter quarters. Once

establish a boundary between the flagstones and Bala
beds, and the whole of this district is as plain as an
oolitic or chalk and greensand one.   I was glad to hear
that you have had calcareous sandstone at the bottom
of the Wenlock in your country, as I think it will prove
to be so here.   I am inclined to suspect, moreover, that
the upper limestone of Glyn will not be so much a
Caradoc as a lower Wenlock limestone.   The difference
is not much, to be sure, and it is a trifling thing in
itself; but it looks like the flags got sandy and calcareous.   We are waiting for Forbes to settle the matter.
He will be down, I believe, at the end of the week.   I
expect Mr. and Mrs. Browne, and perhaps my cousin S.,
to-morrow to stay a few days.   I mean to take them up
some of the hills, whence we see Snowdon and Liverpool, the Great Orme's Head and the Wrekin.

Talbot and I have just sported two plaids ; but we
want you to give us a lesson in putting them on.   I
got into nearly an inextricable fold in one this morning.
I was like Laocoon and the ' sarpints.'   You sinner !
if you get married in Scotland, I'll advertise for a wife,
or have a placard, ' Good husband for sale or hire,'
pasted on my back when I walk out.

Are we to have the new numbers of the ' Memoirs'
presented to us ?   I have got a copy down, on condition
I pay for it if not given.   It is a pity it was not divided
into three parts, so as to detach Hooker's and Forbes'

papers. When you come back, halt at Chester and run over here. It is only one hour and a half, and we'll have a consult.—Yours ever,

J. BEETE JUKES.

Llandrillo Corwen, Oct. 22, 1848.

MY DEAR RAMSAY,—Don't start at the size of the paper, but I fear this letter may be a long one. I was detained on the east side of the Berwyns about Llanarmon-Dyffryn-Ceiriog till last Wednesday morning, when I had only a bed of greenstone to trace up Cwm Llawenog towards here. As the snow was, however, at least a foot deep on the hill-sides, this was impossible, and I walked over the Berwyns in the face of as keen a wind and as pelting a snow as I ever contended with. The two next days were fine; but I could do nothing till the snow had run off the hills, which it had not completely done even yesterday. I accordingly went on Thursday to Bala, and back on Friday, taking a look at the Caradoc and the big fault by the way. I quite concur in Aveline's recent alterations of that part, which do not affect the general question, but make it more ship-shape than it was before. Yesterday I was on the hills, and I have traced the lower ash of the Ceiriog, which in that direction becomes syenite, into a great mass of greenstone on this side. This greenstone

is at the base precisely like the traps in the South
Staffordshire coal-field—a regular melted rock, with very
large and perfect crystals of hornblende and feldspar.
It is somewhere about 300 feet thick at least, and it
has over it another 100 feet of greenstone ash (stop a
bit!), which is traceable thence round the hill into the
regular common ash, that runs in a straight line across
the Ceiriog and up to the mountain limestone without
a single break, except where peat-bogs conceal everything from sight.   Below this comes a second band of
greenstone and greenstone ash, which is likewise traceable just parallel to the other over the crest of the
Berwyns and across the Ceiriog, and for about two miles
farther, when it enters a peat-bog, and I suppose is
drowned and smothered, for it never comes out again.

Greenstone Ash.—The ashes associated with the
greenstones are in every case *peculiar*, and distinct from
those which pass into feldspathic trap or porphyry, or the
common ashes, before they approach or pass into greenstone, first change into greenstone ash.   This was the
case over at Penmachno, and south of Bettws-y-Coed;
but Selwyn did not agree with me that his greenstones
were on the same horizon as my ashes.   I am now convinced of it, because the lower ash of the Ceiriog passes
into exactly the same stuff as the lower ash of Penmachno, and then into greenstone.   We can trace it
here unbroken; but the disturbed country about Pen-

machno did not allow of our doing that. . . . . I should
like to know whether you or Selwyn have observed any
peculiarity in the ashes associated with your greenstones. Anything that will throw light on the ash question is, I think, most important. We shall have a grand
fight some day with the rest of the geological world
about it, as it is yet peculiar to the survey, no one else
knowing anything about it. What do you think of
challenging them at the next meeting of the Association
at Birmingham with a paper, diagrams, and a lot of
specimens? I'll back you like a brick. Here goes for
another subject. The great fault that smashes through
the mountain limestone north of Llangollen is the same
that passes by Corwen and Bala; unity of direction,
sameness of throwdown to N.W., and similarity of
action, all prove it—viz. that it is not a clean cut,
but scatters pieces of the rock here and there along its
course by many cross faults branching in all directions,
&c. Its course is N.E. by E., and S.W. by W., say
55° by prism. comp. This is the exact bearing of the
axis of the great Merioneth anticlinal on one side and
of the series of anticlinals and synclinals in the
Meifod and Welchpool countries on the other. Your
series in Caernarvonshire, I think, do not vary more
than 10° from it, being about 45°, or N.E. and S.W.,
which is certainly that of those about Penmachno,
&c. It is most probable, then, that the big fault and

the production of the anticlinals, &c. (in other words, the general strike of the country affected by them) was contemporaneous. But the big fault was posterior to the coal formation, therefore the forces which gave the present strike to much of the country acted after the coal formation (Q.E.D.). All along the escarpment of the mountain limestone, however, and generally over the county of Denbigh, we find all the Silurian rocks striking E. and W. North of Llangollen the flags are perpendicular, and run E. and W. bang up to the O.R.S.* and mountain limestone, while the mountain limestone rests undisturbedly on the edges of these beds, dipping gently to the E. It is clear, therefore, that before the deposition of the mountain limestone, the strike of the rocks was E. and W.; and it is probable that this was their general strike over a wide district at that time. If this were the case, the north flank of the Berwyns and the country north of Cerig-y-Druidion, &c. are the remains of an old system of architecture, as it were, left undisturbed from a date anterior to the carboniferous period, the last vestiges perhaps of mountain chains that ran east and west through North Wales, degraded and destroyed partly during the O.R.S. period perhaps, and which were finally broken up and obliterated after the coal period by being traversed by forces that had a northeast and south-west direction, and gave the rocks of a

* Old Red Sandstone.

great part of the country their present strike. The reason why the old strike should be left near the great
fault may be perhaps that it was a fault, an actual crack
rather than a bend, and that it consequently did not
raise a ridge of hills over its course, but merely dropped
the rocks down on one side, or lifted them bodily on
the other. I should like to know whether there are
in Europe any vestiges of an east and west strike
among Silurians or other rocks that have probably not
been disturbed during or since the carboniferous period.
That the centre of England was occupied by an east
and west ridge of some sort, either under or above
water, is shown by all the Midland coal-fields reposing
on old rocks without much or any O.R.S. or mountain
limestone, and that the old rocks were disturbed before
the carboniferous period. It must have run nearly east
and west, because the mountain limestone sea deepened
both north and south. The present strike of these old
rocks won't show what was their strike before the coal
because it may have been re-arranged—indeed, must
have been—by the post-carboniferous forces. I'm half
inclined to fire off a little paper on this point at the
Geological this year. You will perhaps have heard from
Aveline that he has got a little thin bit of Caradoc about
Corwen, dipping north-west and north, and that it ends
to the eastward, thinning out, and letting the Wenlocks
and Bala come together. This seems to me to be im-

portant, and to throw light on the Llanfyllin country,
and the other places to the southward, where the Caradoc ends so queerly. I now begin to be almost certain
that the Meifod beds are the true and regular Bala beds
under a different form, and to think it absolutely essential that their calcareous bands should be traced, or at
least some of them, and that I should go and look at
the limestones of Mathyrafal. If one of these assumed
the true Bala character, it would settle the question.
Had I not better go there next year? And indeed, if I
had a look at all the rocks below the Wenlocks in the
Welchpool sheet, I think it would be a good dodge. I
could get a horse at Welchpool perhaps, and do it in
three days or a week. I should then have seen all the
beds that either do or could represent the Bala beds
near the parallel of Bala; and I think it is important
that the same group should be seen by the same pair
of eyes throughout its extension, if possible.

Well, I fear you will begin to look on me as a second
T—. The fact is, it rains like fury, and my *Athenæum*
has never arrived, *hinc illæ lacrymæ.* However, I have
now poured over you all my bottled-up information,
speculation, and argumentation.

I don't know where this will find you, as Talbot
tells me you are scampering off with Sir Henry somewhere. However, I direct to Llanberris. I have four
or five fine days' work here before I can join Talbot.

We were thinking then of going, first to Pentre Foelas and polishing-off that corner, and then coming back along the Caradoc to the mountain-limestone patches south of Ruthin. However, we'll keep you advised. O the weather, the weather!

Llandrillo, Oct. 27, 1848.

MY DEAR RAMSAY,—You rather miss the point of one of the subjects of my last letter. It was not the denudation I was thinking of, and have no wish to interfere with you on that subject. My notion was, that, mountain chains or not, the old strike of the Silurians was east and west; that, either under water or above, there were elevations running east and west through central England; and that this east and west strike had been obliterated or interfered with, by forces producing a north-east or south-west strike after the carboniferous time. No doubt denudation accompanied, as a matter of course; but what I was thinking of was the *direction* of the *forces of elevation*. It would be very curious and quite new, if we could prove the same district to have been acted on by forces having different directions at different times. *A priori,* one would suppose that whatever elevating force acted on a country would almost necessarily act along the old lines of dislocation, &c., and that therefore, whatever change of elevation, depression, denudation, &c. might take place, the strike of a district must always remain the same, when once it had a strongly-marked strike impressed on it. Now if North Wales or any district once had a decided strike in one direction, and became acted on by forces having a different direction, it would be the very thing to produce all the fractures and undulations we find. Is it possible that if a country has an east and west strike, and is acted on by forces that would alone produce a north and south strike, the resulting strike will be a north-east and south-west, or a north-west and south-east, according to circumstances? If so, the forces acting on the carboniferous rocks of the north of England (producing the Pennine chain), and giving it a north and south strike, may have been at the same time acting on North Wales, but, in consequence of its previous east and west strike, have given it a north-east and south-west one. We should then have the curious circumstance of the same force at the same time producing elevations *not parallel* to each other. My obliteration, then, of mountain chains, you see, does not depend on denudation simply or even primarily, but on a new set of elevations and depressions. All the rocks may be there, or nearly so, but in a new arrangement; all the materials, but a new set of buildings. For instance, suppose a ridge connecting the Berwyns and Snowdon (neither of them, perhaps, so

high as at present), with the Bala beds all over it, and
the Caradoc and Wenlock flanking it, broken up by
fresh forces, and the traps poked through, &c. into the
form we now have them, denudation either contemporaneous or subsequent, or both. I should like to hear
a *discussion* on the subject at the Geological. That was
my notion. Let us show them there are some questions
of more importance and of a higher generality than dis-
puted points of greensand and tertiary clays. I should like
to hear the question of the forces affecting the structure of
the crust of the globe more frequently brought into play.

The greenstone ash here, I recollect Aveline said,
you called ' snake' somewhere down in Radnorshire.
. . . If you could manage to join us at Pentre Foelas
when Selwyn goes to Bettws-y-Coed, it would be splendid. We could have a grand consult and a day or two
together in a very knotty district. I should particularly
like to look at that country with you, as both Caradoc
and ashes and traps are most curious.

Corwen, Nov. 1, 1848.

I at length got here last night, coming in a car
after dark, having twice crossed the Berwyns during
the day. Those bothers are done, I'm happy to say.
I think I have got the heart of their mystery in two
whacking faults, which radiate from their north-west

corner, and cut off all the broken ends of beds as they go along for at least ten miles each, one of them likewise coinciding with a mineral vein of Smyth's, and cutting off (in its extension) the end of the Caradoc and Wenlock, that bothered Aveline towards Llanfyllin. At all events, if they are not *vrais*, they are *bien vraisemblables*. Several cross faults coinciding with Smyth's mineral veins help to disentangle the confusion ; and if I cannot explain all the minute details, there is no block of country half a mile square that I cannot assign its right position in the series, and say how it got where it is. So that's satisfactory. . . . I should say by the end of next week the Corwen sheet will be finished. We shall then only have the north-west corner of the Llangollen sheet, and it will go hard if we can't complete that by the end of the month. So do you and Selwyn hammer away and knock off your sheets; you may consider this one safe. And a *werry* pretty one it'll be ; only scored over with faults in all manner of directions. Got a lot of letters to send, so farewell.—Yours ever,   J. B. JUKES.

Note by Professor Ramsay. In 1866, a volume explanatory of the Silurian geology of North Wales was published by Mr. Ramsay, forming vol. iii. of the *Memoirs of the Geological Survey of Great Britain.* In this book, Mr. Jukes contributed nearly the whole of the chapter on the Berwyn Hills, and many paragraphs elsewhere, especially in the descriptions of the Bala country.

Penn Fields, Wolverhampton, Dec. 19, 1848.

DEAR RAMSAY,—I was detained all day on Monday in Birmingham, waiting to be examined, and had to go again on Tuesday. Horrid bore! I told them my opinion was, that the ' swell' at the bottom of the thick coal was an original structure, caused probably by an accumulation of sand or mud in a ridge; that it existed previously to the deposition of the coal which rests upon it, and consequently previously to all the beds above that; while the fault was a rent, fissure, or line of fracture which happened subsequently to the formation of all the beds, as it traverses them all; *ergo* the ' swell' and the 'fault' were different both in nature and origin, and therefore could have no connection with each other. They seemed well satisfied, and rather interested with my evidence. It is curious, and I shall pay more attention to these things; it appears the gubbin ironstone rose with the ' swell' as they cut through it in driving the gateway. To-day has consequently been the first day I could set to work at the maps, but I have finished all except the Corwen sheet, and hope to knock off that this week, and go *pace tuo* to Birmingham on Saturday for a week's holiday, and on January 1 to move to Dudley. . . .

It strikes me that, concerning our Cambridge expedition, it would not be a bad dodge to go down with a day-ticket on Saturday afternoon and spend Sunday in Cambridge. We could talk with Father Adam at all events, and perhaps look over his rocks. . . . I was in hopes we could have had a talk with Selwyn about *my faults* that run into his work. I fear I shall not feel certain enough about them to insert them of my own authority.

LETTERS TO DR. INGLEBY AND OTHER RELATIVES.

Explanation as to the selection and arrangement of the series, comprising letters on philosophical, religious, political, and literary subjects.

A CORRESPONDENCE of a very different nature from the preceding took place about this time between Mr. Jukes and his cousin Dr. Ingleby, who is now well known to the literary world as a metaphysician and Shakespearian.

Dr. Ingleby was many years younger than his cousin, and having lately left Trinity College, Cambridge, was residing in London for the study of the law. He has placed these letters in the hands of the Editor, and has kindly furnished some notes in explanation, which will add to their interest. It is to be regretted that this correspondence was not more extended, since it served to elicit an expression of Mr. Jukes' views on some subjects which rarely formed with him a topic of conversation. That he took a very high view of science, those who knew him best can testify, and many passages in his works show that Nature was not to him a cold abstraction, but a *revelation*, as he says in one of these letters :

'I believe that creation, with all its wonders and

BB

mysteries, and its thousand minute occurrences,·is, as it were, the necessary result of the existence of God.'

The great aim and object of his life was, that knowledge might 'grow from more to more;' but it was never separated from the desire that more of 'reverence' should 'in us dwell.'

Among letters that have been collected with the object of portraying the character and opinions of their writer, it is fitting that those which in any measure convey his views on 'divine philosophy' should find a place. The Editor has therefore added to the letters addressed to Dr. Ingleby portions of others on kindred subjects, written in the year 1843 to her husband chiefly from Australia, at the time when the acquaintance which had previously existed between them became a relationship, together with a few of a later date. Nor would she add any comment of her own, except this one, that whatever might have been her brother's doubts on theoretical points or dogmas of Christianity (and as he himself says, they were '*only* doubts'), his practice was of no doubtful nature. More than the generality of men did he excel in that 'most excellent gift of charity, the very bond of peace and of all virtues,' the fairest of the three queens in the 'dusky barge,' which conveys the soul to the regions of eternal light and eternal love.*

* Professor Morley on the Arthurian poems.

Among the graver letters addressed to Dr. Ingleby
will be found a few on lighter subjects—one on a passage from Shelley, and another on the style of Macaulay
—which may serve to show how many-sided was Mr.
Jukes' mind, and how capable he was of enjoying the
beauties of poetry and literature, as well as of working
out the hard facts of science.

It may be noted also that one point touched upon
in these letters, and to which their writer ever adhered,
is that of the gradual moral improvement of the human
race, and their ultimate realisation of a high ideal. In
contemplating this golden age, however, he threw his
mental glance as far forward into the future as he was
accustomed to look back into the past physical history
of the globe, and there to trace 'processes carried on
through periods indefinitely great before man was
brought into being.' And as from his conception of
those processes were excluded the violent cataclysms
and convulsions which formerly were deemed necessary
operations of Nature, so he had no belief in rapid
growths, or sudden interpositions, in the moral government of the world; but trusted rather to the silent and
almost imperceptible progress which, through age after
age, should raise man ever higher and higher.

> ' For Nature, only loud when she destroys,
>   Is silent when she fashions:
>   She will crowd

> The work of her destruction, transient, loud,
> Into an hour, and then long peace enjoys.
> Yea, every power that fashions and upholds
> Works silently.
>   All things whose life is sure,
> Their life is calm.'

Note by C. M. Ingleby, introductory to the collection of Letters addressed to him.

The letters on the moral government of the world, and on conditions of human well-being (given at pages 386, 389, and 395), were elicited by me on an occasion of great private trouble, which for a time completely jaundiced my mental sight. Looking upon my cousin as a practical man of sound sense, I called upon him to unravel the web of life by which I was then sorely perplexed, insomuch that I perfectly realised the words of the Psalmist, 'My sight failed me for very trouble, my feet had well-nigh slipped.' The help I sought at his hands was promptly, studiously, frankly, and generously extended to me. In the event, it seemed to me, that he himself found a healthy reaction from my appeal to his experience and intellect, that he 'felt the circuit of thought close' (as his and my late friend Sir W. R. Hamilton expresses it), so that his views were at once organised into the form fitted for the *litera scripta.* I have observed that men of action stand in need of some 'plane of offence' whereon to break; and I think I did him that good turn. If so, there was mutual service. He was enabled, as it were, to elaborate and study his own thoughts as reflected in mine. The result was at length the splendid letter of August 16, 1848, one of the finest specimens of epistolary argument I ever met with. In fact, all the letters of the series of July 16 and 23, and August 2 and 16, are very striking. [One can imagine that he may, on rare occasions, have thus shown his innermost mind to his friend Edward Forbes; but unfortunately Jukes' letters to Forbes have not been preserved.] These four letters are the outspeech of manly good sense. Surely it is very noteworthy that Jukes, like a thorough-paced philosopher breaking from the Aufklärung (the school of eighteenth-century enlightenment), rightly objects to the conventional separation of intellect from morality. None ever saw more clearly than he did, that the purest morality must be an *intelligence*. With him, as with Kant, emotions may be serviceable to morals, but morality must be rational rather than emotional. We must own, however, that the intelligence which Jukes had in mind here was the understanding, the faculty which finds cause and effect in particulars, and legislates prudentially for human conduct. In fact, his view on the essential excellence of our appetites—bad only by perversion and excess—is identical with that of the Stoics (see Kant's *Religion in-nerhalb*, Book ii. Exordium). Individual knowledge and social influence were, in his view, competent to keep us from wrong-doing and wrong-suffering. Prudential rules gathered from experience (as laws of Nature are obtained by induction) were to serve us for 'moral laws.' His views on these difficult questions are not for a moment to be confounded with those of the *Revulsion*, whose last and most vapid outcome was Positivism. On the contrary, his belief in God, freedom, and immortality, as regulative ideas, remained unshaken his whole life long, and his deliberate actions were fraught with a religious element.

C. M. INGLEBY.

This series of letters has been inserted here more in accordance with the subjects upon which they touch than as following in order of time.

Penn Fields, Wolverhampton, Nov. 7, 1847.

DEAR MANSFIELD,—I received your last here, where I arrived on Wednesday with Selwyn, a brother officer, to commence the survey of the South Staffordshire coal-fields. . . . I also have had one or two semi-controversies with some old Cambridge friends of mine, who have the same opinions, I fancy, with Mr. ——; but I found it better to let them drop. Where the matter in dispute is not clearly founded either on facts, experiments, or self-evident truths, a controversy is apt to become a mere verbal firework display, in which each is attending to the ingenuity and brilliance of his own pyrotechnic contrivances, and each is occasionally burnt or heated by his opponent's squibs. In matters of religion especially, where opinions are almost as numerous as men themselves, it is very difficult to get common ideas on which to argue, and it would be absolutely necessary in the first place to settle the exact meaning of every word used. Take such words as 'faith,' 'revelation.' I have no doubt if you were to dive into your minds, and fish up the real thoughts, and feelings, and motives, and beliefs, which each of you attach to those words, they would be found to be perfectly distinct, and I am sure that no two men have attached precisely identical meanings to them—I mean, of course, men who have thought upon them. It always appears to me that one man's religion (that is to say, his real, internal, religious feeling and belief; not his printed creed or formula, but the totality of his religious ideas) must almost necessarily be as different from every other man's as the totality of all the small muscles, fibres, and skin that make his face, compose a different countenance from every other man's. It is so in a great measure even in science, wherever opinion—called in science metaphysics—comes into play, and it must necessarily

be so in politics and religion. Nevertheless, it is pretty
certain that the whole mass of human opinion on these
points *includes the truth;* it is difficult, however, in
descending to particulars, to say which bit is part of
the truth, and which not; and it is evidently most
arrogant for any one man to say that he is in possession
of all the truth, or possesses nothing but what is part
of the truth on those matters. . . .

I hope you will get the Burney Prize; the mere
fact of writing an essay of fifty printed pages deserves a
prize of some sort. I certainly always thought the old-
fashioned essay a very repulsive affair, but I can't recollect that I ever said, ' I hate being taught,' though I
am afraid the expression fathers itself on me by the
character of its features. . . .

I do not altogether agree with you in your opinion
of reviewers. An honest and capable reviewer is a very
useful and important personage in the world of literature, and there are now many such. Their principal
fault, taken as a mass, is cliqueism. . . .

J. Beete Jukes.

Llangynnog, Llanrhaiadr, Oswestry,
May 22, 1848.

DEAR MANSFIELD,—I've got a geological song in my
head for our next anniversary dinner, but don't know
what shape to lick it into.   I want an easy rowley-pow-
ley, hey-derry-down sort of a tune, adapted to my limited
powers in the vocal line.   I recollect the chorus of an
old drinking song :

> ' Then troll, troll
>   The bonnie brown bowl,
> A friend and a glass and a lass for me !
>   This is the toast
>   Good fellows boast,
> Whether of high or low degree.'

But I can't recollect the beginning of the tune, nor the
measure of the verse.   Do you know it, or what sort of
verse would fit it ?   If you do, please to note me down
the air, or the measure of the verse.   Make a nonsense
verse to fit it.   I could *petrify* that chorus thus :

> ' Then roll, roll
>   The rocks like a scroll.
> Let them be " flat," or " upright" let them stand ;
>   By " strike" and by " dip,"
>   Through "fault" and through " slip,"
> We'll map them and section them right through the land.'

Again what tune would fit this ?

> ' Free o'er the hills our feet shall roam,
>   We'll breathe the mountain air, sir ;
> Care shall not ever dare to come,
>   Nor grief pursue us there, sir.

> Joyous in Nature's wildest scene,
>   Where rocks lie topsy-turvy,
> And falling waters flash between,
>   We'll prosecute the survey.
> O the survey, the geological survey!
> Health and good humour shall be queen
> Of the geological survey!'

You perceive they are intended to be semi-comic, social, and occasional, and only adapted to the select and initiated. I got into a rhyming fit the other day on the hills, and apply to you as my Handel or Haydn. They, the last especially, had often quite as bad, if not worse, verses to operate on. Don't you think a pedestrian excursion across the Berwyns into the vale of Bala would do you good this splendid weather? If so, come and spend a week with me, and I'll teach you Welsh or geology, which you like.—Yours ever truly,

                                    J. BEETE JUKES.

Llangynnog, Llanrhaiadr, June 1, 1848.

DEAR MANSFIELD,—I am very much obliged to you for setting my verses, and wish I had a piano here to see what your tune is like, as I can't tell merely from seeing it. I only fear it is too good for me, and that I shall spoil it if I attempt to sing it myself. I did not mean to impose the task of original composition on you, but only to select a tune from any you know. However, I am all the more obliged. I shall certainly look for some ferns for you; but I cannot undertake to preserve them. I am continually on the move bag and baggage, and generally in small rooms, crowded with various matters, so should only spoil them. If, however, you will give me your town address, I'll send them by post, as I get them. I have almost forgotten Shelley, as my life lately has been more hard and practical than formerly. Still I can relish his singular beauty at times. He is, however, always mystic, and his admirers must be content with the number of beautiful images, thoughts, and emotions he creates, without attempting to define them too accurately, or to give to each a local habitation and a name. In the passage you quote,* as in most others, you must, after reading the whole, reflect on the total idea produced by the whole in order to get at what is intended, rather than try to separate each passage, and understand them successively. You may

---

* The passage referred to is in Shelley's *Prometheus Unbound*, act. ii. sc. 4. Asia asks,

> 'Who made that sense, which, when the winds of spring
> In rarest visitation, or the voice
> Of one beloved, heard in youth alone,
> Fills the faint eyes with falling tears, which dim
> The radiant looks of unbewailing flowers,
> And leaves this peopled earth a solitude
> When it returns no more?'

then paraphrase it thus : ' Who made that fine, delicate, and exquisite sense which sometimes thrills us in earlier youth, so as with a deep and mingled feeling of pleasure and pain to bring tears to our eyes, when in some lovely scene we feel the first mild breath of spring awakening Nature around us, and hear it breathing over the earth ; or when, alone in some flowery valley, we listen to the voice of one beloved, tears such as flowers (happier, yet ah, how far less happy !) never shed, tears which obscure our sight of them (or perhaps, which, falling on them, dim their appearance), that sense which, tearful as it is, is yet so exquisite that, when in maturer age it is lost, the world seems a desert without it—full of beings, indeed, but beings with whom we have no longer sympathy.'

I don't know Becker's *Analysis,** but should fancy from its title that it could have no possible connection with Shelley.   No matter-of-fact person, no one without some imagination or who is not facile of suggestion, no one without the poetic temperament—mind, I do not mean poetic power—can appreciate Shelley. He will always remain the poet of poets, that is, of those only of that temperament, and thus can never be popular, or at all events, stir the multitude.   I believe this to be a

* Dr. Karl Ferdinand Becker, author of the *Organism of Language*, 1827. My reference was to his German grammar. Fancy parsing Shelley under the rollers of Becker!   C. M. I.

defect—beautiful as are his poems—a great defect, and that all great poets must pass through the process which Shelley did to a higher one. He remained satisfied with the dreams and reveries of his imagination, it appears to me, instead of strongly mastering them, compressing them into form and substance, and setting them before the world as a ' thing of beauty;' to be ' a joy for ever,' *and to all.* He is thus, too, more peculiarly the poet of youth, when the emotions are somewhat undefined; in older age I fancy we require something more pronounced and decided. I do not know which party has the advantage there. Formerly I should have said the young, certainly; now I hesitate; hereafter, I suppose, I shall say the old.

Should you feel the want of some fresh air, after a certain time in London, by all means run down to have a week with me. The railway comes to Ruabon, and you can be there in six or seven hours, and it is only five or six miles from Llangollen, at the north end of the Berwyn range, which I am surveying. I hope my aunt and uncle are quite well. Give my best love to them; and believe me, as ever, yours most truly,

J. BEETE JUKES.

Llanrhaiadr, Oswestry, June 13, 1848.

DEAR MANSFIELD,—As it *rains* like *blazes* (congruous simile !) I am writing letters for lack of something else to do, so can reply to your note.   I daresay your analysis is quite right, as ' subjects,' ' predicates,' ' attributes,' &c. are not at all in my line.   I think, however, the sentence can easily be reduced to the strict rules of grammar by the supplying the little word ' is' or ' are,' or both.   It is a mere case of ellipsis.   The word ' heard' is not the participle, but the third person singular or plural of the present tense, the sign of the tense or auxiliary verb being omitted by poetical license ; thus,

> ' When the winds of spring
> In rarest visitation [are], or the voice
> Of one beloved in youth [is] heard in youth alone.'

I apply ' heard in youth alone' to the sound of ' the winds,' as well as that of ' the voice.'   The difficulty and obscurity of the sentence to my mind is the ' dim the radiant looks,' not your X. and Y.

As to ferns, you must wait for fine weather at all events.   There is no going on the mountains in such rain as this, unless with the full intention of getting soaked to the bones, and perhaps losing oneself in a bog into the bargain.   You address your letter to the Society of Ec. Geology.   It is not a society, but a Government department, a branch of the Woods and Forests. The Museum of Ec. Geology, or the Geological Survey

Office, is the proper title. On looking over your analysis again, I see that, translated into your language, 'heard' would be a predicate of the 'winds' and the 'voice,' instead of an attribute of the latter only. Translated into Latin—canine rather—it would be 'Cum, solus et juvenis, venti veris raro venientes, vel vox amatæ auditur.'*

Give my best love to my aunt and uncle, if in London, and believe me ever truly yours,

J. BEETE JUKES.

Llanrhaiadr, Oswestry, July 12, 1848.

DEAR MANSFIELD,—I have delayed answering your last till I could get some ferns for you. I have now only one, a lump of mountain parsley. It seems curious, but whether it will be valuable in your eyes I know not. I know where you are now. In the year 1835 I went to see my uncle Henry near where you now live. Oxford and Cambridge Terraces were then half finished, and ended in fields near the Grand Junction Water-

---

* Acting on Jukes' hint of paraphrase in a foreign language, I may venture to explain the meaning of ' dim the radiant looks of unbewailing flowers,' by rendering it thus :

' Ch' empie l' occhio debole cotte lagrime che *offuscono*
L' aspetto raggiante del non-sympatico fior,'

or, *del scompiangente fior.* The flowers are not taken as unfeeling, but as being out of relation to human suffering, and themselves untouched by sorrow. C. M. I.

works. In these fields there was a row of small cottages, called Spring-terrace, near the back of the waterworks, looking over a kind of common towards Bayswater, to which there was a footpath, leading between a number of small tenements and gardens, inhabited principally by washerwomen. It was ornamented by lines of clothes, waving in the wind, and poor uncle Henry said he called it 'Washington.' He lived in Spring-terrace, which, I have no doubt, is near the site of the present Spring-street. There are a few remains of the gardens and tenements, I recollect, just round the public-house on the Bayswater-road, unless some lath-and-plaster palaces have swallowed them up since last winter.

What do you mean by a *moral* government of the world ? I have never studied these questions, and so am not up to the technicalities which may be concealed in them. Do you mean merely that there are *moral laws* in the same sense that there are *physical ones* ? If so, it appears to me that, just as the existence of matter being granted, it follows that there are laws which regulate that existence, so the existence of soul—immaterial entity—being granted, it follows that there must be laws regulating that also ; *a fortiori*, if beings exist compounded of the two, there must be laws regulating not only both principles, but the method, &c. of their association.

I fancy, however, this was not all you meant; if so, explain. John Bull won't fight unless something happens to put his blood up; and just at present he'll take a great deal more to get it up than he formerly would. I have no particular personal objection to war myself, and I don't know whether I should not enjoy it if personally engaged, but as a matter of prudence and policy its absurdity becomes every day more evident.

I think the Whigs will stay in till Peel feels or fancies himself strong enough to come in and take their place, and carry Hume's four points, or the Charter, and thus put himself at the head of a new party. I hope some time to see a strong, firm, moderate, democratic party, that, backed by an immense majority of the people, shall come in and sweep away all the relics and dregs of feudalism in which we are now living, reduce the army and navy to a skeleton, remodel the law, the Church, and the whole system of government, abolish all but direct taxation, and, in short, commence a new era in the world's history. How do you manage to exist in London? To be sure, your black mare is an alleviation. It is absolutely hot here, but the hills and country are splendid, and the whole valley redolent of hay.

The new book on sea margins is by R. Chambers. I have not read it. It is, I expect, crotchety, like his *Vestiges.*

Well, I think I have returned your compliment, and feeling tired, knock off.—Ever yours,

J. BEETE JUKES.

DEAR MANSFIELD,—You have set me a job, and I am sure I can't do it. The fact is, I have no defined religious opinions; that I believe to be the true state of the case. I never tried to work them out in any way, and I think it probable I never shall, since I don't see how my solving a riddle, or fancying I have solved it, can alter the real state of the case, or even affect my own condition. I shall merely therefore reply to your letter, that I certainly have no belief in a special Providence, according to the *ordinary acceptation* of the word. If a vessel very narrowly escapes a dreadful shipwreck, it is called a great mercy, and solemn thanksgivings are made; if a wine-glass tumbles, and does not break, a man would say, ' That's a bit of luck.' Now, there must be just as much of providence in the one case as in the other, and according to my idea of the Deity, He would be just as much affected by one of these expressions as the other. I believe that He has impressed upon His creation an *infinite* number of laws, some of which are known to us and some not; and that those laws act permanently and invariably, although their action is either unknown to us, or disregarded by us. What we call chance or accident is merely an expression of our ignorance of the causes of any fact, or the conditions under which it took place. At the same time, it is not a blind destiny that rules everything; because, by discovering natural laws and obeying them, we can rule events to a certain extent. We do so in lifting a chair, and every other action. Special providence is only a pious name for chance.

Llanyedwyn, July 23.

I was interrupted here, and since then have been moving to this place, and much occupied. I do not know that my illustrations above express my meaning exactly, but perhaps you will be able to discover it. I mean that there is a providence in the highest and most general sense of the term, but ' special providence,' in the sense of interposition according to circumstances, seems to me an idea not only unworthy of God, but absolutely contradictory of the best notion we can form of Deity with our imperfect faculties. From our very constitution, it must be that our ideas of Deity are after all but an exalted anthropomorphism; but still, though very incomplete and imperfect, they may contain much truth. Now one necessary attribute seems to me to be unchangeableness. I cannot conceive of God waiting for events in order to act, or interposing to alter the order or consequences of any event; but I look rather upon all beings and all events as the result of invariable laws, the emanation, as it were, of God.

That with a perfect foreseeing and an absolute invariability of law, and therefore predestination, there should still be liberty of will to human beings, is one of those mysteries that we are utterly incapable of understanding, but which I yet firmly believe. I feel that I am free to walk across the room, or not; and yet I know that if I do so, I do it not only according to invariable laws, but that my doing so has been a foreseen and predestined event from all eternity. Should a beam fall and crush me, or miss me, it is equally the result of laws predetermined, although unknown to me. If it miss me, I may, and ought to, be grateful; but I must not on that account suppose it was supernaturally deflected. Moreover it appears to me that, if I make it a subject of special thanksgiving, I am rather presumptuous than devout, and the question occurs, why was the danger allowed to approach me at all?

After all, this subject, like all its kindred ones, when pursued far, leads into an entangled maze of mysteries and arguments, from which differently constituted minds will draw all kinds of different results. I do not feel called upon to pursue it so far for my own satisfaction, since I feel I have not the leisure, and I think not the capacity, for it. I rest satisfied, therefore, with the belief in a God 'in whom is no shadow of changing;' and I believe that creation, with all its wonders and mysteries, and its thousand minute occurrences, is, as it were, the necessary result of the existence of God; the thousand invariable laws that cause and govern them, flowing likewise directly and necessarily from that existence. Does this answer your question?—Believe me, as ever, yours truly,

J. Beete Jukes.

Llanyedwyn, Aug. 2, 1848.

Dear Mansfield,—I believe I now understand you fully, and will answer in this way. It follows from my —and your—idea of Deity, that everything He created is *most perfect* to answer the end intended. What the *object* of creation can be, as far as regards God, we can form no conception. No doubt He intended the greatest happiness of His creatures as a collateral result, and I can't help believing, though I can't understand it, that their greatest happiness is in some way worked out by the existing arrangements. I cannot believe, then, that injustice can form part of those arrangements, or be the result of them, and must conclude, that what appears to be so, would be strict justice, could we *know all its causes* and *antecedents.* I do not believe in a system of rewards and punishments in a judicial sense, as it

were, as if God judged of the action and then assigned
its consequence. This would be nothing else but special interposition, which we both give up. But I am
forced to believe that every action has its necessary
consequence, and that that consequence is a just one;
and we know from our own feelings that happiness
is the result of good actions, and *vice versâ*. The
happiness may not be apparent to others, &c. &c.
I do believe, therefore, that in the original scheme of
creation, it was ordained (as a necessary consequence of
the power, wisdom, and justice of God) that our happiness and misery should be exactly proportioned to our
virtue and vice: in other words, that precisely as we
understood the laws of Nature and acted in accordance
with them, so would be our condition. I see no reason
to suppose that there is any difference between the
moral and physical laws in this respect; but that just
as we discover and make use of the physical laws to increase our physical powers and benefit our physical or
material condition, so we may, by discovering and obeying the moral laws, increase our moral powers and benefit our moral condition. I believe that in each case
early training and exercise (education) is necessary,
but I know no region of the world in which a truly
moral education has yet been attempted, and *therefore*,
in my opinion, is there no region where man has attained to anything like the happiness of which he is

capable, because this happiness depends on the *general spread* of morality. The best man, the man who most truly obeys the laws, must, if placed among the immoral and ignorant, be constantly injured by their actions, and must constantly have his own obedience to the laws rendered more difficult and often impossible. I look upon pain not so much as a punishment as a warning. If you put your finger into the fire and get burnt, you don't call it a punishment—it is a warning to prevent your doing that which would destroy the part. Both the moral and physical pain and misery of the world, are constant warnings to avoid things hurtful and destructive, and incentives to strive for those which are beneficial. I do not know anything which it is pleasant to do, which is not also proper to be done, provided it is done in obedience to the moral laws, under proper restrictions as to time, place, quantity, &c. &c. The grand thing which is wanted is knowledge—that is, knowledge common to all. Were every one fully instructed and trained in the moral laws—many of which probably have yet to be discovered, or, at all events, to be plainly stated and illustrated—the vast majority would act in obedience to them *from choice,* and often with a wise self-denial of immediate gratifications, knowing their bad consequences ; the vast majority do so now, as far as they have been taught, and the small minority who, from weakness of any part of their physical or moral constitution, were liable to make continual ' mistakes,' we will call them, would have all public and private opinion against them—instead of for them, as often now—and thus be not only instructed in the right way, but helped to pursue it, and, as it were, kept in it by force. Our happiness is infinitely increased by its being our own work. This we all feel; indeed if it is not our own work, it is not felt to be happiness till perhaps it is lost. The happiness of the human race, therefore, is meant to be worked out by themselves, by discovering and obeying the physical and moral laws of Nature. This does not necessarily sacrifice the first workers, because much is easily discovered and obeyed instinctively, and it is probable that, even at the beginning, the happiness of the *whole* world is greater so than if it were a mere paradise, with no room for, or occasion for, exertion.* It may seem paradoxical, but I believe it is not untrue in one sense, that ' evil' is the source of ' good,' that without the existence of the one we could not have the other. At all events, the good and happiness to be attained is greatly increased and heightened by the very fact of there being an evil and a misery to be avoided. Evil, pain, misery, therefore, are necessary for the production of the *greatest* good, enjoyment, and happiness.

* Almost an echo of Haller's lines:

' This world, with all its faults,
Is better than a realm of will-less angels.'

Moreover, it appears to me that, for the attainment of the greatest happiness, there should be a great variety of faculties capable of appreciating it, and therefore an almost infinite variety of circumstances under which they can be placed, and in which they can be exercised. We are therefore perpetually liable to do things that shall cause pain or pleasure either to ourselves or to others. That the consequences of my actions should be undeserved injury to another, is what you call injustice (so it is *per se*), but—

First, it is a necessary consequence of the existence of evil and the variety above mentioned, both of which are necessary for the production of the greatest happiness.

Secondly, if done voluntarily, it is punished by remorse, which ought to teach me not to do the like again, and may lead me to atone for it, and thus produce more happiness to both than if it were not done, or its example may, at least, deter others, &c. &c.

Thirdly, if involuntary, it is, if known, atoned for, and if not known, still in a thousand ways a greater good may arise from it to the injured person from myself or others.

Fourthly, if it be a pure injury, never reversed or ameliorated in this life—for I confine all remarks to that—it is a rare exception, and its permanency must almost necessarily be the result of some weakness or incapacity or fault of the recipient, the observation of

which may still be useful to many others, and thus produce a greater sum of happiness than if it had never
been done. You may spin a thousand cases, but the
most difficult one to imagine is, the one from which *evil
only* arises.

A recipient of evil only becomes one of your small
expiators, but I can't say I ever met with one. Without partial evil and injustice, what becomes of the pleasure of pity, compassion, generosity, sympathy, &c. ?
There would be no room for these faculties, and thus a
great sum of happiness lost. I thus arrive finally at
the conclusion, that 'justice is sooner or later done; so
much pleasure for so much virtue'—and a great deal
more—' so much pain for so much vice'—and no more
than is necessary to warn mankind to avoid it ; only I
define virtue and vice as obedience and disobedience to
the moral and physical laws of God, and I believe obedience to be the necessary consequence of the knowledge
of them ; and therefore I conclude that *pain or evil is
the result of ignorance only,* but as we can never have
perfect knowledge in this world, evil must necessarily
exist, but may be diminished to almost any extent.

I don't understand you about free will, but can't
discuss it now. It is two o'clock in the morning. I
have had three hard days' work, and must have another
to-morrow, having already had a nap on the sofa.—Ever
yours,   J. BEETE JUKES.

Llangollen, Aug. 16, 1848.

DEAR MANSFIELD,—I can easily conceive, that much which I have written to you is not strictly logical or coherent, because I have never thought deeply or consecutively on the subject, never reasoned it out. I find, when I take up my pen, the ideas or notions in my mind; but how they got there, or when, I have no certain knowledge. As to mine being 'a theory of compensations like that of the modern sensationalists, such as John Stuart Mill,' as I never read any of their writings, I cannot say. Our minds certainly seem either differently constituted or differently trained, inasmuch as we have some difficulty in understanding each other. I do not think that my belief in God's justice can be rightly termed faith. It is, to my mind, an attribute of the Deity, inseparable, from His existence, so that I could not possibly believe in a God at all, if I did not consider Him a perfect being, all-wise, all-good, as well as all-powerful. The existence of free will with necessity, and of evil with good, is one of those mysteries which we may perhaps be able to understand in a future state, but which I do not think the mind of man adequate to comprehend in our present state, with our limited faculties. Neither can I understand the connection of spirit with matter, nor the method of the production of thought, nor life itself. If I do but move my finger, it is an action the nature

of which I cannot comprehend. I wish to do so, and
it is done; but how or why I can never know. Not
only is the connection of the material with the immaterial obscure, but the very existence of matter is a
puzzle which, as you know, has led some men to deny
it. I am content, then, to rest satisfied with the belief
—or faith, if you like—that matter exists, and also the
immaterial, and the connection between the two, although
the clear and certain knowledge of it is unattainable by
me, and, I believe, by every one else. When I said that
knowledge is the one thing needful, I meant a perfect
knowledge, which is, of course, unattainable by man.
If a man could foresee every consequence of each action—if it were all present to his mind—it seems to
me impossible but that he must choose the one which
should be productive of the greatest happiness in the
totality; and I think it is evident that that could only
be produced by *doing right*. Now, although a perfect
knowledge be unattainable, yet we may approach it by
a quantity less than any that can be assigned, and I believe we are approaching to it, though slowly. As an
instance, take drunkenness. Could society ever go back
to what it was in the last century? Is not the abandon-
ment of it by all the instructed classes, the direct result
of the increased knowledge of its consequences. I know
that that knowledge does not enable a drunkard to re-
form, but the general spread of that knowledge produces

a *public opinion adverse to the vice,* instead of in favour
of it, renders it disreputable, and therefore not only
diminishes the number of drunkards, but increases the
number of the sober; *i.e.* not only frees those disin-
clined to it, from the necessity of complying, but re-
strains those who are naturally inclined to it. No
doubt some men voluntarily do what is right *because it
is believed to be right;* and some men have a frame of
as much grace, strength, and agility as the human body
is capable of; but the mass of men fall far short of this
type of human goodness, moral or physical, and they
require aids and appliances to effect that, of which the
first are capable singly. Now if the existence and the
possibility of using these aids and appliances may be
used as proofs of design in the physical world, why not
in the moral? Are not the passions, the sentiments,
the affections, the relations of mind with mind, the
wonderful variety of intellectual powers and abilities,
their mutual adaptations to themselves and the physical
world around them, capable of affording as many proofs
of benevolent design as any adaptations of the physical
world? Taking your case of a person suffering consti-
tutional disease for the sins of his ancestors. There
are many men careless of themselves, but feeling for
others. To them the knowledge of the probability of
the transmission of disease to children may be a stronger
check against the act that shall produce it, than the

probability of disease or death to themselves; or a man may find himself condemned to celibacy from this circumstance, and thus be a warning to others, &c. &c. I believe that even hereditary insanity will in this way be eventually eradicated from the world.

I think one great source of error in these subjects arises from the practice of separating the *moral* from the *intellectual.* I do not believe in any *essential distinction* between the two. At all events, the distinction is a mere arbitrary one, for the purposes of our nomenclature and classification rather than in the nature of things. I believe man to consist of body and soul (what is called 'mind,' is the action of the soul through the organs of the body); but the union of body and soul is so intimate, that it is impossible to refer any action or thought wholly either to the one or the other. Still, our (1) actions, (2) feelings, or (3) thoughts may be classified as (1) sensual, (2) moral, or (3) intellectual, according as we feel (1) the body, (2) the body plus the soul, or (3) the soul to be more directly concerned in them; but yet there can be no voluntary action in which the soul is not concerned, nor any thought which is not modified by the body. (Whether the souls of all men are equal and similar, and the diversity among them caused wholly by varieties of body, or whether they ever have been equal and similar in a pre-existent state, or ever will be so in a future one, I do not even care to know.) It is evident,

then, that there can be nothing sensual in which there
is not also something intellectual; so the connection between the moral and intellectual must be much more intimate, and their boundaries often insensible. Education
(including training) is necessary for each assemblage
of faculties; but while man can educate and train and
exercise his senses and bodily powers, or his intellects,
to a great degree singly and in solitude, his moral
powers especially require to be educated, exercised, and
trained in conjunction with those of others. They can't
be exercised in solitude, for they are then merely intellectual ideas. Now the utter absence of this moral
training has been the cause of almost all the evil and
misery in the world. Its general diffusion over the globe
some 10,000 years hence, say, is that which shall regenerate the whole earth. *Then cometh the Millennium.*

I find, on reading over, that I have been giving
you my own notions rather than trying to enter into
yours; but perhaps that is the best way, as enabling
you to judge and compare. You seem to me to think
that the balance in this life is in favour of evil, of
unhappiness, that there is more of both than of good
and happiness; to my mind the balance is the other
way, so that there is no wretch so miserable, or wicked,
or depraved, but has more pleasure than pain, and
more of good in him than evil. In all cases pain and
wickedness are the small exceptions, or they would

cease to be felt or remarked. There is certainly more than enough of both; but pleasure, goodness, and happiness are *infinitely* more abundant. That they are the rule, and not the exception, is shown by the fact that they are not felt or remarked till they are lost. We are so accustomed to them, that the exceptions only strike us as wonderful; and then it is more often the absence of a certain portion of them than the presence of their opposites. That we can always imagine a higher degree of felicity than that in which we exist, while it tends to make us dissatisfied with our present state, and is so far evil, yet gives us always hope, and always urges us on to endeavour to attain it; and hope and action are great goods. I cannot at all understand why you fail to see proofs of benevolent design in the moral world; they appear to me quite as abundant, and more striking than those that are purely physical. What think you of family affection, of the pleasure derived from generosity, making benevolence pleasing to the doer, as well as useful to the receiver; nay, attaching pleasure to the exercise of every quality, such as courage, and even rewarding endurance of pain or self-denial with a pleasing consciousness in the very act of their exercise; so that a man, by denying himself a gratification he is tempted to take, shall eventually experience more pleasure than if he had taken it—a pleasure springing up unawares and unforeseen in his

own heart, and utterly independent of all external circumstances. The pleasures of hope, memory, and
imagination are the very themes of poets. In short, I
know no moral faculty which is not the source of pleasure, and the more so the more it is exercised, while
the reverse is the case with bodily pleasures. Now all
this seems to me strong proof of benevolent design.
You might fill volumes with instances and examples.
Well, I have occupied a wet morning with the above.
I hope they have a better harvest in America than we
are likely to have here.—Yours ever,

J. Beete Jukes.

Dudley, May 6, 1849.

Dear Mansfield,—I have not been able to read
either Layard or Macaulay yet. I have read some of
Macaulay's essays, and do not wholly agree with you as
to his style. Speaking of style generally, you are right.
I like a plain, homely, condensed Saxon style,—nervous,
strong, picturesque, and expressive; but I should hardly
wish Macaulay to change his style. It appeared to me
to be not *diffuse*, not *laboured*, not *puffed up*, not
padded, nor trained into a march, with a regular pom-
pous cadence and measured swing like Gibbon's; but to
be clear—very clear—striking, nervous, perhaps a little

DD

too trenchant, glittering, and brilliant, but only a little,
and its brilliancy set out its beauty with a graceful air,
for which one quite forgave it. (One pardons a really
beautiful woman a few airs of coquetry and pretension
which would be insufferable in a dowdy.) For such a
style the pure Saxon is not alone sufficient, and we
must recollect that *English* is not *pure Saxon.* When
we have good Saxon words with the *full* meaning of
those of foreign origin, it is better not to use the latter;
but there are many ideas, and many more shades of
meaning, for which the Saxon has no expression, and
by taking words from foreign languages we have extended and enriched the Saxon into English. Every
good writer, therefore, ought to take full advantage of
all the resources of the instrument placed at his disposal. I send you this small scrap of criticism as a
kind of observation across the table on reading your
note; and am yours most truly,

J. BEETE JUKES.

Dudley, May 9, 1849.

DEAR MANSFIELD,—So far from finding fault with
the three sentences you first quote from Macaulay, I
maintain the repetition of the word is not only proper,
but proper in a high degree, as adding clearness, force,

and strength to the expression.* Your formulas I hold
to be bosh, and don't understand what you mean by the
' universe.' Your reference to page 50 does not fit the
edition I have, which is the fourth. 'Noisy or tumultuous' tribune is by no means fully equal to ' stormy.' He
might be noisy or tumultuous, and not succeed in
raising a storm after all.

'Who had wrung money out of dissenters.' You propose ' extorted ;' exactly the same meaning—twisted
out ; but ' wrung' is Saxon, and ' extorted' is Latin.

'The conflicts of the royal mind did not escape the
eye of.' You say, ' The king's indecision was observed;'
feeble, since 'indecision' falls far short of ' conflicts,' and
' not escaping the eye of' conveys the idea that they
were attempted to be concealed, and therefore might
have escaped, if he had not looked sharp, while simply
' observing' them would only convey that they were obvious to observation.

* The passages criticised in this letter are in vol. ii. of Macaulay's *History of England.* I refer to the third edition, 1849.
Possibly the passages on p. 50 were those containing the repetition
of the words ' hypocrisy,' ' vice.' That concerning the feelings of
the Anglican clergy towards James II. is on p. 307.

Jukes' apt comparison of a good style to female beauty might
have been suggested by Macaulay's reflection on the dowdy Cathe-
rine Sedley ' affecting all the graces of eighteen.' Evidently the his-
torian, unlike the majority of his sex, found more attractions in *form*
than in *person*, or he would not have been astonished at the king's
preference for the fond and sprightly matron.   C. M. I.

'The clergy now regarded the king with those feelings which injustice aggravated by ingratitude naturally excites.' You substitute for that 'indignation;' but Macaulay shows you here the cause, and also the kind of indignation; he gives you the reason and the measure of it, and at the same time accuses the king and excuses the clergy. Moreover, there are more feelings involved than simple indignation. The sentence last given seems to me to be a capital one, pregnant with meaning, clearly but most tersely and concisely expressed, not one word superfluous. I should be at a loss how to express so much in so few words. I have not yet been able to read more than a few pages, and formed my opinion of his style from his essays.

Hard at work, sectionising all day.—Yours ever,

J. BEETE JUKES.

---

*Parts of letters to A. H. Browne from on board
H.M.S. Fly.*

North-east coast of Australia, April 10, 1843.

. . . You and I are totally opposed in our religious and political opinions. You may not have perceived how totally, because these with me are serious matters, and serious matters are more rarely on my tongue than in my brain. As to religion, I give no farther weight to authority of any kind, whether written or oral, than my own judgment accords to it. In politics my heart is with the mass of men. I have never yet taken a part in public affairs, because I hate strife and detest party spirit and clannish feelings; but should I ever feel myself called on to act, my voice, my hand, or my pen would be at the service of what is called the mob, Radicals, Chartists, and all. When I say at the service, I do not mean to flatter or pander to them, but to render them true real service, to correct their faults, and to elevate their condition if I could. We may, then, be publicly opposed to each other; will that raise any barrier to our friendship in private life? On my side, no. Some of my oldest and most valued friends are as much opposed to me on these points as yourself. In private conversation they are sealed subjects with us, but we are not the less friends on that account. I should detest the thought of letting a conscientious difference of opinion on any subject interrupt my friendship with a man who was worthy in other respects to be my friend.*

---

* It is almost needless to state that this generous sentiment was heartily reciprocated, nor was it found necessary that religion and politics should remain 'sealed subjects.' Mr. Jukes had, indeed, at this time an imperfect knowledge of his brother-in-law's political opinions, which were independent of party, and were as much opposed to oppression and wrong as his own. Still a great difference of opinion on both subjects existed, and continued to

On reading over my letter, I find I have spoken with
great plainness.   The fact is, I was lying awake in the
morning-watch thinking of these things, and as soon as
I could see, I got up and dashed off what you have read.
Will you take it in all the plainness, sincerity, and
good feeling with which it was meant?   Loving A. as I
do, it was hardly possible that my feelings should not
be agitated when I thought of what so nearly concerned
her happiness, and words in that case flow from one's
mouth or one's pen somewhat without control.   Enough;
you will, I hope, understand me.

H.M.S. Fly, Oct. 11, 1844.

As to religion, you are right when you say I have
not devoted much time to the study of it, if by study
you mean the reading of many theological or polemical
works.   How many years would it take even to read
through the works published by the various Christian
sects, so as to arrive at a knowledge of their various
opinions, much less to balance the evidence for and
against each opinion, and to arrive at an honest and
conscientious judgment on them?   If, however, you

---

the end; but the ties of friendship and brotherhood were never
broken, and Mr. Jukes was ever regarded by his brother-in-law
as the embodiment of manliness, generosity, and intellectual
power.

throw these aside, and come merely to the evidence for and against revelation itself—that the Bible is the word of God—my doubts (for they are only doubts) rather hang upon *à priori* considerations.* What do you mean strictly by ' revelation'? what do you mean by the

---

* So appropriate are the following noble words of the Bishop of Carlisle to the frame of mind here described, that he will pardon their being quoted from a lecture delivered on May 12, 1871, in connection with the Christian Evidence Society :

' Be it ever remembered that the word *sceptic* is derived from a word which means to *look* or to *see*. It is the same word which forms the root of the word "bishop" or "overseer," and, accordingly, there is nothing radically reproachful in the name of " sceptic." It implies that a man is determined to look into matters for himself ; not to trust every assertion, not to repeat a parrot creed ; and, so far as this determination is concerned, it is high and noble, and is, in fact, the very root and spring of all human knowledge. But who can wonder if looking should lead to doubting, and that so the name of sceptic should popularly imply, not the man who looks and believes, but the man who looks and doubts ? And I am not ashamed to confess that I have much sympathy with this sceptical frame of mind. Not only is it closely connected with a noble instinct of inquiry and search for truth, which God has implanted in the human mind, but also, as I believe, it is well-nigh impossible that an inquiring mind should deal seriously with religious subjects, and remain entirely free from doubt. In my opinion, the amount of scepticism which has, during some period of his life, occupied the mind of each thoughtful earnest man will be merely a question of degree ; while at the same time, I most sincerely believe that scepticism ought not to be, and need not be, the lasting condition of the human soul, and that all doubts may be made to vanish in the light which God has given to lighten every man who is born into the world.'

'word of God'? I do not ask what is the nature of God, nor even of spirit. I do not and cannot understand what my own soul really is, and how it is connected with the body. I seek not for a solution of mysteries; I only deprecate the use of mystifications. Have I any other faculty than reason, or any other means than the use of it, to arrive at a knowledge of anything, to understand even the commonest abstract term? God Himself, if He addressed me, must address me through my reason, since He has given me no other or higher faculty by which to receive His communications. *A fortiori*, therefore, I have no other means of comprehending man or the works of man. But the Bible was written by man. Now, my chief difficulty lies here—what is, or was, *the nature of revelation?* how did it act upon or govern the reason of those who received and imparted it? There is something of human in the Bible; there is much of what we call imagery, metaphor, poetry in it, there is difference of style. Each person who received this revelation imparted to the portion he transmitted, the tinge, the colouring of his own mind. Can we be sure there is not also something of the imperfection, the faults, the errors, the prejudices of his own mind also? Before, therefore, I enter on the subject of external evidence, I must have the internal nature of the work or works explained to me. I have never yet seen this done, and should be glad if you could tell me in what book I may find it. In short, let any one explain or define revelation to me, and I will at once tell him whether I believe in it or not. I do not now heartily believe it, solely because I am at a loss to understand what it is. Give up the common and stricter interpretation of it; let it guarantee not the absolute truth of any and every part of the Bible, not every text, but the main and general doctrines which are to be deduced from the whole—such as the immortality of the soul, universal charity, faith, hope, love, forbearance, long-suffering, peace—and but little external evidence would be required to convince every man.

Priests, sects, creeds—these have been in my mind the grand enemies of Christianity.

I suppose, according to the usual course, we should neither make much change in the opinion of the other on these points. . . . The less you can approximate to certainty and demonstration on any subject, the more eager and determined are men's minds upon it. The cause is plain. Reason admits not of dispute—it is one; passions, feelings, prejudices are infinitely various, they are consequently necessarily opposed. The one is a straight line capable of being divided into parts, and many parts of which are unknown to one man or to all; the others are an infinite number of lines of all sizes, and meeting in all directions and at every possible angle.   There is a magazine-article sentence for you.

Toryism, under whatever new guise it may now please or be forced to disguise itself, is essentially this —the mass of the people are to be kept in subjection, in ignorance, and consequently without power ; they are to be kept at the minimum of power possible in each state of society, in order that *the few* whom accident of position, natural genius, or other circumstances may elevate above them, shall govern them quietly, and divide riches, honours, and power among themselves. The more benevolent among them would supply ' the poor' with material goods—food, clothing, and shelter, and think they go to the extreme of philanthropy when they make them ' *happy and contented ;*' but they have a pious horror of the ' turbulent many,' the ' many-headed monster,' &c.   The fundamental fact is, that they cease to regard them as fellow-men, as *equal in the sight of God.*   Now, I am no leveller.   Political and social equality is a dream.   Men must be as various in station and condition as they are in face ; but I do protest against the contempt of the *lower orders* innate in the bosom of what are called the higher classes, both Tories and Whigs, however fashionable it may be to disguise it at the present day.   Of this habitual contempt I have difficulty in even divesting myself, so inveterate are the prejudices of early education and asso-

ciation. It is mixed up with all our thoughts and all
our deeds concerning our fellow-men. No land, no
nation is yet free from it, has hardly made an effort to
be free. It is as rife in democratic America as in aris-
tocratic England. For forms, institutions, and parties
I care nothing, except as means to an end. You must
use them as instruments. Human nature is yet in its
infancy, and requires restraints and encouragements,
whippings or sugar-plums, according to circumstances.
The time will come when every man, whatever his station in society, shall stand erect to his work, shall act
in it without fear of his neighbour, shall be conscious
of his own soul, and possess it with perfect freedom.
Such is an imperfect hint of the foundation of my political creed. It is as yet an abstract faith merely, without hope that either I or any one can in this age act
up to it. But shall we fear, hide, or disguise the truth
because we ourselves are not at all times true to it?

Llanwddyn, Welchpool, April 19, 1848.

DEAR ALFRED,—Will you enlighten me on the obscure subject of the land tax—how much it is per acre,
who pays it, &c.? Government will now, if they are
wise, themselves propose some of the points of the
Charter, or something similar : a very large extension of
the suffrage at the very least, and a kind of electoral-

district arrangement. I would amalgamate localities and populations, so that, while I left no large district, however thinly peopled, unrepresented, I would, at all events, in thickly peopled districts, give a proportional number of representatives. Then the present aristocracy and their followers must succumb, and a very large infusion of democratic principles and action and feeling must be poured into the government, and, if possible, into the tone of society, and thoughts and habits of the people. I don't want to smash things all at once—that is the very thing I want to avoid; and in order to do so, I would have men who should acknowledge they wanted altering, and avow they were ready to set about it. Then let it be done with all solemnity, deliberation, and discussion. I believe if you were to propose to the present House of Commons just now to make the monarchy absolute, to double the army and navy, and rule by divine right, they would vote them all by a majority of 170 ! There is a temper and feeling abroad in the country of which the Ministry and the House seem little aware, and which it would be salutary to them to know.

Llanwddyn, April 29, 1848.

DEAR ALFRED,—Much obliged for your abstract of the land tax, which I never understood before. It is a very good tax if fairly rated and collected, and I should then be glad to see it increased tenfold.

Forbes and I are having a political consultation. We agree in:

1. Educational suffrage—every man that can read and write easily so as to be understood.

2. Electoral districts, and representation equal to population.

3. No property qualification.

4. Taxation strictly according to property.

5. Simplification of law, public prosecution of criminals, and no private costs but in private suits.

He also proposes:

6. The constitution of a *clerisy*, consisting of the clergy and the medical and legal bodies, and the restoration of church property to this clerisy, and its including the entire instruction of the people.

7. Landholders to be made trustees of land for people, and to be made to do their duty by the state.

8. Every citizen has a right to either food or work from the state.

9. Class representation. Electors to be registered in classes according to their pursuits, and only to elect members of those classes.

10. Limitation of titles.

Of these, I think No. 6 impracticable and objectionable; 7 unjust and absurd; 8 I don't see my way into;

9 not only absurd, but socially most pernicious; 10 I don't care much about.

We neither of us care for the ballot, and I rather object to it, as not requisite in so wide a suffrage to protect the voter, and opening a door to trickery, as now it appears in France.

I am not at all willing to wait the result of the French experiment because, first, I don't want to wait at all, but principally because France has nothing on earth to do with the matter. Suppose they utterly fail and sink under a military despotism, it could not alter my views in the least as to the best policy for England.

I advocate the points mentioned above simply because they are just and right in themselves, and most expedient in the present circumstances of the case. It is true, the continental movement exercises some influence, since it makes us demand at once what else we should have been content to wait for, in the certain expectation of getting it at last. We feel now, that other people have had their demands granted them, and why should not we? This is all the effect the continent has, and does not in my opinion alter the nature of the demands a bit. However, our post is going. We move on Monday to Llanfyllin.—Ever yours,

J. BEETE JUKES.

The following letter may here find a place. It was written before physical geography had received the attention now given to it.

Llangollen, Aug. 15, 1848.

MIA CARA,—I am sorry to hear the *Athenæum* never arrived. I suppose the post in the vale of Llanyedwyn is just as badly managed as everything else. Had I known it earlier, I would have made a row about it, but now I cannot enjoy even that satisfaction. Aveline and I ran over to Liverpool on Saturday, to dine with my old friend Archer; spent Sunday in Birkenhead and Chester, going to the cathedral. I am glad to hear of the study arrangement, though when you are to finish Alison, at the rate of two hours a week, would require a long calculation to make out. You will find my maps of great use to you, and to arrive at any comprehensive notion of the relation of states and nations, the intercourse of commerce, and the art and science of war, you should make a frequent and thorough use of them. There wants, however, a few maps on a new plan—a set devoted to the physical geography of countries, instead of their political geography. For instance, a map of England, which, instead of being crossed and scribbled and confused by county boundaries and names and roads, should show the natural features of the country, its river-basins, and the hills or ridges which bound them. The principal basins of drainage are three: first, the

Severn; secondly, the Thames; thirdly, the Humber.
To these may be added that of the Wash of Lincoln, the
Solway Frith, that of the Mersey and Dee, and the
watershed of the south coast of England.   The ranges
of high land are : first, the two parallel ranges of the
chalk and the oolite, running from the south coast to
the north-east—Dorset to York—of which one the chalk
sends two spurs to the south-east, ending respectively
at Dover and Beachy Head. Secondly, the Pennine
chain, running from Derby to the Cheviots. Thirdly,
the three knots or groups of hills of *Cornwall and Devon*,
Wales, and the *Lakes*.   Now each of these three have
quite peculiar characters, differing from the other two
in external features, in soil and mineral and agricultural productions, and in their inhabitants. The ways,
manners, and customs of the people of Cornwall, Wales,
and Cumberland more nearly resemble each other—notwithstanding the difference of race—than they do those
of the carboniferous Pennine chain, or the agricultural
ridges of the oolites and the chalk. The details of the
features of the country determine the great routes across
England, also nearly in the same directions, whether
for purposes of war or commerce. The Romans made a
great military road from Dover, by Daventry, to Tamworth, and thence to Chester and Shrewsbury; the
moderns took first a great canal and then a line of railways along exactly the same line of country, and in

many places they now run as close side by side as is
possible. Were our island to be invaded again, the
hostile camps, if the war were protracted, would be found
posted nearly in the same spots as the old camps of the
Romans, the Danes, and the battles of the Middle Ages
and civil wars ; and this notwithstanding the changes
in the art of war. To understand the plan of a cam-
paign, and the reasons for it, you must comprehend the
physical geography of the country,—more especially one
of a great military genius, like Napoleon, who would
seize at once on the great line of communication that
penetrated the vitals of the land, and then work his detached parties and corps round that in such a way, that
all the strong points should be at his command, and yet
that no corps should ever have its communications with
the rest intercepted, and be always ready to retreat or
advance as was most beneficial not only for itself, but for
the whole body to which it belonged. Thus managed, a
large army becomes a great, wide-spread, flexible, and delicate machine, obeying the intelligence of one man, and
giving him, in the right of intellectual ability, a power
that no amount or array of brute force, no mere multitude, however courageous, no mere wealth, however enormous, could attain to. I never read Alison, but from
what I have seen and heard of his book, I should doubt
whether he had sufficient of the stuff of which a general
or a statesman is made, to set their reasons or movements

E E

before a reader. Arnold of Rugby, had he not died, could have done it, as may be seen in his imperfect *History of Rome*. Napier, in his *History of the Peninsular War*, does it admirably, but I do not recollect any other history that even attempts it ; but then I have not read much. Historians hitherto have attended principally to style and narration, and made pictures rather than represented facts. Almost all history wants re-writing. Nevertheless, it is most useful to read what we have. There's a fragment of an essay for you, written as hard as pen can go, while we were waiting to see whether it meant to rain or not. We are now going out. By the bye, I think we shall not be very long at Llangollen : there is less to do than I thought. Could you come before September if the weather is fine ?

Best love to mother and Alfred from your affectionate brother,

J. BEETE JUKES.

LETTERS TO PROFESSOR RAMSAY. 1849, 1850.

Written by Mr. Jukes (prior and subsequent to his marriage) from Dudley, Halesowen, Worcester, Harborne (near Birmingham), Walsall, Aber, Conway, Bala, Ffestiniog, Llangollen, and Rugeley, on the progress and details of the survey, and on geological questions arising in the course of his work.

Dudley, Feb. 11, 1849.

DEAR RAMSAY,—Thanks for your kind note received this morning. I had a capital day yesterday with Blackwell and Beckett, who had got hold of a cunning man in the fault line. They are sinking a very interesting pit in the new red sandstone near Westbromwich, of which I give you a section. It shows clearly that we have really a lower new red here conformable to, and graduating into, the coal measure. You may call it the top of the coal measures if you like, but it is red soft sandstone, with some red marl, and with nothing but new red above it. . . . In the red rock are many fragments of plants, fossil-trees apparently, with imperfect markings—mere longitudinal striæ, such as have been found in the higher parts of the new red in the Midland counties. Both the plants and the rock are precisely the same as those found by Sedgwick in the north under the mag. limestone, and between it and the coal measures. Half a mile north of this place they have got thick coal at a depth of 980 feet, after passing through about 200 feet of new red, but I have not got the details exactly. There are some whacking faults, which must at present remain rather obscure, till the ground is more tried. About two miles south of this, on the up side of a fault, they have been sinking down in the Silurian shales, expecting to come to coal; and even the intelligent ground bailiff with us would hardly believe me when I said they were below the coal, and approaching the Dudley limestone. I dine to-day with Lord Ward's agent, and hope to get access to his plans and sections this week. If I only had the people under command, I could complete the district in a month—as it is, I lost two whole days this last week dancing attendance at Lord Ward's office for nothing. The agent at last asked me to dinner with many apologies, and said he could then get time to have a quiet talk with me. I have written to Reeks to get me Logan's paper on under clays. I am nearly convinced it won't do in this district. The coals are well shown in several cuttings now, but I can find no stigmariæ underneath them, or at least no more than elsewhere. I believe the state of the case to have been this : stigmaria was a subaqueous plant for which clays (especially fire clay) was the best soil. Fire clay is also often the floor of a bed of coal (though there are many fire clays without coal, and some coals without fire-clay), because it is evident that coal was *tranquilly* deposited, and the *clays* also must have been so from their fine texture. The existence of clays, therefore, or other beds containing stigmariæ always or frequently below coal, do not militate against the drift origin of coal, or prove it to have been *entirely* formed peat-wise by plants growing *in situ*. All I contend against is the terrestrial or quasi-terrestrial formation of coal. If subaqueous, it may have all grown, for what I care. I think I had now better stop here, and take the folks while in the humour.

Dudley, March 14, 1849.

I have been down the Congreaves pit to-day, and have got many interesting facts. . . . This district ought to have a *resident surveyor* (to take advantage of the opportunities constantly turning up) with a large map, &c. William found marine shells in the ironstone while I was below, so they are spread over a much wider district than I at first thought. I sent Forbes word of our, or rather William's, finding a smashed echinus; we'll pack and send a lot up to-morrow. I went three-quarters of a mile in one gate road in thick coal to-day, rather a magnificent affair. There were one or two ' swells' of the floor also, in which the lower measures seemed to have been actually bent up into a

ridge before the thick coal deposit.   They may, however, have been formed thus over some 'ridge of deposition' below, without any actual movement having
taken place.   The mine was rather sulphury in some
places, owing to the high wind; however, we had no
blow up.

I've lost a good many days lately in hunting ground
bailiffs—a species of sport not particularly exciting.
Some were gone to London on law business; some were
in a state of fluency or catarrh; some so busy that they
could only be seen before seven o'clock in the morning,
or after seven o'clock at night; periods at which I have
not yet summoned up courage to go in quest of them.

Dudley, March 17, 1849.

I've read over carefully Hooker's papers, and also
Sir Henry on his rocks of South Wales, the part relating to coal, &c.; and, sinner that I am, I'm not convinced, nor even shaken in my infidelity, as to the
terrestrial growth of coal *anywhere.*   I can see no evidence for the growth *in situ* of any other plant than
sigillaria and stigmaria, and no sufficiently strong evidence to show that it was not a subaqueous plant.
Mere vague relations to Lycopodaceæ—a desert and
arid-plain loving set of plants—in some particular bits
of structure, don't go far with me.   How many relations

might not be established between fragments of a whale and an elephant! My argument is this: you find in a great mass of stratified materials of vast thickness and many beds, all of which are clearly the result of the deposition in water of materials drifted from a distance, certain beds of coal. These beds of coal alternate with, and are interlaced among, the other beds in every possible way—thickening and thinning out as they do; sometimes many layers coming together, and forming a thick bed or set of beds; *those same layers* in other places separated by layers of shale or sand, just as layers of sandstone or shale are separated by other layers in one place, and come together in another. The sole difference in the method of occurrence between coal and other rocks being that it is of finer grain; it is one of the extremes (coarse sandstone the other) in the scale of minuteness of division of the particles composing it; therefore its consequent more wide and general distribution. Beds of coarse sandstone are most irregular and local and partial. Beds of fine sandstone less so; sandy shale less; argillaceous shale less; carbonaceous shale less; and coal most regular and widely spread. There is every gradation in mineral character, in method of deposition and stratification, &c., between a coarse quartzose sandstone and the finest cannel coal. Now to prove that they were not all formed by the same causes—*i. e.* were not all formed by the deposition of

sediment in water—requires, to my mind, an over-
whelming amount of evidence, compared with which
all I have seen, read, or heard, is but a trifle. Indeed,
to confess the truth, I am at a loss to imagine what
possible evidence could be discovered. Direct seeming
evidence—if such could be—of the terrestrial growth of
coal in any one locality, instead of showing that all
coal was so produced, would, to my mind, only remain
as a puzzling exception, to be explained, perhaps, hereafter. Indeed, I never hitherto much cared to examine
such evidence as has been given, because I have always
looked upon the notion as one of those fashionable
fancies which spring up sometimes; and I have been
waiting patiently for it to blow over, as Sydney Smith
said. If I can give it a helping puff, however, so much
the better.

There are many special instances which are easy
and obvious to explanation on the drift theory, which
the terrestrial growth and subsidence dodge (it's like a
legerdemain trick—hocus pocus conjurocus) renders
very difficult to understand. Take, for instance, the
two following bits of sections, which are about half a
mile from each other. Now, according to dodge afore-
said, the thick coal subsided at Shut End ten feet,
while half a mile off, it went down 130 more ; then
came the flying reed coal at a level, and this time the
Shut End part went down eighty-four feet, while the

K. Swinford only descended twenty-nine, to make the level for the Herring coal. See-saw, Margery Daw!

I'm going to Birmingham to-night, to spend Sunday with Matthews, and have to mark in some faults, so farewell. I thought the question above sufficiently important to devote two or three morning hours to its careful consideration.*

Dudley, April 1, 1849.

DEAR RAMSAY,—I have been considering the question of this coal-field quietly and impartially. I really, however, have no partiality about it, for I should be puzzled to say which I should prefer—finishing it by myself, or working with you in Snowdonia. I have, however, come to the conclusion that, as a *matter of policy* for the survey, not only I should stay and work here this summer, but that more force should be put upon it, and either you yourself stay some time—which would be best—or Aveline come and help me to run sections. Mind you, in these times of economy, we shall be expected to exhibit some *practical work*, and, at all events, should be ready and prepared to show that we are working to some *practical* end. Now if we could

---

* *Note by Professor Ramsay.* Mr. Jukes' matured views on the points discussed in the preceding letters will be found in his *Memoir on the South Staffordshire Coal-field* and in his *Manual.*

exhibit this at the meeting of the British Association
in September, it would become a matter of public notoriety, and do much to give us a *stability* in *public
opinion.*   I could get Matthews and Blackwell to speak
in the section of the great want of a survey of this district, of its importance in a practical point of view, and
several influential and moneyed and landed men to cry,
'Hear, hear!' Matthews has been speaking to Sir Henry
about it, I find, and there is really beginning to be an
anxiety among the coal- and iron-masters here to get our
maps and sections of the district.   A great lot of longitudinal sections, to exhibit together with the map, would
be the thing. There need not be much of a paper, merely
a few points stated, but with ' a wish to exhibit to the
section the progress made in the survey of so important,' &c. &c., ' hope to make out the relations of the
new red sandstones to the C.M. in adjoining parts,' &c.
When we have done this and shown our intentions, we
can then go on with North Wales, and knock it out of
hand entirely with our whole force upon Snowdonia, &c.
We want notoriety, and public standing as a public
body; and it appears to me that this opportunity should
not be lost of getting some of it, because we can get
more of it here and more easily than at any other place
where the Association is likely to meet in future.   We
should gain also more credit with the higher powers
by such an exhibition than we should lose by our delay

in North Wales. The people here are very favourably disposed to us, and quite ready to make a public manifestation in our favour. Now don't you think it is a matter of sufficient moment to make a push for, and could not you devote some part of the summer, if not the whole of it, to completing the district? All the central and important part of the district is ready for sectionising. The mapping still wanting in this sheet could be done in a couple of weeks easily, and two more would enable one to complete the mere mapping of the whole coal-field. The longitudinal sections are the heaviest business now, but my pit sections will give plenty of data. We only want the shape of the ground. There wants one north and south section about twenty miles long, one ditto about ten miles, and about eight or ten east and west sections of eight to ten miles each. Country all pretty easy and often level. If we could get those done by September, we should have the materials for a simple and easily-constructed skeleton model, which I will explain to you when you come down. The more I think of it, the more does the good policy of the move on behalf of the survey appear to my mind. . . . When you can find leisure, please turn the above over in 'that thing you call your mind.' I think we should be able to state three important practical points by this summer's work, viz. under one or two districts of new red, Silurian would probably be met with instead of C.M. In one untried district of C.M., the place of the productive beds is over 200 yards deep, and the beds *very poor*. In other districts of new red, the thick coal and the productive measures would be certainly found at a depth varying from 350 to 550 yards, and would be worth working either now or in a very few years' time. It would make a difference in the market value of land of from 300*l*. to 1000*l*. per acre. People would look upon us with some respect if they began to have a notion that we might have such a power as that. By George, sir, it appears to me almost worth while to clap the whole disposable force of the survey on this district this summer, and just show them what we could do if we had the means. You and I, with Selwyn and Aveline, could dig the extra grits of the coal-field out before September. However, I remit it all to your judgment, and settle it all when you come. At the Shut End colliery they have a section across their property of forty feet to one inch. It looks very pretty, and makes one long to continue it.—Yours ever,

J. BEETE JUKES.

In September of this year, Mr. Jukes married Georgina Augusta, the eldest daughter of Mr. John Meredith of Harborne, near Birmingham. As may be readily supposed, the engagement and marriage caused some diminution in the length and frequency of his correspondence with his friend, and only a few hasty letters were written during the summer, from which the following are extracts. In July, the position of geological surveyor to the colony of New South Wales was offered to him; but he was somewhat unwilling to accept it, and preferred to remain in England, if the authorities would consent to raise his salary to the not very munificent sum of 300*l.*, being half that offered by the colonial government. This addition was conceded, and he continued his work in North Wales and Staffordshire until his appointment to the Irish directorship at the end of 1850.

Dear Ramsay,   *Halesowen, June* 21, 1849.

. . . I have contrived through all to do good work (if the pole dodge turn out well), having run a long section over the Clent Hills nearly up to Dudley, and shall bring it up to the castle to-morrow. The people are excessively civil about our going through grass and corn.

Halesowen, June 26.

I have three of yours to acknowledge—a circumstance most unexampled in our correspondence, but then circumstances alter cases; however, I must now take up the dative, or you'll be using the vocative and accusative cases. Having proved that I can work at

Harborne, I shall have the better heart in undertaking
to draw sections there, especially as I can have a work-
room all to myself, and work from seven to one o'clock
(with one hour for breakfast), and sundry half hours at
other times in the afternoon or evening.

I went up to Harborne on Friday to dinner, and in
the evening I turned to at the logarithms. To my delight and surprise, Gus, who professed herself more
stupid at figures than anything else, learnt in about
five minutes how to look out logarithmic series, and
with her assistance I got on swimmingly; so rapidly,
indeed, that we set to work again at seven o'clock on
Monday morning, and by half-past twelve o'clock had
filled several sheets of calculations that I find most
useful. She had each quantity ready for me when I
wanted it, and once or twice corrected my mistakes
when I got puzzle-headed. I did more by that time
than I should in the whole day by myself.

I have a better opinion of the pole dodge now, as I
find a minute's error does not make much difference
till we get to angles less than $1°$; at $1°$ an error of $1'$
makes only an error of twenty-one and a half links. . . .
Old Gov.'s foot improves, but I think he had better take
up permanent quarters at my sister's when they have
discharged Forester. I knocked up William to-day,
so that he said he never was so tired—the first time on
the survey, I think, but it was *werry* warm.

July 2.

In re verses. I am rather fond of changes of rhythm, but I daresay I introduce them too abruptly and profusely. I should be much obliged, when you have time to bestow on trifles, for any criticisms on my productions, as they may as well be done as well as I can. Be as carping and as savage as you like. I have just written them out again and can't discover the change of measure you talk of, except in the rhythm of the second line of the third, and the fourth of the fourth verse, but these were designed. The lengthening pause on certain syllables occurs often as a beauty in good writers. They might all be sung to the same tune without violence to the words, I think. . . . I've just written to Sir Henry and Forbes, and told them I'm going to be spliced.

Halesowen, July 10.

I of course mean to complete the mapping of all this quarter-sheet; my only doubt was whether I could run and finish three or four more sections across that country. I think now of drawing these sections in outline, only then enlarging them on cartridge paper to three or four times the scale roughly, and colouring them for exhibition. Your bad weather surprises me. It has been a succession of the most exquisite weather here, and how lovely Harborne is, I feel most painfully now I am down here and alone. However, duty! duty! duty!

Worcester, July something.

I was brought here yesterday by a subpœna, and kept in court to no purpose all afternoon. However, the trial comes on this morning the first thing, so I hope to get back to Halesowen'to-night. I got yours on Sunday. As to sonnets, I have read all Shake-speare's of course, Wordsworth's, Shelley's, Coleridge's, and some of Keats', besides Milton's and lots of others, but never yet saw one I could take any delight in. They never seem to me to be spontaneous, but always something artificial — exercises, toys ; things done merely to try if one can do them; poetical puzzles or tricks, like a man dancing on eggs. I don't deny but that there may be fine sonnets, but then it is in spite of the measure, and one always feels inclined to split them up, to cut the laces of their stays and let them go free.

I know I shall have a tough job of it to get ready for the Association, and begin almost to despair at times. However, I'll do as much as I can. I always looked forward to having to enlarge not only the sec-tions done and many more, but also to making an en-larged map so as to show the faults, &c. What I wished to see done was a complete inspection of the whole coal-field, as it were, for the whole section. (The Geological Section at the British Association meeting.) That is, of course, now impossible. Our maps are only useful for some of the knowing ones to look at afterwards, and the sections won't be visible unless enlarged ten times; however, we must do what we can, not what we would.

Halesowen, July 19.

I yesterday came from Worcester; read Sir Henry's note, wrote to Augusta about it, and then started off and finished the Clent Hills, and did not return till 10 P.M. . . . As to the New South Wales business, all my notion is that 600*l.* per annum is very tempting, and the work would be very pleasant. There are, however, many reasons against it : such as uncertainty of colonial affairs, its leading to nothing beyond, want of maps and proper assistance; but the great thing is Augusta's health. Could she stand the voyage and the climate? I can come to no decision till I see them. I should be most reluctant to leave you and the survey, and if I were not going to be married, I don't think 600*l.* per annum would induce me to do it; but you see it seems like the very thing I was sighing for, dropping at my feet. However, I must discuss it with Augusta

FF

and her mamma before I form my own decision. I should be very glad, indeed, to have your full and free opinion. Let me have it by Sunday at Harborne.

Harborne, July 22.

I have been waiting for yours (received this morning) as a help to making up my mind. It coincides, as far as it goes, with my own notions. I have, however, just written to Sir H. to say that I have not actually decided, telling him exactly our state and condition here, and telling him also that 300*l*. per annum on our survey would at once decide me to remain. I have not asked him directly to give me 300*l*., but simply told him that the feeling in my mind is, that with 300*l*. on our survey I should at once reject the Australian, even with 1000*l*., and that if he felt authorised or justified in offering it to me, the thing is settled; but if my pay is to remain at 200*l*., then I must take the Australian plan into consideration. Mrs. Meredith leaves the matter entirely in my hands; Augusta hopes we shall not have to go, but if I say we must, she will pack up without a word. . . . I feel sure that prospects are brightening for all of us, and that practical geologists are 'looking up.'

Harborne, July 25, 1849.

I get up at half-past five every morning, work from six to half-past eight, and from half-past nine to half-past one, by which means I have already drawn out the outline of all the sections but one. That I shall do to-morrow, and it will not then take me long to fill them up. I have not got any more sections north of Dudley, but if I finish these in time, and the Walsall mapping, I should like to run one or two over there, *nous verrons.* I have not heard again from Sir Henry, but wrote him yesterday a kind of penitential note, explaining my former one, and doing away with any bad impression it might have caused. I finished by saying that I almost entirely gave myself up into your and his hands to do with me as you would. The sections come out very fairly. There is a slight discrepancy in the lengths sometimes, but not to any serious amount. In one it amounted to about 0·1 of a mile in eight miles, as compared with the map; in another the length was slightly in excess, which is all right. The heights of known points come out very nearly, which is the great test. . . . I have been writing this at eleven at night, after they are all gone to bed; and as I have now been up eighteen and a half hours, I shall turn in.

I should, or ought, to thank you for your good wishes and exertions in my favour as to the future, but between *us* there is no need of *phrases.*

Harborne, July 28.

I have heard again from Sir Henry. He was not
at all offended at my application, but with his usual
caution says increase of salary depends on others, does
not know whether the money will allow it, &c. He has
heard no more of the Australian affair, and does not
know whether it will be necessary or not to hear again
from the colony before the appointment takes place. So
now it is in abeyance, but I have almost entirely given
up the idea of accepting it. What made me write about
lodgings was this—that I had to wait about two months
for lodgings at Bettws-y-Coed the year before last.
Llanrwst would do very well, but there are no lodgings
to be had there. For a poking room leading out of a
kitchen, which you had to cross to go upstairs, they
asked me thirty shillings per week, and it was the best
room that was ever let out in the town. Selwyn asks
me if I have any work he can do for me while he is
laid up, so I think of sending him down the sections to
enlarge.

Walsall, Sept. 1.

I've just heard from Selwyn that he can't finish the
sections. I have not an hour to spare even to show
any one how to do them, and hardly know how I shall
get time to write my papers for the British Association.
What a relief it will be to cut off with Augusta into

Wales! I've stuck stoutly at it this week, and am going to meet a baillie this afternoon at Darlaston. I'm in a horrid funk about your success in the lodging department. Indeed, I have no hopes of Aber at all, and expect we shall have to go to Bangor or Conway, or that miserable spot Llanrwst. Hotels are out of the question. Bettws-y-Coed is out of the way of the work.

Harborne, Sept. 2.

Thanks, my dear fellow, for your exertions in our behalf, of the successful issue of which I was delighted to hear this morning. Augusta desires her especial thanks also, and is quite charmed with the idea of her future residence. . . .

The British Association met in Birmingham in September, Mr. Jukes acting as one of the local secretaries to the geological section. Shortly after the meeting, his marriage was celebrated at Harborne, from whence he and his wife departed for North Wales; but ere many weeks had passed, he was summoned to the death-bed of his mother. After an illness of ten years, she expired peacefully at the house of her son-in-law, at Penn Fields, near Wolverhampton, and was interred at Bushbury (a few miles from the same town), where her husband and two daughters had been laid before her. Near to this village Mr. Jukes had inherited a small property—part of an estate that had belonged to his great-grandfather, Mr. Mansfield of Bushbury-hill.

DEAR RAMSAY,   *Aber, Oct.* 28, 1849.

. . . By the way, I have been thinking of firing-off a letter to the *Daily News* on the Woods and Forests, with a suggestion for remodelling that department and dividing it into two—No. 1, the surface department, or whatever you like to call it; No. 2, the mines and minerals department—and giving up to *us* the control of the latter—of course, *us* on a greatly-enlarged scale ; so that we should have not only the geological survey, but the entire management of all mines and minerals held from the crown, with a geological lawyer and a staff of clerks and financiers, treasurers, or whatever you may like to call them.   Officers, perhaps, to be paid partly by salary, partly by a percentage on the revenue they raise and hand over to the state.   This, with a school of mines, would indeed put us on a good permanent footing, and be a great saving to the nation, as well as a vast benefit.   It is a raw hint; tell me what you think of it, and let us see if we cannot concoct a scheme. When will you come and see us ?   We've brought down Toby, who seems to approve of Wales very much.— Wife's kind regards, and those of yours ever,

J. BEETE JUKES.

Merchlyn, Conway, March 21, 1850.

MY DEAR RAMSAY,—Yours of the 19th received yesterday, but no maps made their appearance yet. I will take care about William, and get him some wine. I sent to Bangor for some the other day, and if it turn out good, he shall go by the rail and get three bottles more of the same sort. Aveline and I had a grand day on Monday at those bothering rocks. He says they are like the Bishop's Castle ones, and that you decided they were ash with showered crystals of feldspar. There is a small district up there in which there is a different rock every five yards; I must indicate them somehow. One of the rocks represents, I believe, the Bala limestone; it is equally calcareous, and there are fossils in the slates in good quantity about it. It occurs in two places, with a big mass of injected greenstone intervening, the slates near the greenstone converted into hornstone still retaining in some places the casts of fossils (I think). These last two days we have devoted to finishing the Caradoc. It runs right up to the top of the hill south of Conway Castle, and there jumps up into the air and finishes as usual, nobody knowing what has become of the end of it. One thing is very remarkable, that, just under it, we get the 'pale slates' all along here.* Now these pale slates always occur at the top

---

* *Notes by Professor Ramsay.* These 'pale slates' have since been shown to belong to the upper Silurian strata, and have been

of the Bala, wherever that is seen in all the Bala and
Cerrig-y-Druidion country, and in all the Llangollen
country, being overlaid in one case by Caradocs, and
in others by Wenlock shales.   They are therefore truly
part of the Bala beds, and their appearance at the junction of the Bala and Caradoc, all the way from Bala to
Conway shows that in all that country there is a general conformability of Caradoc on Bala, and that the
faulty-looking places where the pale slates are not seen
are really faults, and not overlaps.   All hereabouts the
Caradoc* is quite conformable to the lower beds, which
are black slates just below the pale, and gray slates just
below the black.   Conway Castle stands nearly on the
very top of the Bala beds, somewhere about where the
Hirnant limestone† ought to be.   This is unexpected—
to me, at all events.   The nearest trap to the Caradoc
hereabouts is exactly like the nearest trap to it in the

called Tarannon shale.   They were afterwards remapped by Mr.
Aveline, who first showed their true stratigraphical relations,
when Sir R. Murchison became Director-General after the death
of Sir H. de la Beche.

* This so-called Caradoc sandstone was afterwards proved to
belong to the Wenlock series, and consists of the Denbighshire
flags of Professor Sedgwick.   The supposed conformity is de-
ceptive and accidental.

† The beds mentioned at Conway are now believed to be the
equivalents of the Bala limestone.   Mr. Jukes himself came to
this conclusion.

Bettws-y-Coed country, but the calcareous beds here are, I think, below that a good way. Another curious thing is, that if there be conformability from Caradoc to purple slates, the thickness will be greatly less than we supposed about Bala, as here, notwithstanding many undulations and very much trap, the distance across the strike is not more than ten miles, while about Bala it must be twenty.

Concerning the finishing of this district, it will not be done quite so soon as I expected, as there seem more elements of botheration than I thought for. To-morrow, however, we start for the far country to the south, which I hope will be finished on Monday, and then all the ground is within easy distance, and unless it turn very crinkly, the rocks must yield their souls up shortly. My wife thanks you for your note, and will write shortly. Aveline will write when he has finished the dry proof of Denbighshire. He wants to know whether he is to put in the dry proof the boundary of the 'upper flag beds.'—Yours in the body and the spirit,

J. BEETE JUKES.

———

Merchlyn, April 2, 1850.

DEAR RAMSAY,—Aveline and I finished yesterday all that was essential for us to do together, although another day would perhaps be beneficial. I made up my mind yesterday about those rocks I spoke of before; they are certainly aqueous and stratified, although half formed of slightly worn crystals of feldspar. I saw a block yesterday as big as a cottage, with angular lumps of slate and pieces of all sorts of rocks. It is just an old submarine-volcanic conglomerate. I saw in Teneriffe this section [section given]. Now these were clearly deposited beneath the sea in the neighbourhood of the old volcano, and since uplifted. If they were baked hard, they would be just our sandstone. But there is nothing to show that they synchronised with any eruption or protrusion of igneous matter; on the contrary, they probably marked a period of repose in the igneous forces, when the aqueous ones resumed their sway, and acted on the old igneous rocks. It is therefore hardly right to give them an igneous colour. I acknowledge, however, the great difficulty of neatly separating these rocks here from the greenstone porphyries with which they are often surrounded. Put that in your pipe and smoke it! O, I forgot you had given up smoking; for the *seventh time,* I believe!

Bala, June 1, 1850.

Wet weather has kept me in two days, and it raineth villanously. I've got the map, and am going to set to work at it. Shall I send it to you, or had I not better send it to Aveline, for his work and dips to be put on it? I want his opinion on one or two points rather. Six fine days will enable me to take the section to the ridge beyond Penmachno, where my knowledge ends; but I want to know *exactly* where you mean to carry it on there, in order that I may bend it accordingly. I've brought it across the lake, and measured a base for the width of the same, as I cross it obliquely. Comes out very satisfactorily. Pen-y-boncyn where I begin is just 1600 feet above the lake. Ought I not to get a boat and take soundings across the lake, in order to get the shape of the bottom? I could do it some quiet evening, only I may have to buy line.

Bala, June 9.

. . . So they are making a row in town, eh? All the better perhaps; but I hope Sir Henry will take a high stand at once—say it is impossible to work by the square mile, and that if the survey is not to be done thoroughly, *carefully*, and *deliberately*, it had better not be done at all, and that we'll all strike work rather than be responsible for defective work. Send Lord Seymour down to examine us a bit; and let him come and do a week's work.

On Tuesday, having had a most sweltering day on
Monday, I set out in a car with my bag and baggage,
the ladies on horseback, for Yspytty Efan.   Having
proceeded to a point where I had sent on some men, I
descended and became no longer 'car-borne,' and we
proceeded by a wildish mountain-path under Carnedd
Filiast and Gylchedd, forming quite a winding array up
the valleys of horsewomen and footmen, and dogs and
carpet-bags and instruments. Arrived at the Brolch, the
ladies and the baggage descended to Yspytty, while the
surveying department climbed Gylchedd at a vast expense of breath and sweat, and continued the section
down to the Conway, getting down to dinner at 8 o'clock
nearly.   The next day we ran it over to the Machno ;
and here I carried on my straight line a good deal
farther than it was marked previously, and then diverged
at a right-angle in order to cut a place where the limestone is well shown, *reposing on trap,* and thus prepare
people's minds for the limestone trap of Snowdon, &c.
Next day breakfast at 7 o'clock, over to Penmachno,
and run the section across the ridge down to Dolwyddelan. Now I had had neither of your letters, but I took
it at a venture in a straight line across the ashy beds
down to a house in the flat, called by the natives thereof
Ty-isaf. I left the section on the doorstep of that house,
which looketh to the south-east ; but your worship will

doubtless be able to extract him up the chimney by a long shot across the valley. The ladies this day set out on horseback, accompanied by a car, to bring us way-farers home. . . . I think I can certainly finish the line going west of Cerrig-y-Druidion, of which more than half is done, and the part of No. 6 between Caeran Crwyni and the Arenig top, and perhaps the part of No. 4 be-tween Llandrillo and Pentre Foelas before the end of July. Then, if we return by August 14, I believe I could complete the remainder by the end of September up to Llangollen. After this, I suppose it would be necessary that I should go and complete the South Staffordshire coal-field. Anything you order of course I shall go and do, though that is superfluous to say. How do you get on with your men now? Selwyn had to chain up the Aran himself, leading the chain, and then coming down for the theodolite. His men, however, seem to do very fairly, and I think could keep a good reckoning if each had a book. . . . I shall have to work the cars a little bit about Bala to get to the ends of my lines; but time is money, and we must use one or the other.

Bala, July 8, 1850.

I have this morning sent off to Selwyn two tables: one of them the oblique section table, which I am very glad I have had the perseverance to go on with, as I

think it will be very useful. The results are exceedingly
curious in some respects, and not what I should have
previously expected. For instance, in crossing a dip,
or any other inclined plain, obliquely, until the vertical
plane of the section crosses the vertical plane of the dip,
&c., at a greater angle than 30°, the diminution of the
angle in the section will not be greater than about one-
eleventh of the angle of dip. For instance, a dip of 45°
crossed at an angle of 30° becomes 40° 55'; when crossed
at 45° it becomes 35° 9'; and it is not till the angle of
crossing becomes greater than 50° that it diminishes
very rapidly. Now, your honour, these tables and
others. I think it is a pity they should be lost, which
in all probability they will be, if left in MS. What
think you of getting a lot of tables useful in geological
surveying, and publishing them in the next volume of
the ' Memoirs' ? We could then have a hundred copies
or so struck off for our own private and particular
uses, for each of us to paste into our note-books, &c.
By the bye, concerning the chain-man. He is a very
decent old fellow, but very poor. Now, we only had one
day's work for him last week. Notwithstanding, he was
at my beck and call, and therefore could not accept any
other permanent engagement, every day. Half-a-crown
for this was so little that I gave him five shillings.
Ought we not to allow some sort of retaining-fee under
such circumstances ?

My wife wants to know about William : how he will be supported if his pay ceases—I think it would be a great shame it should cease—but she wants to know whether we should not all subscribe to send him something. I shall be most happy to do so. The best way will be to draw up a subscription-list, and send it round to all those he has worked for, would it not? . . .

Bala, July 14.

. . . Father Adam is to be at Edinburgh. I heard from him the other day. He says he's got a book coming out about 'geology, psychology, theology, Deism, Atheism, Pantheism, procreation, transmutation, parthenogenesis, academic training, Popery, and tomfoolery.' I long to see it. I hope he will speak out his whole mind—come out somewhat in the Carlyle style, and drive a good broad harpoon deep into the sweltering sides of the floating carcass of Humbug. I am beginning to feel savage with the world in general. Here's the mail come in ;· the bags pass down the street under one's nose, containing all our letters and papers, &c., and yet they are to lie in the post-office till 8 o'clock to-morrow morning.

Llangollen, Sept. 22.

I am anxious to know when I shall be likely to be wanted in Ireland ; and indeed you must try and ascertain for survey reasons, because it appears probable to my mind—maturely meditating on these matters during matutinal hours—that instead of going to Wem, I ought to proceed at once to Rugeley, and knock off all the northern end of the South Staffordshire coal-field, and thus make sure of finishing that district, at all events. Any one could do Wem, but no one could do Staffordshire so well as me, because they would have to get up my back-work. This is our wedding-day. We are celebrating it alone in *our hearts*, but our friends may keep it with as noisy a festivity as they like.

Llangollen, Oct. 8, 1850.

I have now finished my Welsh work in the field, but shall protract the last two sections, and overhaul the whole before I start for Staffordshire, which will keep me till Monday next. Mr. T. of Pool Park is a capital fellow. He took me a deuce of a tramp through a lot of wet woods, to verify Aveline's work as to a narrow band of old red below the mountain limestone there; and was much pleased when I found it for him in the bed of a brook and in the bank of a lane. We got out a rustic with a spade to dig for red sandstone rock, who had never heard of such a thing thereabouts, and was much surprised when the first stroke of his spade showed it him. He is greatly interested, as an estate-agent, in

the question of the boundary of the coal-field about Rugeley, and says that all the landowners there are anxious
on the point. He has written to several agents, &c. to
give me information when I arrive there; so I expect I
shall get on swimmingly.

Concerning the Irish matter,* I have got a temper-
ately-expressed, but very strong, testimonial from Sedgwick, of which I am going to send Oldham a copy; and
I am going to send notes to all those who I think will
come forward on my behalf. Among others, I must of
course look to you for one, if you have no objection to
write me out a formal document. . . . I think it very
likely that Howell would be very useful in South Staffordshire; he could, at all events, hunt for calcareous
beds in the Young Red.† I will let you know soon after
I get there.

DEAR RAMSAY, *Rugeley, Oct.* 28.

I've just returned from an expedition to the north,
into the middle of 72 S.E. sheet, and I have found a
great sprawl of lias in the middle of *Staffordshire,* between Rugeley, Uttoxeter, and Tutbury, where nobody
ever dreamed of such a thing before. The other day at
dinner at Mr. Landor's, T. Turnor asked me about
some beds above the red marls, which he said were

* This relates to an application for the Chair of Geology,
Trinity College, Dublin.

† This Young Red is now known to be Permean. A. C. R.

blue and black shale, with sandstone bands and ironstone balls. 'Why,' I said, ' that must be Lias !' ' Well,'
he said, ' Pickering (a mine agent) has written to me to
say he thinks so too, but wants me to get your unbiassed
opinion on the point.' So to-day we drove over together, taking Hull and Howell, and Lias sure enough it
is. All over the high ground of Needwood Forest
there's a capping of twenty-five, thirty, or forty feet
thick of the lower Lias, with bands of excellent limestone, like the celebrated Barrow-on-Soar limestone. I
only found some obscure marks of shells, which I shall
send to Forbes to-morrow. I saw it at two places a
mile apart; and according to Turnor's account, it must
form a capping on the high ground there four or five
miles long, by two or three wide.* Now what shall
I do with it? I half think that sheet had better be
knocked off out of hand. If any part of the Ashby
coal-field comes into it S.E. of Burton-on-Trent, it must
be a very *leetle* piece indeed; but I could cut down there
some day and see; all the rest would be red marls and
gypsum, with lias in the centre and portions of red-
sandstone on the N. and S. borders. Hull goes to-
morrow to Colwich, a very pleasant place between here
and Stafford, on the edge of the red marls, whence he
can work a large part of that sheet very conveniently.

---

* *Note by Professor Ramsay.* These are now known as the
Rhetic beds, and lie at the base of the Lias.

I heard from Oldham on Saturday. He says he shall certainly give up the survey, so I must go to Ireland anyhow.

Rugeley, Nov. 1, 1850.

My doing that Lias is out of the question. I shall barely be able to finish the Staffordshire coal-field alone, and certainly not able to draw any of the sections. If I can run them and sketch them, I shall be well off. I shall sectionise this end before leaving here; but there is a broad track between Cannock and Walsall untouched yet, and there are several knotty points in the field to the S. about Westbromwich and other places, besides tracing the cross of the beds above the thick coal all over the field. . . .

Brereton, Rugeley, Nov. 15.

. . . I hardly lost a day at Stourbridge, as I fettled a point about the Clent Hills, and did good work both going and coming back. I've just sent a sort of report on the waterworks question. The surveyor proposes to sink through the New Red and C.M. near Hagley obelisk till he comes to a water-bearing stratum, and then drive out right and left, and tap the trap rocks of the Clent Hills; pump up the water into a reservoir, and then distribute over low district. I've called it an *ex-*

periment, but one worth trying, as likely to give water of good quality at a high level. I'm going to steal a day to-morrow to shoot. The Marquis of Anglesea is out, and I'm going to breakfast with his physician at Beaudesert, and go out on to the Chase with him afterwards. Blackwood got me the leave through Lord Clarence Paget.

Rugeley, Nov. 17.

It was such a wretched day yesterday that I shot nothing. Heaps of game, but as wild as hawks. Oldham says, ' At last I think I may say that I go to India certainly.' He wants to be relieved at the end of this month, and wishes me to go over then, or before if I can. Now I commence sectionising to-morrow, and finish on Wednesday; on Thursday I have an appointment with a cunning old ground bailiff, who can enlighten me on obscure points; then if I can get a few days' horseback work between here and Walsall, I shall have finished all the essential parts of this coal-field. There will then be two sections to be run, and a few little points of detail to be settled, and the thing is done; perhaps one more section, but it is not very necessary. Hull is doing excellent work in his district. He has got another fault downthrow to W., which, with the one he had before, shows an elevated tract of lower red sandstone running from this coal-field towards that of North Staffordshire, bounded by downthrow faults, just as the coal-fields are bounded by downthrow faults ; so that though the faults are not continuous, there is an indication of the same movement in the same general direction. I believe Selwyn will find similar facts to the S., in the Droitwich sheet.

Rugeley, Nov. 20, 1850.

I have this morning had my official appointment for Ireland. I must be there on the 30th. Now I can leave here some time on Thursday, and have a talk with you on Friday at Holyhead. . . . I have asked Sir Henry to let me come back into Staffordshire for a week to pick up scattered threads, and then go to London to see him and arrange about our Welsh sections, &c. I have had two wet days' sectionising, but to-day was too bad. I think Howell safe for the Lias ; there's no marl stone, only the Lower Lias limestone beds, like Barrow-on-Soar. . . . By the way, ought there not to be a coloured copy of as much of our map as is finished in the Great Exhibition ? Seriously, I think there ought ; one copy mounted, and another table of sheets under it. I'm sick of writing.

Rugeley, Nov. 23, 1850.

. . . I shall be in Holyhead on Thursday evening by express train. I had a very kind letter and testimonial from Sir Roderick this morning; ditto from Darwin. . . Have you any extra-sharp razors? I am to borrow one of yours to shave with at Holyhead. I think we must have a solemn feast on the occasion, and offer up my beard as a sacrifice to the Infernal Deities, to propitiate their favour — making due libation of wine, and all things in order.*

* *Note by Professor Ramsay.* Jukes came to Holyhead on the 30th, just before dinner. Selwyn and I were at the hotel. Jukes was well bearded, but hesitated about shaving. However, we insisted that his beard would injure his prospects of getting the Chair of Geology at Trinity College, and next morning I lent him my razor. He appeared at breakfast amidst great laughter, and the waiters thought he was a stranger.

GEOLOGICAL SURVEY OF THE UNITED KINGDOM.

IRISH SURVEY. 1851-1869.

GEOLOGICAL SURVEY OF THE UNITED KINGDOM.

IRISH SURVEY. 1851–1855.

Acceptance of the directorship a turning-point in Mr. Jukes' life
—Official worry—Sense of order and duty—Letters to Professor Ramsay from Dublin, Dungarven, Monkstown, Cork,
London, Macroom, and Rathdrum, in regard to the working
of the survey, structure of the Irish rocks, and other geological
questions—Experiences and views of Ireland and the Irish—
Geological science as applied to mining—Presidency of the
Geological Society of Dublin—Red tape—Letters to his sister—Full work and scanty rest—Journey to Auvergne.

MR. JUKES' acceptance of the directorship of the
Irish survey seems to have been a turning-point in his
life, unfavourable in the end to mental ease and bodily
health. His letters after this time assume a different
tone ; the light-heartedness which could laugh at difficulties gradually fades away. He shows the heavy
sense of responsibility, and the bodily toil and mental
anxiety begin to tell upon his health and spirits. These
letters, now read by the sad light of after years, show

this to have been the case long ere his friends were
aware of it. In one of them he speaks of the climate
as depressing, and though the proximity of Ireland to
the 'melancholy ocean,' of which we have lately heard
so much, might from the consequent dampness have
had some lowering effect, yet the greater humidity of
Newfoundland had failed to elicit from him any com-
plaint. But far more than from anything outward or
climatic was he oppressed by the causes indicated in
the following painful words, which were written to a
friend a few years later : ' It is this perpetual strife and
atmosphere of contention in which one is obliged to
live here that, fight against it as one will, laugh at it
and bear up against it with good humour and courage
in public as one may, yet, from the perpetual strain
upon one's nerves and the gradual sap of one's heart
and spirits, ultimately breaks one down.'

Many minor cares, which he would formerly have
easily thrown off, now began to fret and annoy him.
Among these was his inability to carry out his own
ideas of orderly arrangement with respect to the survey
and the maps and specimens connected with it. Those
who possess this faculty can fully understand how ha-
rassing it is to be thwarted in attempts to institute a
perfect system, whether in the classification of a museum, the work of a survey, or the arrangement of a
library. To many persons this is a ' missing sense,'

and they cannot in any degree sympathise with the feelings of those who are worried and wearied by vain efforts to get or keep things *straight*. Mr. Jukes possessed this sense of *order*. Not that he was particularly *neat*—in fact, he was rather careless than otherwise in outside matters—but the order was of a deep-seated nature, and denoted among other things that he liked to lay his hand on every book or paper he possessed in an instant; that he wished to have every specimen in the museum so classified that it could be at once referred to, every map and document in its right place.

The writer of these words well remembers the pain she felt, when visiting him in Dublin, at seeing his annoyance on the receipt of some mandate from South Kensington to ' turn out the Geological Museum into the streets,' as he expressed it. Well might even this slight irritation impress her in one whom she had always known as so remarkable for placability and evenness of disposition, that she could say with truth she never remembered his being in younger days what is commonly called ' out of temper.'

It seemed, as he says in a letter to Professor Ramsay, ' as if life had lost its charm, and only duty remained'—*duty*, that word so full of power and meaning, which in the life of an Englishman seems to occupy the place of *la gloire* in that of a Frenchman. Both ideas may lead on to noble deeds and acts of self-sacrifice; but duty needs no stimulus from admiring spectators, being content with the quiet approving voice of conscience, and trusting to the 'perfect witness' of Him who is ' greater than our hearts,' and who ' pronounces lastly on each deed.' The following letters will be found to confirm these observations.

Dublin, March 3, 1851.

My dear Ramsay,—I finished my book of calculations on Saturday. . . . I can by no means agree with you about the O.R. north of Llangollen. By ' thinning out' I did not mean overlap, but thinning out to nothing, and I do believe that where I cross the O.R. in my section is its very thinnest part in all North Wales. Take a pair of compasses with an opening of two or three miles, and put one leg on Ty-u-chaf, and with the other you will describe the boundary of the O.R. *under the M.L.,** *i. e.* beyond that space there is no O.R. under the M.L., which rests directly on the Silurians. It is a mere patch of O.R., which in the north of England (that is, north of Shropshire) is nowhere a continuous formation, but a mere thing of shreds and patches, little spots of sand and gravel accumulated here and there in hollows of the other rocks. I go the full extent of saying *there is none* and *never was any.*

* Magnesian Limestone.

You cannot speak of it as its ' outcrop;' it is not an *it*.
If we could see deep enough, the section might cut into
another patch of it, but of that we can know nothing.
Now for another subject. Sir H. more than hinted
when I was in London at our giving up the six-inch
map here, and taking to work on those footy little
county maps of half or quarter of an inch to the mile
he has now revived, and, in fact, ordered me to do so.
At first I thought of at once resigning the appointment; on consideration, however, I perceive that would
be a foolish and also a cowardly way of acting. The
interests of the survey here being committed to my
charge, I am bound to stick by it and fight for it to the
last. Sir H., therefore, gets a letter from me this
morning, telling him the six-inch map was one of my
greatest inducements to come here; that I was disgusted and disappointed when I found they were not
published geologically, and that I am resolved not to
rest till they are so published; that before long you in
England will have the six-inch maps to deal with, and
that of course you will publish them; that at all events
then, if not before, we shall publish them in Ireland;
that I am prepared to show to his satisfaction, or any
other reasonable person's, that the six-inch map is the
smallest scale on which good geological work can be
done, and ' that my geological existence here depends
upon it; take that away, and I have neither interest in

nor capacity for the work.' I close with a hope that he will continue to get us the six-inch maps to work on as before. Taking a strictly 'service' view of the matter, I can show the absurdity of paying 1500*l.* per annum out of the public purse for merely a slightly-amended edition of Griffiths' map. The real question is the amount of *detail* and *accuracy* of work.

I heard from Hooker the other day. His letter did me good; it smacked of the sea.

Dungarven, June 14, 1851.

MY DEAR RAMSAY,—I am very sorry I have grown so matter-of-fact. I can hardly tell you why it is, but I feel that it is so. I hardly know whether it is the air of Ireland, or the nature of the work here, or what; but certainly much of the zest of life has departed, and nothing but duty and business remain. . . . As to age of valleys, my work in those days (having no ordnance maps) went in the old style, somewhere near the truth, there or thereabouts. I fancy I suspected the existence of N.R.S. in the mouths of several of the smaller valleys, and was confirmed in the idea by finding it in that of Leek. Besides, you find the old rocks, don't you, peeping out here and there through the New Red between Ashbourne and Derby? What do you make of the Cheadle coal-field? are its boundaries simple

faults, cliffs, or denuded slopes? Smyth writes about
some red marls, supposed to be Coal-measures, between
Newcastle and Rentham; what are they? There *are*
red marls in the Coal-measures of North Staffordshire.
You'll see when you go to Walsall the difficulty of tracing
the boundary north of that town. By the way, I think
of coming to town to see the big show at the end of July.
Could you not get a requisition to stop me in Staffordshire for a week on my return? I could join you at
Walsall, and we'd go over that ground together, and
settle all the important points in that time. . . . I am
glad Percy is made metallurgist; he'll do good work.
I am glad also to hear the educational dodge is coming
out. I would there was a prospect here of something
of the sort; I am almost tired of perpetual field-work,
and long to read and to lecture. . . .

Monkstown, Cork, July 26, 1851.

But for my having sent accidentally to the Cork
post-office, your letter addressed there would have been
returned to Dead-letter Office. We are nine miles from
Cork. I think you have done rightly, but it is a case
in point as to pushing and driving. The memoir on
North Wales ought to have appeared simultaneously
with the maps; you have now other occupations coming thick upon you, and I see very clearly it stands a

great chance of never being written at all. I'm sure all my recollection of my work is fast going out of my head. As to money, by Jove, sir, this country is awful! It may be cheap to live in if you can settle down anywhere; but for travellers like us, I assure you it is at least as expensive as England, without half the comfort. They all look upon me as a government officer and fair game, and they naturally, without previous concert, combine together to impose on us. They are like the Welsh in their determination to have a bargaining match, and always commence by asking four times what they mean to take; but as they fight every step, we are reduced either to give what they ask, or go without.

Monkstown, Cork, Oct. 15, 1851.

My dear Ramsay,—I wonder whether you have really left Llan &c. (too long a word that to write oftener than is absolutely necessary; not but what we have some queer names here; for instance, Garranckincfeake and Ballykickabog), and I also wonder whether you have left it a free man or a captive. I am delighted to hear that Selwyn's work is coming out so well. I am deep in the memorial of South Staffordshire, and have as yet steered clear of King Charles' head. . . . On looking at —— 's work, I certainly did

'The marvellous work behold amazed.'

It exhibits an audacity which is either admirable or exe-
crable, and for the life and soul of me I can't say which.

Oct. 28, 1851.

... Went to a lecture on electro-biology last night,
and never laughed so much in my life. Some queer
things, though, as the people acted on were well-known
in Cork. He set one young fellow whistling, and would
not let him stop—a perpetual stream of weugh!—one
dull note for a quarter of an hour, while the fellow was
begging him, by signs and pointing to his mouth, to let
him stop. Another very good-looking young fellow, but
rather dreamy-looking, he made imagine himself Father
Matthew, and he gave a solemn religious discourse on
temperance, amidst the shouts of the audience. He
made him, moreover, come forward as Mr. Wilson and
sing a song, in the middle of which he disenchanted
him, and the fellow stopped short, looking about him
in a most bewildered manner, blushed, and rushed to
his seat. . . .

During Mr. and Mrs. Jukes' residence at Monkstown, Forbes paid them a short visit, of which the
following account was furnished to Mr. Geikie, and is
taken by permission from his *Life of Edward Forbes :*

'My wife and I were living at that time in the old

H H

vicarage of Monkstown near Cork, and Forbes came
down to me to examine the fossils of Cork Harbour
and the neighbourhood. He was then busy with his
great work on the British Mollusca, and especially on
that part of it relating to the Limacidæ. At that time
he always carried a tin box in his pocket with half-a-
dozen fat slugs in it, on the description of which he
was then engaged. The vicarage, being an old damp
house, was well adapted for a slug preserve : the kitchen and pantry were every night baited with pieces of
turnip or other delicacies suited to the limaceous appetite, and the prey diligently secured in the morning.
Our Irish servant was horribly disgusted at these *slug-
gish* propensities of our guest.'

Forbes also inspected some remarkable fossils which
had been found near Ballyhale in the county Kilkenny,
respecting which Mr. Jukes writes : ' They were certainly a very remarkable group : fronds two feet across
of a large fern, since called by Forbes (provisionally) a
Cyclopteris, with the specific name of *Hibernica* ; several stems of *Stigmaria Calamites* and *Lepidodendron*,
or other similar plants ; and a shell like our *Anodon*,
three inches across, to which Forbes afterwards, in
spite of my remonstrances, insisted on affixing my own
inharmonious name as a specific designation.'

Dublin, Dec. 18, 1851.

May it please your honour!—The South Stafford-shire coal-field already looks seedy, not only in coal, but more especially in ironstone. Most of the best beds of ironstone are being rapidly worked out; there is a large importation from other places, not solely for mining. In great part of the district the thick (coal) is all gone, or else is so drowned by water, in consequence of the faults, &c. being all worked through by gate-roads, that its draining is next to impossible. I believe that the publication of our maps, with the Permian marked on them, will have a marked effect in raising the value of the parts so coloured. This is one reason why I wish to be so especially cautious in doing it ; an unsuccess-ful experiment on our authority will damage us most seriously, a successful one would establish us for the next half century. Imperfect trials have been made in abundance ; lots of people will be ready to make a great one on our recommendation—expend 20 or 30,000*l.* perhaps.

Sedgwick, on a visit to Sir H. Halford in south part of Leicestershire, saw an engine, &c. at the top of a hill a few miles off, and was told it was a coal-pit. He na-turally went to look, and on going up the hill found Lias shale with Lias fossils, &c. 'Why,' says he, 'this is Lias !' ' Lias !' says the proprietor, coming down on him ; ' you're a liar, and you're all liars together !' &c.

Naturally, the man ruined himself. I recollect a shaft being sunk on the Oolite close to the town of North-ampton. It was by a company, the shares of which were principally purchased by poor shopkeepers, servants, &c. at the instance of some 'practical coal-miner.' They sank for a long way till they had got all the money they could, when one fine morning the projector cut his stick. Buckland used to tell a story of Lord Oxford sinking on some property in Oxfordshire, on the Kimmeridge or Oxford clay, by the advice of a 'great practical man' and in opposition to all Buckland could say, who offered to be broiled on the first ton of coals raised there, and Conybeare made a caricature of Bucky frizzling in his own fat. Not only these gross errors have still place ; for instance, well-educated Londoners, gentlemen, have assured me there was plenty of coal on Blackheath common, only there was a law against getting it—a very unjust and oppressive law, sir, which ought to be repealed. In South Staffordshire I knew two instances of ground bailiffs, intelligent men, well versed in *coal-getting*, continuing to sink in Silurian shale, and heaps of the fossils of that formation lying on the pit-bank; they were still going down for coal. Not a man in South Staffordshire, except Blackwall and Matthews, knows what a fault is ; they imagine all faults to be contemporaneous with the beds—the coal grows, and so do the faults. Moreover, ' anything is a fault that injures the coal,' ——'s words. I mean to come out strong on this and other points in my memoir. There are several old coal-pits in the Black slate near Cerrig-y-Druidion, and a gentleman there has still 500*l.* at the service of·any man who will undertake to go deep enough to find the coal. The name of these stories is legion. The money wasted *in this century,* for want of the very rudiments of geological knowledge in those who wasted it, would have paid for the whole Survey and Museum since its establishment, and given it an endowment of 2000*l.* per annum for ever.

You may say that on my authority, if you don't like to say it on your own. I mean not only wasted on fruitless trials, but in bad mining where coal was. Even the 'great swamp' in South Staffordshire, where there are many thousand acres of thick coal now under water, each acre of thick coal alone worth from 800*l.* to 1000*l.*, is the result of bad mining. I don't know the exact extent of it, but say it is worth only 500,000*l.*, there is the interest of that, or 25,000*l.* per annum, lost for many years, and great part for ever, as to drain it now will cost an immense sum. Is that enough for your worship ?*

* *Note by Professor Ramsay.* These anecdotes were sent to me for my introductory lecture to the geological class, at the first opening of the School of Mines. I wanted some information of this kind to show the common *practical* uses of geological knowledge.

In farther illustration of this subject, an important
note was added to the account of the Geological Survey,
given by Mr. Jukes in 1866, which is here subjoined:

' Very few people are aware of the enormous amount
of loss which has been incurred, and is even yet of annual occurrence, in fruitless mining enterprises. The
loss from bad mining is, I believe, something frightful;
but I now more especially allude to that utterly useless
expenditure which takes place in the vain search after
mineral veins or coal, in places where their occurrence
is quite hopeless. Even in such rich mineral districts
as Cornwall and Devon, it is said that many of the most
experienced mining men there have strong doubts as to
whether a profit has been realised on the whole amount
of expenditure. Mr. Hunt, in the mineral statistics for
the year 1862, states the whole value of the tin, copper,
and lead raised in that district during that year to be
nearly three millions sterling. The necessary expenditure in raising that amount is doubtless very large; and
when to that is added the money spent in the vain
search after fresh veins, and the continuance of work on
old veins after their riches have been exhausted, it is
quite possible that the profits on the whole transactions
even of that year may have been much smaller than
would at first be supposed.

' Miners and geologists are equally destitute of that
knowledge of the mode of deposition of minerals in veins

which would enable them to avoid so great a loss; but
it is obvious that this knowledge can only be ultimately
attained by the application of the most accurate scientific investigation into the nature and origin of mineral
veins, and that the necessary preliminary to that must
be the exact delineation of them and their attendant
phenomena on maps of a sufficiently large scale to exhibit
them clearly and without distortion.

'When to the loss incurred in rich mineral districts
is added that which is of yearly occurrence in the search
for mineral veins in remote parts of the United Kingdom, where no one ever hears of it, except a few of the
peasantry or a stray geologist, the annual cost of the
Geological Survey is seen to be a trifle in comparison.
But the money wasted in the search for coal is almost
equally great. The foolish expenditure of which, during the thirty years of my geological life, I have been
myself personally cognisant, cannot have been less than
150,000*l*. The expenditure of which I never happened
to hear must, I should suppose, have been at least as
great. The total would make a sum which, capitalised,
would pay the present annual cost of the Geological
Survey *in perpetuity*. As a specimen of bad mining, I
may mention that about three years ago I was consulted
by some connections of my own respecting a coal-pit in
the South Welsh coal-field, which had cost something
like 30,000*l*. It had been sunk chiefly on the advice of

a practical man, who managed an adjacent colliery, but who did not notice that the axis of the synclinal curve in which the beds lay was itself inclined, and that accordingly the coal of which they were in search had cropped out a mile before reaching the place where the pit was being sunk.   This is but one, and by no means the worst, of numerous instances that might be given. The Geological Survey has already checked much of this fruitless expenditure, and I believe that I have been myself the means of putting a stop to as much of it as would repay the nation the amount of my own salary for the five-and-twenty years I have been in her Majesty's service.   The increase and spread of geological knowledge generally, however, to which the survey has contributed, has doubtless produced a still wider effect than is known to any of us.

  'None know more than the really eminent vein-miners how much their profession is infested by quacks, who often conceal their ignorance under a profuse use of technical mining phraseology.   Some of these are doubtless knaves as well as quacks; but the most dangerous class are the honest quacks—men who, having a smattering of mining knowledge, fancy they know everything, and would spend their own money, if they had it, in the enterprises they recommend to their employers. The obvious conviction of the truth of their fancies and crotchets, by which these men are animated produces a necessary effect on the minds of those who suffer themselves to be influenced by them. Moreover, it sometimes happens that their positive and confident assertions cannot be met by any direct proof to the contrary, although the more a man trained to sound logical reasoning actually knew about the matter, the more hesitation he would feel in coming to the conclusions to which they leap at a bound.'

Museum of Practical Geology, London,
June 18, 1852.

MY DEAR RAMSAY,—Concerning the obliteration of the trap of the Clent Hills, I fully coincide in it. I have always doubted them, and if you recollect, the last time we were on the Lickey, going up that steep bit, I said it was just as like trap in shape, &c. as the Clent Hills, and suggested the probability that the latter were, like it, only breccia. Still, the prestige of trap was so strongly on them, that I hesitated to knock it out altogether. However, away it goes now, and although it will involve much alteration in *Memoir* and section, yet it so simplifies matters that I am quite delighted. The getting that horizon in the Permian and the two calcareous bands are capital. What deceived me about that trap breccia, when trying to survey the trap, was, that taking the existence of trap as *undoubted*, I thought the loose angular fragments or breccia were drift, resting on the New Red, and pressing down into it, owing to its surface having been simultaneously washed up. Then, at that place near Clent Hall where it went under the pebble-beds, I suppose this was an old local drift, or accumulation of débris, anterior to the new red times. What still strikes me as wonderful is, that where the pebble-beds abut against it and overlap it on the sides of the hills, there should be so little mingling of the two things. You would see between Clent and Calcott Hill that a boundary could be traced across a ploughed field, not more than ten yards wide, with all angular fragments on one side of it and all rounded quartz pebbles on the other.

Museum of Practical Geology, June 21.

I have already in my description spoken of the great probability of the Bromsgrove fault running by Clent up to Wychbury Hill, if it were only to account for the apparent overlap of new red in Hagley Park. What think you of making the boundary-fault split dottedly at Wychbury Hill, one branch going to south-west and ending in new red, the other going on to Clent, and joining by dots to the Bromsgrove? . . . I have mentioned the fact, that if those two faults join, we get a continuous line of fault running north and south, and throwing down to west for forty miles and more, viz. from east of Droitwich to west of Stone, across four quarter-sheets. I have determined to send you the last fragment of my second chapter for your revision. You will see that I have endeavoured to be as brief as possible in my account of the New Red, and I think I have done justice to every one concerned. All my description of Clent Hill trap is in press, and will have to be cancelled; or do you think I had better let it stand, and insert a correction of it? . . . Whatever be the nature of the junction between the pebble beds and the angular débris, it is still very curious that two things so easily washed about should not be more mingled at the surface. By the way, I ought to call attention to it in the memoir. I have determined to start on Thursday at latest. I have still another chapter to write on odds and ends, but I cannot be any longer absent from my Irish work.

June 24.

Now for a deliberate answer to your letter, and a pack. I think it is a good plan for Howell to go and have a look at that Foxbrake affair. I recollect that knoll you mention, but east of that I do not believe there is a bit of rock seen *in situ*. The Permian there depends on pieces ploughed up in the fields and dug out of the ditches, showing there was ' red rock,' and not C.M., at surface. As to the memoir, I wish you had time to read it all from beginning to end. I fully intended you should do so with the proof-sheets, and revise it; but if you are going abroad, it will be impossible, I fear. I wanted you to look over my Staffordshire Memoir, and I would do the same by your North Welsh one. They would both probably profit by the operation. . . . There are going to be some lectures on gold got up here immediately. . . . Since writing the above, and just as I was on the point of leaving, I was sent for to the council, where it was proposed to me to stop, and, if you had no objection to the proceeding, to stay a week longer, and give the first lecture on the geology of the gold regions of Australia. Sir Henry is to write to you on the subject. I am quite willing. Moreover, I had collected materials, and was going to add a chapter to the unbound copies of my book on physical structure of Australia, about the gold regions, so that it would help me rather than otherwise; but if you think it interloping on your province, of course I drop it. Respond, O king!

DEAR RAMSAY,   *Cork, Sept.* 22, 1852.

Touching the Ludlow and Wenlock, there is a difficulty. It is impossible in Staffordshire to draw a line

between Upper Wenlock shale and Lower Ludlow ditto
—*i.e.* there is no difference either in the rock or the
fossils, so far as is known, all the way from the Am-
yestrey limestone to the Wenlock limestone. Moreover, the Wenlock limestone passes upwards into the
shales above by such insensible gradations of nodules
and bands of limestone, that it is equally impossible to
draw a top to that. . . . What think you, *pro hac vice*,
of introducing a third shade of colour without specify-
ing it, and colouring the shale below the limestone
darker than that between it or just above it?   I have
tried it, that you may see how it looks. The only blotch
is south of Walsall, where no one can tell where the
devil the limestone is, though I think it will hereafter
be found running down to the Delves and the Goodwins.

I have now just finished the proof-sheets of my
memorial. Heigho! If I had only known the work
it was to give me, I would never have asked for the
district. However, it is all finished except a wood-cut
of the map and a reduction of one of the sections, which
are to accompany it. Now there is just time, before I
send it finally to be struck off, for you to look it over,
which I shall be very much obliged if you will do ; and
if you find anything to say, just annotate on a separate
sheet of paper. Criticise as severely as you like. Flog
away. I'll find back as long as you'll find cat. Only,
as it would now be expensive to alter much, I would

only do so in case of gross errors or faults. . . . I am
sorry to say my wife has been laid up ever since our
return. I am almost beginning to fear she will not be
able to stand our wandering life much longer. You will
have heard of Mrs. Playfair's recovery. I hardly hoped
ever to see her again. . . .

Macroom, Nov. 5, 1852.

DEAR RAMSAY,—I have three of yours to answer.
Why should I not draw the basalt? What I object to
is the having to waste my time in draughtsmanship
unless I can do it as fast as I write this. Our time is
too valuable to be wasted on clerks' work—our pay is
head pay, not hand's, or we should shortly lose it. As
to the tadpoles, of course Lowry knows they are meant
for dots. He told me himself, especially in the vertical
sections, that it was no use my bothering myself; so as
soon as the geometrical drawings were done, I finished
them as fast as I could write. If I had stayed to make
them look pretty, they would not have been done now.
As to your memoir: if you really have all the data, I
would advise you to set to work at it in real earnest,
and tell Sir H. that it is impossible for you to do otherwise than to go somewhere and write it. . . . Forbes
and I quite agreed that it was no use going on accumu-
lating data that were never to be reduced, arranged, and

utilised, and that we must have clear time to bring out and publish our results.

It is, in the end, the best policy also. We, that is you and I, must not allow ourselves to be reduced to the position of mere map-colourers; what we do we must explain ourselves to the world, with the assistance of the palæontologists, chemists, &c., if necessary. We must assume the right, and keep it, to be ourselves the judges of the time when and place where this shall be done. . . . There is one thing here I am in doubt about. All the Silurian district of S.E. Ireland is finished without a memorandum by my predecessors. I do not know whether to go over it again at once, or wait till the Devonian district of S.W. Ireland is completed. We are getting some inch maps out; before they are published geologically, the ground must be gone over again. . . .

Macroom, Cork, Nov. 10, 1852.

Your letter of the 7th is refreshing, and I thank you for it. I seem to myself somehow to want a faculty which other people have—that of thin-skinnedness or whatever you call it. If a fellow thinks me an ass, or a fool, or a knave, or any modification of those denominations, it would never anger me that he should call me so (unless I saw he did it, not because he thought

so, but for the pure intention of insulting me). If it
were a downright honest expression of conviction, I
might endeavour, perhaps, to show him it was an er-
roneous one, but it would never occur to me to feel
offended that he entertained it. I can look back to
several instances in my younger days when I resented
such appellations—only after I found it was expected I
should by the lookers on, and not at all from any feeling
of my own. Even now it offends me much more when
a man begs my pardon for intimating that he thinks I
am not exactly right, and asks me to reconsider anything, than if he were at once to say, ' Jukes, you're a
fool or a rascal!' I suppose it is in accordance with
this idiosyncrasy that, when writing to *a friend* about
anything I disagree with him upon, I always feel inclined to put it ten times stronger than I actually think
or feel, simply as a demonstration of friendship and
good will. If I wrote civilly to him he might be quite
sure that I either hated, or feared, or distrusted him.
However, I am beginning to have a glimmering apprehension that people in general either don't or won't
comprehend this, and that I am alone or in a small
minority on this point. The only remedy is, I suppose,
that when I feel strongly tempted to write to a friend
about anything—when it occupies my thoughts day and
night, eating and drinking, walking and talking, and
rattling sentences seem coining themselves in my brain

every five minutes without my being consulted in the matter—is either to hire somebody to blow them at, or to put them all down on paper, and then put the paper into the fire. Because I recollect that it has often occurred to me, after once firing off a missive, to forget all about it, and to be very much puzzled when the answer came back (sometimes from a wounded, sometimes from an angry spirit) to recall to memory what the dickens it was all about. . . .

Dublin, March 19, 1853.

My dear Ramsay,—I have finished my sections, and am now looking them over. I believe that the solid geometry is as perfect as I can make it. The mechanical part I must trust to the draughtsman and engraver. . . A day or two's work in Kildare lately nearly convinces me too of a large Cambrian tract there also, hitherto mapped as Silurian. If not Cambrian, it is the very uttermost base of the bottom of the Silurian, or *somewhere below that*. In addition to the unconformability of these two, another good result comes out of our late work, and that is, that the granite comes up through the other rocks without heaving them up at all, except just close on the borders to a slight extent. On the contrary, the general dip is towards the granite, especially along a band of country running parallel to it, and one or two to four or five miles distant from its present surface boundary. I'll draw a section of what I mean on the other side. I am to have a paper on Wednesday week on this country. I wish it were to the London Society, as it would, I think, tell well there.

I am getting on with my various kinds of work, devoting the morning to one sort, the afternoon to another, and the evening to another, but the progress seems to be slow in each.

In the *Proceedings of the Geological Society of Dublin*, 1853, a paper appeared, ' On the Structure of the N.E. Part of the County of Wicklow, by J. Beete Jukes and Andrew Wyley, Esqrs.' This is probably the paper above referred to ; it contains the following passages : ' The position of the stratified rocks with regard to the granite is very interesting, as tending to modify the ideas with respect to the physical action of this rock, which, if not now prevailing, have only just ceased to be universally entertained. It was always thought that the eruption and elevation of a great range of granite invariably brought up upon its shoulders the lowest formation of the neighbourhood, and flung off the upper ones to a greater distance from its flanks in proportion to their newness.

[*Note by J. B. J.* ' I may here remark that I had

long ago been suspicious of the fallacy of this notion, and been inclined to attribute the elevation of rocks generally to a great widely-acting force, most probably the action of great heat, of which the production of igneous rocks was one of the local symptoms; but to look on the actual outburst of igneous rock as tending to produce depression rather than elevation in its immediate neighbourhood, except so far as the mere puckering and crumpling of the beds directly in contact with them are concerned.]

'In North Wicklow, however, we see, according to our present results, that the granite has in no instance the lowest formation (viz. the Cambrian) in contact with it, in no instance brings any portion of it up upon its flanks; but on the contrary, that the Cambrian rocks either dip towards the granite when they approach within a couple of miles of it, or pay no regard to it at all; and that the Silurian rocks which rest upon the granite have their beds tilted up by it only when very near it, and then at comparatively low angles, while a mile or two off they are almost invariably vertical, much contorted, but seem to have a general tendency to plunge headlong in the direction of the granite. It seems as if the elevation of the granite and its outburst had left a great hollow or cavity, as it were, running parallel to its present general direction, and a little removed from it, and that the rocks had sunk bodily

into this cavity, suffering greatly from lateral pressure,
doubtless, during the process, while rocks still farther
removed had remained wholly or comparatively unaffected.   Although it is beyond the limits of this paper,
we will just add, that the position of the rocks on the
Kildare side of the granite confirms these conclusions.
The lowest rocks are farthest removed from the granite,
and notwithstanding many sharp flexures which are
seen here and there, the general dip of the whole is
towards the granite, except immediately on the flanks
of that rock, where the beds are bent up against it.'

In order to illustrate Mr. Jukes' unfailing desire to
do full justice to his colleagues when compelled to differ
from their conclusions, the opening observations of this
paper are here added :

'In revising the six-inch map of the Geological
Survey, previously to laying down the results of the
work upon the new inch-sheet map shortly to be published, we were led to take a rather different view of the
main features of the structure of the north-eastern part
of the county of Wicklow from that which has previously obtained.   In stating this view, we wish em-
phatically to remark that we impute no deficiencies to
any of those who had previously examined the country,
since, without the benefit of their previous labours, it is
possible that we might not have arrived at our present

results. The country is one of almost unexampled difficulty; difficulty arising in part from the very complicated, and often almost inexplicable, structure of the rocks.'

Rathdrum, June 15, 1853.

MY DEAR RAMSAY,—I can only say that the colour for granite on the published Irish sheets is indistinguishable by any eye from that for Cambrian in the Anglesea ditto. There is an altered Silurian also, which is *werry* nearly the same. When I said green should be kept for feldspar trap, I meant ' quite the reverse ;' that it should be kept for hornblendic, augitic, or pyroxenic trap—confound their mineralogical jargon ! (That is spoken quite aside.)

If ever you look at my little book, you will see I have hooked granite on to the feldspar traps, though I hardly worked the connection out, simply because I had not a piece of rock to look at when I wrote it, and only wrote from a vague impression that I had formed that conclusion once. I am going to get Sullivan down here next week, and we mean to work the very entrails out of all the traps hereabouts, and not leave 'em a single atom of mystery to pride themselves on any longer. Worst of it is, when one has made out the chemistry and mineralogy, one can't see the geology in this confounded rolly-polly wood-covered country.

If I had all back work cleared off here and in the museum—and if I had my own way, there should be no more forward work done till all the back work *was* cleared up—I could, by help of the six-inch map, adopt such a system of work, that with six mappers, two collectors, and one office-man, I could have maps, reports, sections, and memoirs, rocks and fossils all prepared, arranged, and published in one continuous stream, and yet have five months' lecturing, six months in the field, and one month abroad. I could not do it with the inch maps, because there you cannot at once fasten on the data that looks shaky, and go and examine them. Moreover, I must be at perfect liberty to devise and carry out my own system, and must have publication and all that in my own hands. In that case, I would in two years' time have everything as smooth and regular as clockwork; so that the things should do themselves very nearly. It is this seeing my way to what could be done and not being able to do it, that frets and annoys me very often.* . . . One other reason why I want a lecture

* It was in the same year in which this letter was written, that Edward Forbes was also feeling thoroughly disheartened by the inability to carry out his own ideas with respect to survey work, and he thus writes to Mr. Horner from Jermyn-street, October 1853:

'Rest assured there is *no post here, or possibly to be made here,* that would induce me to remain, could I become Professor Jameson's successor in his professorship and keepership. Lyell, with the kindest of intentions, wrote to Mr. Cardwell of his own post is that, though well strung up again now, I know I may break down any day for field-work. However, I can't stop bothering you, when there is an article in the *Literary Gazette* awaiting me. Good-night.

Rathdrum, June 25, 1853.

MY DEAR CARA,——I have had a week's very exhausting work, what with the heat and the nature of the country ; but to-day, being a drizzle, I am indoors, and attempting to knock off some of my back correspondence. I am glad you are pleased with my new book.* It has only been reviewed in the *Literary Gazette* yet. ... I should like very much to come and see you at

---

accord, and did not tell me that he had done so until afterwards. Playfair did the same without informing me. I have assured both, that I cannot accept any office in preference. Lyell does not know how we stand here, and has never felt the horrors of being slowly strangled by red tape—a process of slow torture and eventual death we are undergoing in this place.'—*Life of E. Forbes.*

It surely indicates that something must have been greatly ' out of joint' when the same burden pressed so heavily upon, and probably shortened the lives of, two such men as Professor Forbes and Mr. Jukes—men whose sole aim was thoroughly and conscientiously to perform their work, and who, moreover, knew so much better than the ' officials' by whom they were thus hampered, what that work was and how it could best be accomplished.

* *Popular Physical Geology*, published in 1853 in Lovell Reeve's series.

Compton, but do not see any probability at all of such an event. I am going to try for the appointment of examiner to the Queen's University in geology. If I get it, that will take me a fortnight in Dublin in September. Then Binney talks of summoning me as a witness to a mining trial in Edinburgh, at the end of July or thereabouts. All these will so cut up my time, together with the possibility of having to go up to Dublin to meet Mr. Cardwell when Parliament breaks up, &c. that I shall be entirely used up till winter sets in.

How fearfully fast the years fly round now! I declare we shall come to the end of our days in no time at all; and I already often look forward to it—another reason why one must stick to one's work, or there will be no time to finish it. Hoping you may long enjoy your little nest at Compton, and with best love to Alfred and the boys, believe me, dear Cara, your affectionate brother,

BEETE.

Stephen's Green, Dublin, Oct. 20, 1853.

MY DEAR RAMSAY,—I was bothered so much all morning yesterday, that I put off answering you till afternoon; but afternoon brought me an impertinent letter from a pig-headed lawyer to whom I owed a wipe, and it took me till post-time mixing and compounding the plaster and polishing the lancets with which I meant to operate on him. I took the dignified gentlemanly tone, made everything as suave as possible, and treated him altogether as a matador does a bull in Spain, avoiding his rushes and gracefully sticking an arrow in his flanks. I can imagine him foaming at the mouth at the present moment over my epistle. *N'importe.* If you recollect, I ran a section which very nearly completes that sheet; it is engraved on it, and I thought the sheet was published long ago. There may possibly still be room for a little section across the Lickey, and I have drawn another pencil line, nearly coincident with an old one, only starting from Northfield church, and ending at 'Coppice,' anywhere from marl to marl through the coal-measures of the Lickey will do. For a datum you will require to run a short line from the railway up to Northfield church.

In February 1854 Mr. Jukes, as president, gave the annual address to the Geological Society of Dublin, in which he briefly sums up the contributions that had been made during the two previous years to geological science in the British Islands. He especially notices M. Elie de Beaumont's 'Theory of Mountain Chains,' as explained by Mr. Hopkins; the papers of the latter on the 'Drift;' Professor Ramsay's paper on the 'Sequence of Events during the Glacial Epoch, as evinced

by the Superficial Accumulations of North Wales ;' and
that by Professor Forbes on the Eocene series in the
Isle of Wight.   Mr. Jukes also refers to several papers
of great interest by Professor Sedgwick on the paleozoic
rocks of Devon, North Wales, and Cumberland, and on
the ' Separation of the Caradoc Sandstone into two distinct Groups ;' and to a paper by the Duke of Argyle on
the ' Granite District of Inverary,' the latter being of
special interest to himself, because it tended to confirm
his own views with respect to the action of granite.

This address concludes with the hope that, together
with the scientific and economical benefits which Geology has conferred on the world, ' she has contributed
her share to this moral benefit also—that philosophers
can differ in opinion without loss of temper and without
loss of respect for their opponents, and that, while each
is conscious of his own single aim at the discovery of
Truth, he is ready to give credit for the same singleness
and directness of purpose to any one who may fancy
that she lies in an opposite direction.'

Few letters between Mr. Jukes and Professor Ramsay passed or have been preserved during this year.
The following extracts relate chiefly to the death of
Edward Forbes, conveying some faint idea of the grief
which that sudden and lamentable event caused to his
friend and colleague.

Dublin, Nov. 19, 1854.

MY DEAR RAMSAY,—. . . . In my last address to the Geological Society, I mention briefly the geological *survey,* and say I hope to be able to give a more detailed account of our progress on a future opportunity. I am very busy now, as I have a lot of practical details to get up for to-morrow night; but on Tuesday I shall write my letter, which Forbes quite approves of, Mrs. F. says. I have still good hopes for him, and, in comparison with his state, I care little for this affair.

Nov. 21.

. . . I lectured again last night to a crammed theatre and a most attentive audience, although quarrying is a dry subject enough. The theatre has, indeed, been full every night except one, which was fearfully wet. How it will be when they have to pay, I don't know. I am bothered by not being able to get a draughtsman, though I have 50*l.* at my disposal for diagrams. I am going to send in a demand for heaps of books, geological maps, models of coal-fields and mining districts, &c. I hope that 'no news is good news,' as regards Forbes.

3 P.M. I have received the news. God help us!

Dublin, Nov. 25, 1854.

One seems at last recovering one's senses after a heavy blow. Regret, I think, can only cease with life. Remembrance is forced upon one at every turn, by every circumstance of one's daily occupations; so that, highly as we all thought of him, it is only when he is gone that we find the frequent necessity for his aid, and the large gap of our life that he has dragged away with him. However, it is all useless. Sorrow, regret, disappointment, are equally in vain. I could sometimes rave at fate, and feel a longing for some person or thing to take revenge upon, if grief did not almost prostrate one's energies. Enough said where nothing can be done. . . . I have three men at work upstairs copying mining plans and sections into diagrams ; but they are awfully slow workers. . . . I mean to ask for a fortnight's leave on December 20, but shall not come to town unless sent for.

Heigho ! I must go and draw a diagram myself, so farewell.

In his second 'Address to the Geological Society of Dublin' (1855), in addition to the affectionate lament already quoted, Mr. Jukes speaks thus of the peculiar powers of Forbes' mind :

'Although he possessed all those natural powers of minute and accurate perception and discrimination,

without which no man can become a real observer in
natural history, he was the opposite of a mere species-
maker. . . . He never remained satisfied with the knowledge of a barren fact, where it was possible so to connect
it with others as to make it tell a story. His ultimate
aim and object was the philosophy of natural history in
its widest and loftiest sense.'

In the same address, which was delivered in February 1855, the state and progress of the geological
survey are briefly summed up :

' I would always wish to see a national geological
survey carried on with reference not so much to what
may be thought our immediate requirements, as to the
wants and requirements of the future. It ought to be
in advance of its time, in order that, in a few years, it
may not be found to lag behind it. It ought to contemplate geology and its practical uses, not so much as
they exist now, as foreseeing what they will be in the
next and future generations. You may perhaps be inclined to ask me how far the geological survey of Ireland
will be in accordance with the high aspirations I have
been putting forth. To this question I may answer,
that it is one thing to form a great ideal model, and
another to attain to its perfect execution. I hope that
the future will prove our work to be not greatly below
the standard of merit which I think ought to be attained, though, from a variety of circumstances, I fear

it must in many instances fall short of it. One drawback has hitherto been the want of sufficient means
to do our work properly; but I am happy to say that
this impediment has been to some extent removed
during the past year by the increase of our staff.   Still,
even now we have to choose between apparent slowness
and a hasty imperfect execution.   We are compelled to
overcome our difficulties as best we may with the means
afforded us, hoping that, when a better day comes, and
the value and importance of geological work is more
truly appreciated, we may be found to have done our
best and to have been not altogether useless in our
generation.'

The autumn holiday of 1855 was spent in a journey
to Auvergne, of which the subjoined letter gives some
particulars :

Dublin, Sept. 15, 1855.

MY DEAR CARA,—At length I have got a few moments' leisure, when I can give you a brief account of
our movements. . . . We landed in Dieppe on Monday
evening, August 6, slept there that night, and after a
pleasant stroll in the morning on the beach under the
chalk cliffs, proceeded to Paris.   The journey through
Rouen up the valley of the Seine—from half a mile to
two miles wide, with its steep vine-clad banks about
300 feet high—was very beautiful; and we reached the
Hôtel de Tours in time for a late dinner.   We only

just succeeded in getting a room, having to wait while a
man cleared out. Next day, Dr. W. and his friend Mr. L.
joined us, and we stayed in Paris a week, seeing a good
deal of the sights and the Exposition. On August 14,
we set out by rail for Clermont at ten o'clock A.M.,
arriving there about eight P.M.; and the next day but
one we crossed the mountains to Pont Gibaud. Here
we went up the Puy de Dome, 5800 feet (*i.e.* W. and I
did, but Mr. L. and Augusta contented themselves with
the Puy Parion, a lesser but more perfect crater), and
saw several of the principal lava-streams. On August 19,
we went to the baths of Mont d'Or, stayed three days,
and went up the Pic de Saucy, 6200 feet—W. and I on
foot, and Augusta and Mr. L. on horseback—and to
several other points of that very beautiful and wonderful
district; bathed in the hot baths (old Roman), and
collected specimens. On August 22, we all rode across
the mountains, down into the valley of St. Nectaire,
along the course of a black lava-stream, which the river
has cut through again, and in some places nearly
washed away. Saw the encrusting springs. We slept
at a rural hotel, built near some more hot baths, and
the next day rode across a most interesting district
down to Issoire, and returned by rail to Clermont.
Here, after one day on Mont Gergovia—where Cæsar
was nearly defeated by the ancient Gauls, and about
which the guide had a story of a princess ' who was made

war upon by ' Le Prince César'—W. and L. left us.
We set out by rail to Issoire, intending to proceed by
diligence to Le Puy en Velay, but found every seat in
the diligence engaged, and had to stay all day at a
wretched inn, and travel all night to Le Puy, toiling up
mountain roads at a funereal pace, arriving at Le Puy
about eleven A.M. This was a most singular and picturesque town, in a basin surrounded by hills of 600 feet,
with volcanic mountains in the background. Strange
crags, narrow and quite precipitous (only to be scaled
by narrow steps), rose in the valley, capped by ruins of
castles. One rose out of the suburbs of the town with
a church on the top of it. It is 265 feet high; but as
the stairs are broken, the entrance to them is closed.
Another still loftier crag rose from the middle of the
town, with the ruins of an old castle on the top, the
streets being for the most part mere narrow stairs.

It was still blazing hot, and nothing to be done in
the afternoon. However, we one day drove in a sort of
calash up to the mountains and had a picnic *à deux*
at the edge of a wood among some crags. The coachman thought us mad, and the coach-proprietor equally
so, when we could not tell him where we were going
nor what we wanted to do; however, it was set down to
the remarkable eccentricity of the English. . . .

Aug. 31. We started by diligence for St. Etienne,
with a French officer returning to the Crimea, very

melancholy at going back, for he said last winter was
'horrible.' He was an oldish man, and two or three
stalwart, gray-moustached brother officers came to see
him off. They kissed each other most affectionately
at parting, mingling their gray moustaches and giving
a hearty smack. The porter told us there were four
Russian prisoners on parole in the after part of the
diligence, but at our first halt I discovered them to be
Professor Roemer of Breslau, his brother, and two other
German geologists. Roemer speaks English well, and
is a capital fellow, and it was with no slight regret we
found we had been at Mont d'Or, Issoire, and Le Puy
together without ever meeting till we were leaving the
country. We still had mountainous, almost Alpine
roads, frequently very beautiful, till we descended into
the valley of St. Etienne, the principal coal-field of
France, and found ourselves surrounded by smoky
chimneys and black faces. St. Etienne, 70,000 people,
is made up of pieces of Coventry, Birmingham, Manchester, Bilston, Tipton, and Gornall Wood, all mingled together, the stone houses being all black, and
one wonders what brings all the colliers and work-
people talking French ; for in looks, dress, and manner,
they might be all South Staffordshire. The next day,
after a visit to the Ecole des Mines, we all set off for
Lyons. It rained tremendously, and we found Lyons
in a perfect flood. Taking leave of our German friends,

we started at ten o'clock P.M. for Paris, and arrived by
one o'clock next day. Here we had a final peep at the
Exposition, and on Friday morning were just finishing
our breakfast, and in five minutes we should have been
gone, when who should come in but Prestwich. He
said the Geological Society of France were holding an
extraordinary *séance* and making excursions all round
Paris, and that I must stop; accordingly we did till
Friday, I getting up at five o'clock in the morning and
going off twenty or thirty miles by rail, and not returning till seven or eight at night, leaving poor Augusta
(whose friends had left Paris) all alone for three days
running. On September 7 we returned *viâ* Boulogne
and Folkestone, and reached London at two A.M. on
Saturday. For the first time for nearly six weeks did
we then know what silence was. In France, in town
or country, no matter where, there is always somebody
shouting, singing, stamping in wooden shoes, slamming
doors or windows, always coaches or carts rumbling,
dogs barking, cocks crowing, bells ringing, so that from
mere noise alone, it was only when utterly wearied out
(and not always then) that we were able to get an unbroken night's sleep. I never suffered so much from
heat as in Auvergne, but they said they had never had
such heat before. . . .

I have now a lot of maps to do; then the examination in the Queen's University; then I have to run

down to Killarney; and on November 3 commence lecturing, and, with the exception of a fortnight at Christmas, shall be lecturing every other day till the end of March.

LETTERS ON THE SURVEY

(CONTINUED)

AND ON THE STATE OF IRELAND.

1856–1858.

Letters to Professor Ramsay from Omagh, Galway, Dublin, Bantry, on surveying work—*Manual of Geology*—Letters to the *Times* on the state of Ireland: No. 1. Universal feeling of estrangement from the English government — No. 2. Social condition of the people—No. 3. Evils of governing by vicemasters.

Omagh, Jan. 1, 1856.

MY DEAR RAMSAY,—A happy New-year to you and yours. . . . I am lecturing two hours a night here, with all this business and other matters, quarter's accounts, &c.; so you can easily imagine I have enough to do. Drove thirty miles before daylight on Saturday, over a country like that about Trawsfynnydd, in a gale of wind and rain-storms. Back again on Monday morning.

Jan. 2. What do you think of the triple division of the Lower Paleozoic which I have proposed? I think it would make a nice symmetrical classification, and would be true to Nature and avoid all systematising and theorising:

PALEOZOIC.

Upper.	*Lower.*
f. Permian.	*c.* Upper Silurian.
e. Carboniferous.	*b.* Lower Silurian.
d. Devonian.	*a.* Cambrian.

Having got *six* letters and *six* colours to represent *six* things, we can leave the exact demarcation of any of them to work themselves out hereafter.   The precise boundary between *c* and *b* would often be as difficult to hit as between *a* and *b*, or as that between *c* and *d*, perhaps, in Shropshire, or *d* and *e* *certainly* in the southwest of Ireland, and *e* and *f* perhaps in other localities. Still, the main masses would be distinct enough. . . .

————

Leenane, Killeny Harbour, west coast of Galway,
Sept. 7, 1856.

MY DEAR RAMSAY,—Yours of 3d received here last night.   Post only three times a week.   Last year no post-office nearer than twenty-five miles. . . . If you want another odd section, I'll give you one [section]. . .

We did not discuss the Lower Silurian or Cambrian question at all.   I had other things to think of, being more in the Old Red Sandstone line.   Dunoyer and Salter now suppose part of that great red series at Dingle to be Old Red Sandstone, with other Old Red

Sandstone conglomerate, *unconformable*, at top of it. I doubt.*

Certainly the Silurians there altogether are lithologically more like old red sandstone than anything else, as they consist largely of bright red sandstone with large beds of brown conglomerate; but interstratified with these are beds full of Silurian fossils. It will all come right in the end, I have no doubt.

6 P.M. Beaten back from the pass of Bandarra by rain.

The great conglomerate here in the Silurian rocks is certainly rum—perfectly rounded boulders of granite and greenstone, at least eighteen inches across, so firmly set in a hard gray grit, that you can rarely knock them out. It all breaks together; but when they do come out, the surface of the pebbles is like billiard-balls for smoothness—quite regularly bedded, and interstratified with beds of very fine hard greenish-gray grit, like Cambrian rock. . . .

Dublin, Oct. 9, 1856.

Your letter enlightened me a good deal. I was aware that Bala = Caradoc and did not = Llandeilos —nay, I maintained the former on physical grounds

* *Note by Professor Ramsay.* He ceased to doubt, and proved the unconformity. See *Manual.*

alone, in my address to the Dublin Geological Society
three years ago—and that the Pentamerus beds = May-
hill sandstone; but I did not know that the Craig-y-
Glyn limestone = Llandeilos, and not Bala. I suppose,
then, the Gorwyllt limestone is the same. I never
even saw the sections across Craig-y-Glyn that I recollect, and certainly did not know they were published;
neither have I had any maps of any sort sent me for
about three years, nor the last decade, nor anything
else. I merely hear of these things or see them in
advertisements, and then forget to write for them; but
I will, and also have them sent without. I had no
idea that the ' pale slates' (a name I think I first used
to Aveline) were coming out so importantly, or were
so closely related to Pentamerus beds, or that they
had fossils at all. As regards a name for the Pentamerus beds, Sedgwick's of *May-hill sandstone* has clearly
the priority and is in use. I rather think Sharpe
was the first to state boldly that Bala = Caradoc.
Sedgwick calls your Montgomeryshire grits by the name
' Denbighshire grits,' if I mistake not; and it is not a
bad name, since they sweep all through Denbighshire
in great force and very typical form from Conway, where
they have the pale slate underneath them to Pentre
Voelas, and round north of Cerrig-y-Druidion, down to
the Druid, lying quite comfortably all the way between
the pale slate and the Wenlock shale or flag. They

cover a larger area north of Pentre Voelas and make
bigger hills than anywhere else. You must clearly go
carefully through the *Geological Journal*, &c., reading all
the papers on North Wales, and giving every one his
due—mentioning all the points where each man is right,
and adopting his names, saying nothing of where he is
wrong unless absolutely necessary. You must assume
the office of *judge*, and be impartial above all things.
I am not quite clear that 'May-hill sandstone' is not
the proper name for these, including the Pentamerus
beds and the pale slate. The latter is nowhere overlapped by what we call Caradoc that I know, unless at
Garn Brys, near Pentre Voelas. On the contrary, supposing it to be the top of the Bala, I always maintained
its constancy, and that therefore there was no unconformity between Bala and the beds above in North
Wales. I do not think either that our old Caradoc is
so much overlapped by the Wenlock as that it *dies out*
to the east. It is evidently made of the waste of the
old feldspathic traps on the west, which must have been
out of water, and their sand was not washed beyond a
certain line to the east. If you look at our maps, the
several places where our yellow colour ends are all
nearly in a straight line north and south. I believe
this ending could now be distinctly traced south of
Llangollen. There are a lot of sandstones coming
down from Corwen towards the Ceiriog, that Aveline

was strongly disposed at one time to put in as Caradoc, but decided against them on mineral character alone. I think they must be above the pale slate, the importance of which we were not aware of till some time after that—when, indeed, we were leaving the Llangollen country. (That was the reason of my mistake north of Llangollen; and it was only the very last day I was out there, that I saw the pale slate there and its bearing.) Now I think that just north of Tomen-y-Meiru there are just a few feet of fine-grained sandstone between the pale slate and the *Wenlock flag*, and that these are the dying-out of the sandstone we called Caradoc. I had always a misgiving about that place, and the boundary north of Moel Ferna seemed to me to go across the strike of the beds. However, I never worked the ground, and may be all wrong; but had we then known of the ' pale slates,' it would have saved us a world of trouble and prevented these mistakes. I hardly know what to say to that ground and the limestones of Glyn Ceiriog. It ought to be gone over again with our new lights; for, just looking at the maps now, I am afraid we all got all wrong somehow, and that the pale slates as the base of the Upper Silurian may go unconformably in many places over the Bala beds, and that may be the cause both there and elsewhere—about Conway, for instance—of much that puzzled me when on the ground. Heaven! if I had not to lecture, I would ask

leave to go and dash over the borders of all my old work
this winter. Indeed, I hardly see how you can get out
of shifting the boundary of the Bala beds to the base
of the 'pale slate,' and making all above that Upper,
and all below it Lower Silurian. Recollect Sedgwick's
name for the pale slates is the 'Pasty Rock.' I am
going ahead with the Berwyns. Forty pages already.

Oct. 10. My birthday. Forty-five I am ! . . .

Dublin, Nov. 5. 1857.

My DEAR RAMSAY,—I was very glad to hear of your
safe return. You seem to have seen much and well.*
I feel greatly in want of some such a run even now at
the beginning of the winter. That confounded book of
mine hangs over me like an incubus.† I am now at

* Alluding to a journey made by Professor Ramsay to Canada
and the United States, to attend a meeting of the American Association for the Advancement of Science.

† The book referred to above was his *Manual of Geology,* the
first edition of which was published in 1857, and is inscribed to
the memory of Edward Forbes. In the preface to this edition, he
says:

'Early in the year 1854, the late Professor E. Forbes asked
me to be his fellow-labourer in writing the article on geology in
the new edition of the *Encyclopædia Britannica,* and a text-book
to be founded on it. At the meeting of the British Association in
Liverpool, we had agreed each to sketch out a plan and to submit
it to the other; but before even that could be done, death deprived
the world of his services. When, after some time had elapsed,

work at an alphabetical list of all genera, with a reference to every page in which each is mentioned. I begin it at six o'clock, A.M. It is an awful task, and has already taken me a month, and not near the end.

I am convinced the old red sandstone of South Wales is two things, but it would require very close and detailed work to separate them. . . . The upper Old Red leaves the lower behind at Caermarthen (as you know) and also at Usk, going on between the Usk Silurian and

---

the publishers decided to intrust the work to my hands, I immediately commenced it; but as I could only devote to it occasional hours not occupied by my official duties, I found myself unable to complete the article in time to come in its proper place in the Encyclopædia. It was necessary, then, to defer it till the publication of the letter M made it possible to bring it in under the term Mineralogical Science. In the mean time I had felt in my own lectures the want of a text-book which should treat the subject of geology more systematically and more succinctly than any yet published, and the same want had been expressed to me by others.

'The *Principles of Geology*, by Sir C. Lyell, must be read and re-read by every one aspiring to be a geologist, and the perusal of his excellent *Elementary Manual* is almost equally necessary. They are, however, more adapted for the advanced student than the mere beginner, and presuppose the possession of much knowledge of collateral subjects, some of which I have here endeavoured to supply. Instead, then, of wishing to supersede, my object is rather to lead up to the study of those works. . . . Neither do I aim at supplanting the excellent treatises of Phillips, of De la Beche, of Ansted, of Portlock, or of Page. I have wished to enable the student to arrange in his mind and digest the knowledge he may acquire, either from the books above mentioned, or from the great works of Murchison and others.'

Pontypool *without* the lower or Cornstone group. The two are of course conformable about Abergavenny, being both nearly horizontal; but off towards Llandeilo there is either total unconformity or a lot of faults along the strike. Doubtless, at Castell Craig Cennen the junction between upper and lower old red is a fault. Altogether I had only time to see that something was to be done, and had no time to do it. Not a ghost of a conglomerate in all South Wales, so far as I saw; but I conclude there must be some, as everybody says so. I saw a few pebbles about as big as peas in a sandstone on the Vans of Brecon, but could hardly call that a conglomerate, unless very hard up for one. No time for more. Shall I send you my maps with the dips on?—Ever yours,

J. B. JUKES.

Dublin, March 27, 1858.

MY DEAR RAMSAY,—I'll look after Turk's Head.* I know Mr. Bennett, and will be glad to do anything I can for him. As to the *livableness* of the place, a man can live anywhere if he can build a house, and take plenty to eat and drink, and send a boat for some more when he has done. . . . I, this morning, working from seven o'clock to two *sans* intermission, finished the revision of my South Staffordshire Memoir. There are 280 pages of replacing and additional MS. I divide it

* In Newfoundland.

now into fourteen chapters, with appropriate headings. What I want you to do is to write the chapter giving the general description of the new red, to revise that on the Permian; and then, after I have described the boundary faults of the coal-fields, to describe the position and lie of all the rocks that don't come within a margin (of say half a mile) of any part of the coal-field.

You may say as much or as little as you like; or, if you like, I will cut out everything that does not come within that limit, and refer to some memoir of yours to be published in the year ——! for an account of the same. When do you finish your lectures? Could you run over here for a week, and stay quietly with us, while we had a good talk, and work over these and other things? I find it tremendously hard to do any work in London, while here I have got everything *en train* for hard work.

I find I have only a very imperfect set of the Staffordshire horizontal sections. Could you send me a complete set, uncoloured, if you have them, and also a copy of Hull's? I think I could improve them now, with all the ideas fresh in my head.

If I had had any notion of the amount of that Staffordshire work, I would have let it alone. It will have taken five weeks' hard work away from my other labours here, where there is botheration enough.

We have tumbled into some controversial ground, and I have the greatest difficulty to steer clear of a row.

———

Dublin, March 29, 1858.

. . . I can't find the least mention of Turk's Head in any chart, map, or book I have, nor have I any recollection of the name. I have been told of copper mines in Placentia Bay, and daresay it is there. If Mr. Clements reads over my *Excursions in Newfoundland*, he will see the kind of life he will have—a rough fisherman sort of life. All intercourse carried on by fishing-boats or trading schooners; no agriculture beyond a potato-garden; no fresh meat, except on great occasions; fresh and salt cod in abundance, salt pork, salt butter, ship's biscuit, any amount of rum, wine, or beer he likes to send to the merchant's store for. That will perhaps be in another store twenty, thirty, or fifty miles away. If he gets into an English harbour, *i. e.* a place held by English, Scotch, or Jersey men, he will be jolly enough. If into one held by South Irelanders, he will find it not so pleasant, and will have to keep on good terms with the priest. Only the sea-board inhabited; the interior is a rough trackless wilderness, rocky, barren, and overgrown with *scrub.* . . .

Castletown, Bantry, Sept. 2, 1858.

MY DEAR RAMSAY,—. . . As to publications, Sir R.,
when over here, told me distinctly he did not wish these
' data' to be ' memoirs,' but something distinct.   They
certainly ought not to be memoirs, or you would have
two sets of memoirs going over the same ground.   Each
sheet should be issued with its description—a short,
cheap, sketchy thing, intended to explain the sheet and
how the work was done, the data on which it was founded,
where people were to go to see the data, sections, quarries, &c.—for the use of the multitude.   The memoirs
should be the condensed extract of these descriptions,
addressed to the scientific geologist only.   In the mean
time the Irish stationery office have put a price of two
shillings each on these, instead of sixpence, threepence,
or gratis.   They say they can only do what they are
ordered—charge what they cost.   I suppose it is of no
use writing to Sir R. now to get this altered, but I shall
do so as soon as he comes back within the region of
civilisation.   I have heard nothing about the new bone-
caves or the fossil man, but have not the least objection
to believe the human race as many millions of years
old as may be warranted by the evidence. . . .

The following letters by Mr. Jukes appeared in the *Times* in December 1858; and a perusal of them will show that they are by no means inapplicable at the present time :

THE STATE OF IRELAND.
No. 1.
To the Editor of the ' Times.'

Sir,—I believe that some remarks on the state of Ireland, coming from a competent and impartial observer, might be acceptable just at the present time. In order to show that I ought to be able to observe truly and judge impartially, I will state in the first place that I am an Englishman born and bred, a graduate of an English university, and that my pursuits led me in early life to move much about England, and to mix on familiar and intimate terms with all ranks and classes of the people ; that I have since visited many countries, including some of our own colonies, as well as those of other nations ; and that for the last eight years I have resided in Ireland, living in different parts of it for weeks and months at a time, and here again associating with people of all ranks, of all sects, and of all parties, on terms of the utmost frankness and cordiality. I ought, therefore, to be as free from local and national prejudice as most men, and have at least had the means of comparing a good many different races of people under many different circumstances. I will only farther

premise, that personally I never had reason to complain
of any one in Ireland, never, that I know of, made an
enemy, but many warm friends, and that I have no personal feeling to gratify nor any personal aim in what I
now write.

I would first speak of the feeling of the Irish people
towards the English government. Englishmen in general are, perhaps, hardly aware of the very widely-spread,
if not very deeply-rooted or well-grounded, feeling of
estrangement towards the English government that
exists in Ireland. I speak now of the middle-class
gentry, persons of education, people 'one meets in society,' as well as the tradesmen and shopkeeping class.
The peasantry have, I think, very little fixed opinion
on this point either one way or the other. This feeling
of estrangement may be gathered even from the conversation of those who would think themselves affronted
if you spoke of them as any but loyal subjects of the
English crown. They talk of the blunders of the English government, or the occasional blunders of the English arms or diplomacy, as if they were things foreign
to themselves, of which they had no share and for which
they felt no responsibility.

But there is a very large class, especially among
young men, who would not be at all offended if spoken
of as wanting in loyalty; some who merely leave their
sentiments to be inferred, others who will openly avow

that they are not loyal subjects. Of these, perhaps the majority are Catholics and Young Irelanders; some, however, are Protestants, and some even avowed Orangemen—men who will speak of themselves as Orange Republicans or Republican Orangemen, whichever way you like to arrange the incongruous terms.

This feeling does not ordinarily go beyond a vague sentimentality, such as one occasionally meets even in English society, which would never under any circumstances lead to any positive effects. But when its expression is freely and widely indulged in, even in Dublin, one cannot wonder at its more exaggerated form assuming a tangible shape, 'a local habitation and a name,' in some parts of the country among young enthusiasts, ignorant alike of the things they profess to hate and those that they suppose they admire.

I well recollect my surprise when, on my first coming to Ireland, I heard a gentleman by birth and education, a young man, an associate and friend of my own, avow, in evidently the full strength of sincere conviction, his belief that 'the English government was the worst in the world,' that he 'would rather live under any other government;' and on my asking why, I found that he was perfectly convinced that, among other delinquencies, the famine of 1848 was *caused* by the English government, or, if not actually originated, was at least encouraged by them; that it was their fault en-

tirely that so many people died; and that they might,
if they chose, have taken such measures as would have
prevented a single person from perishing of starvation.

I need scarcely add, perhaps, that at that time my
friend had never been out of Ireland; but I mention it
to show the feeling with which he had been imbued by
every one with whom he had hitherto associated. Indeed, he seemed both surprised and offended at my
venturing to oppose his assertions or to combat his opinions, never before apparently having met with any one
who thought differently from himself upon the subject,
although he moved much about Ireland.

Such ideas originate principally in ignorance—ignorance alike of the spirit of the British government
and its power. Irishmen are too greatly imbued with
the continental notion of personifying a government.
They look upon it as an individual animated with a
peculiar intelligence and feeling, as a ruler acting from
his own impulses, and expect it to guide, direct, and
govern everything. Englishmen regard a government
as merely a temporary committee, a national vestry selected to carry on the national business for a time, in
accordance with the national feeling that may then
happen to prevail, until the national will or caprice
shall substitute another set of men, who shall be more
in harmony with the new national feeling. Irishmen
and Scotchmen have, or at least they may have if they

choose, just as much share in making the selection and contributing to the feeling as mere Englishmen, and if they do not choose to take their share and be satisfied, it is their own fault. If they have their own grievances, let them bring them forward, grumble and growl as much as they like; and if they are substantial grievances, they require nothing but united action and perseverance to get them righted. Any section of the people of the United Kingdom who do not know and feel this, and are not always ready to act upon it, are ignorant of the spirit of the constitution of the empire of which they form a part.

But how much more ignorant of the power of the British government must be any set of men who in these days conspire to overthrow it by force! These foolish fellows of the Phœnix Society, for instance: how utterly unaware they must have been of the magnitude of the task they proposed to themselves! Give them a halfpenny squib and tell them to go and blow up Carran Tuohill and the Reeks, and they would have laughed at you. And yet that would really be a more reasonable proposal than one to shake the all-pervading power of the British Empire (which embraces the whole earth, against which a union of all nations of the earth would not ultimately prevail) by means of a Phœnix Society and a regiment of American filibusters.

Would it not be well, for the sake even of these poor

foolish fellows themselves, that some more visible and tangible representation of the mere physical power of the empire should be placed here and there about Ireland, than the few constabulary and coastguard stations which at present do that duty? We are spending vast sums of money in fortifying some points of the coast of England. A little of that expenditure extended to Ireland would be both an immediate benefit to the industrious part of the population in the employment afforded, and its results would be permanently beneficial, in rendering the very notion of revolt manifestly ridiculous. There might, indeed, come a time when a few strongholds might be useful against the foreigner; but in the mean time they would be most advantageous in giving an alteration to the tone of the Irish mind, by showing that Ireland was neither neglected, nor despised, nor treated with less favour than England; and the idea that she is so is very prevalent.

You will, indeed, hardly meet an Irishman who in his heart does not believe, not only that Ireland was unfairly and unjustly treated formerly, but that the English government and people have still a spite towards Ireland, and that she is unfairly and unjustly treated down to the present time, although they do not find it easy to ‘condescend to particulars,’ as the Scotch say.

Now, if any unfairness has been practised during the last twenty or thirty years, it has been manifestly the other way, Ireland having been rather unfairly petted, perhaps, by the English government, though no Englishman would, I am sure, complain of that, if it would in any way contribute to obliterate the remembrance of the past and to produce cordiality for the future. No one can read the history of Ireland and wonder that such a feeling of alienation from England should exist among the Irish people. We are, however, too apt to suppose that, because the operating cause of such feeling has been during the present generation withdrawn, therefore the effects of its past action should immediately cease. 'The sins of the fathers shall be visited on the children, even to the third and fourth generation,' may be taken as a great political axiom ; and three or four generations must yet pass by before the national wounds, now barely skinned over, shall become completely healed, nor should we be surprised if the scar is never wholly effaced. I cannot help looking upon the viceroyalty of Ireland as one at least of the continuing causes that serve to keep up the memory of the past, to perpetuate the feeling of distinct nationality, and thus act as irritants to retard the perfect healing of the old sores. This, however, is a part of the subject which would lead me to trespass on your space to too great an extent. My object in this letter was simply to lay before you the result of my observations on the state of feeling of the Irish people towards

the English government. If you will allow me to trespass on you again, I propose to confine my remarks to
the social state of the people among themselves, and in
the mean time am, sir, yours,

COSMOPOLITE.

Dec. 28, 1858.

THE STATE OF IRELAND.

No. 2.

To the Editor of the ' Times.'

SIR,—I wish now to turn your attention and that
of your readers to the social condition of the people,
and especially to the relations between landlord and
tenant.

Now, sir, no civilised man can do otherwise than
execrate the cowardly and treacherous conduct of that
part of the people which instigates secret assassination,
and combines to protect the assassin. I say *cowardly*
advisedly, because I believe a base and abject cowardice
and want of manly feeling is at the root of it. There
is a system of secret terrorism which could not exist
if even one man had the courage and manliness to come
openly and frankly forward, and at all risks to himself
refuse to submit to it any longer, and, if needs be,
publicly denounce those who attempted to enforce it.

Not one man, it seems, dares do this. How seldom,
indeed, one hears of a man in Ireland, of any rank or

class, who dares to act independently and do what he thinks right, in defiance of the opinion of those around him !

This fear of public opinion, this slavery to ' Mrs. Grundy,' is one of the greatest defects of the Irish character; and it is to that feeling of moral cowardice that is to be attributed the possibility of secret associations, whether Ribbon, or Orange, or Phœnix. While, then, we must abhor as well as deplore the feelings and the acts of the class of Irish tenants towards the class of landlords, I must say that I have arrived at the conviction that it is in great part, if not wholly, the fault of the landlords themselves. That action and reaction are equal and in opposite directions, is an axiom equally good in morals as in physics. Let us briefly examine the action of landlords as a class. My own personal observations were chiefly made in the south and centre of Ireland, but the description will, I believe, be more or less applicable throughout the country. I would premise here, that I do not impute blame to individuals nor to the present race of men. I speak of the system, the ordinary practice and custom, in which the present race, and their fathers and grandfathers, have been brought up. I speak also of the rule and custom, to which I am aware that exceptions are beginning to spring up more and more frequently; but if we wish to ascertain the causes of things now generally in action,

we must look to the past rather than the present, since
time is required for local action to have a general effect.
Now the rule and custom in Ireland is, that the landlord lets to the tenant, whether by the year or by lease,
for a term of years *the bare land,* the tenant building
his house as well as his barn and offices, such as he
has, and not receiving one sixpence from the landlord
towards these very necessary parts of a farm.   The
landlord, I repeat, as a rule, merely allows the tenant
to take possession of so much land, in the condition it
may happen to be at the time, not only not building
anything, but not even thinking of repairing anything
that may be out of repair, or doing anything towards
draining, fencing, manuring, or putting the farm into
proper condition.   Nothing struck me with more surprise, when I first came into Ireland, than this entire
apathy on the part of the landowner.   I recollect speaking of it to an old gentleman in the county of Cork,
with whom I was spending an evening in 1851.   He
was a retired solicitor, having a small landed property,
which had been his father's and grandfather's before
him.   An average kind-hearted man he seemed, neither
better nor worse than his neighbours, and appeared to
express merely the common sentiments of his class
when, on my mentioning my surprise as above stated,
he replied, ' O, tenants ! ―― them, the scoundrels !
never do anything for them; if you do, put up four

mud walls for them. Let that be the outside you do ;
they will only wreck and ruin your property.' Subsequent inquiry and observation, continued down to the
present time, have shown me that this sentiment,
though not often perhaps so openly expressed, has
been, and to much too great an extent still remains,
the prevailing one among the landlord class towards
the class of tenants. I have heard from good authority,
and have even seen in print, statements as to the management of large estates, belonging to noblemen and
others, in remote parts of the country that seem almost
incredible, but which have, I fear, too large a foundation in truth.

These statements go to the extent of saying, that
it has been the practice to keep whole populations in
complete ignorance as to their debts to their landlords,
no tenant ever knowing how much was still owing, nor
having anything to show how much was paid, occasional demands being sent out, seizures made, and
cattle sold if the demands were not complied with ; but
no balance ever struck, no receipts given, and no accounts rendered. Such things have been told me by
respectable persons, who had no interest either one way
or other in deceiving me. Can we wonder at a feeling
of hostility existing between class and class when such
things—whether actually true or not—are believed to
be possible ? Even if there be a certain amount of

exaggeration and misstatement in such extreme accounts, the general fact remains undoubted, that *the
rights* of the landowner are generally used to the uttermost, both in the amount of rent demanded and the
method of exacting it, while *his duties* are seldom even
thought of. I have myself known of two or three instances with respect to persons in a rank of life con-
siderably above the mere peasant farmer—persons of
small property, country practitioners, and the like—
men, too, who were at least on speaking terms with the
landlords of the houses and grounds they rented. These
men, trusting to verbal promises and assurances, or
relying vainly, as it has happened, on their knowledge
of the landlord's supposed good-feeling towards them,
have expended money in improving and beautifying
their holdings; and as soon as their improvements
were complete, notice to quit was served on them, and
either the rent raised, or the places taken and given to
some friend of the landlord *or his agent.* Such instances
may of course occur in England or anywhere else, but
as exceptions only. Here in Ireland they seem to be
the rule, and were spoken of by others as a matter of
course, the sufferer only pleading in excuse for his folly,
' O, I thought he would not have done it to me, after
what he said when I,' &c. &c. Now, if such things are
unblushingly done to men who could, if they chose,
make their complaints heard, have we not too much

reason to put faith in the statements of systematic
oppression and cruelty of which we have only the
rumour from a class cut off, by their ignorance and their
want of confidence, from all intimate communication
from those above them? I have no wish to use high-
flown phrases about the ' finest pisantry on earth,' and
so on—I know their faults as well as their good qualities:
their want of truth, their want of honesty and candour,
the absence of all trustworthiness either in their word
or their deed; but are not these faults the very results
we should expect to find in any race of men who have
for centuries been treated as an inferior race, as a lower
order for whom any treatment is good enough, with
whom no friendly communication or intercourse is to
be held, who are regarded as a *servum pecus*, whose
only use in life is to minister to the wants and pleasures of their masters? You will observe that, when
speaking of the ' landlord class,' I make no distinction
of creed, or party, or race. So far as I can make out,
the Catholic landlords are as bad as, if not worse than,
the Protestant—the old families certainly worse than the
new—worse absolutely in their treatment of their tenants and dependents, and still worse relatively, because
the people submit to them more tamely and completely,
and take worse treatment from them without murmur-
ing, than they would from others, and become therefore
more degenerate and degraded.

I do not pause here to inquire into the many and various causes which have led to this utter estrangement between class and class. I merely wish to describe the actual fact, and to show to your readers the extent to which that estrangement has proceeded, and the kind of feelings which the superior classes entertain towards the inferior, and the treatment which they ordinarily give them. The feeling is universal. It may be detected even in the treatment adopted and the tone used towards domestic servants and other dependents; but its most baneful influence is felt in the relations between landlords and tenants. Can this alienation exist only on one side? If the landlords as a class—even those who reside on their property—stand aloof from the tenantry, and if their only connection with them be as hard exactors of rent—of rent so high as often to leave only the most miserable chance of existence to their tenantry—can we wonder that the latter look upon the former as a hostile class; that a common feeling, an *esprit de corps*, unites the peasantry on the one side, as well as their lords and masters on the other; and that the peasant feels bound to 'stand by his order,' even in acts which he individually disapproves? Can we wonder, too, that men, individually kind and estimable, fall victims to the feelings of hostility with which their class is regarded? When two hostile armies meet in the field, how is it possible for each to weigh the individual merits of their enemies? The *esprit de corps*, when once prevalent, overrules all individual considerations even among the best of us.

There is nothing anomalous or exceptional in the unfortunate state of feeling among the Irish farmers and peasantry (there is no distinction in Ireland between these two classes); it is simply human nature. Place any other race of men in the same circumstances, and you will get similar results, varying only according to minor variations in the constitution of the race.

Those circumstances are not the offspring of laws, nor can the evil be cured by legislation. The only cure must be an alteration in the habits and manners of the people, and that alteration must commence with the landlord class. I have often been puzzled to make out how the resident gentry in the south of Ireland pass their time. They inhabit spacious mansions, secluded in wide and beautiful demesnes, that are encircled by walls ten and twelve feet high, with overhanging coping-stones. They seem to live in a sort of strenuous idleness within these walls, and only pass beyond them in pursuit of field-sports, or when visiting the neighbouring gentry, or going a journey. There is no school-superintendence, no visiting of sick or distressed neighbours, no hearty chats with farmers or their families, no persons coming to ' the house' for assistance or advice, no parish business—in short, no intercourse or com-

munion with the people outside the wall and the gate,
except casually and accidentally.   There are no county
cricket-matches as in England, no curling-matches as
in Scotland, where all ranks mix for the time on equal
terms.   If the young resident gentlemen happen to
have any intercourse with any of the neighbouring
lads, it is only as patron on the one side and as fawn-
ing parasitical flatterers and hangers-on on the other.
Not unfrequently may be seen, up to the very gate-lodge
of the domain, a line, on each side of the road, of the
most wretched tumble-down houses or reeking cabins,
such as no English proprietor would allow on the most
secluded part of his estate, if but for their very unsight-
liness.   Even with the lesser gentry, whose houses are
not so fenced-in from the commonalty, the intercourse
between them and the surrounding peasantry is equally
stiff, formal, and casual, with no cordiality on either
side, but with a careless air of superiority on the one,
and something approaching to servility on the other.
Go where you will, you find not only this complete
separation between the class of the landed proprietors
and the rest of the community, but you look in vain for
any effort on the part of the gentry to assist or to improve the condition of the peasantry, to sympathise
with them in any way whatever, to alleviate their dis-
tresses, or even to pretend to notice them and be sorry
for them.   So used are the peasantry to this entire

absence of all sympathy on the part of their superiors, that if any one in the garb of a gentleman or lady enters their cabins, either for charitable purposes or for mere conversation, they seem puzzled and confounded, and rather inclined to be suspicious of the motives of their visitor.

I could expatiate to a much greater length on the social condition of the people, and give many curious anecdotes in illustration of it; but I forbear out of regard for your space, and will in my next letter endeavour to point to one at least of the causes of this deplorable state of things.—I am, sir, yours,

COSMOPOLITE.

Dec. 30, 1858.

THE STATE OF IRELAND.

No. 3.

To the Editor of the ' Times.'

SIR,— Any one reading my last letter might naturally conceive of the Irish landed gentry and their class that they were a hard, severe, haughty, and repulsive set of people. He could not possibly make a greater mistake. As a class they are remarkable for their kind-heartedness, their courtesy towards any one who is entitled in any way to be considered their equal, their indulgence towards their inferiors, their genial cordiality and good-nature. You rarely meet among them any of

M M

that 'landed manner' which Sydney Smith spoke of as characterising the landed gentry of England—that cold exclusiveness and repulsive condescension which, at a country dinner-party in England often makes you long to get two or three of the squires and baronets in the backwoods, and teach them their own value. Personally and individually the Irish gentry, if not quite so conscientious as they might be, are at least a more lovable race, who, if they had not been brought up to think much of their pleasures and little of their duties, would at least be personally incapable of cruelty and harshness. How is it, then, you may ask, that so much harshness and so much cruelty exist in their general treatment of the tenant class as accounts for, if it does not palliate, the innate and organised hostility of the one class to the other? This mutual feeling of repulsion arises, doubtless, from many circumstances, which must be sought in the past history of that country—a long chain of events reaching back to a period even long before the Norman invasion, when its real history commences. There is, however, one cause still in existence which is a most efficient one, if not in originating, at least in continuing, this repulsion. This cause is the system of agency which is universal in Ireland. Ireland itself is governed by a viceroy; the land of Ireland is *managed by vice-owners* and *vice-masters*. I believe there is now in Ireland no class of persons more pernicious as a class than the agents who manage and administer and are masters of the land, though they do not own it. The acres in Ireland which are not in the hand of an agent are few and far between. These agents are younger sons or nephews of landlords, retired or half-pay officers, briefless barristers, solicitors, broken gentry who, having lost their own estates, are of course the fittest persons to manage those of others, or lastly, persons who have been bred and brought up as agents, and have never had the pretence of any other occupation. They are not mere law agents, to look after the deeds, the leases, or the encumbrances of the estate; still less are they mere stewards or accountants, who at stated times receive the rents, and account for them to the landlord, reporting any deductions that ought to be made for repairs, losses, &c.

They are not represented by any class of men that exist in England. They are virtually the masters of the properties for which they are called agents, removable individually doubtless at the will of the owner, but to be succeeded only by other masters or vice-masters of exactly the same kind. The agent acts as if the property were his own, except that he hands over a certain balance of rent to the real owner. He grants leases or not, as he likes; he alone visits the property, he alone sees the tenantry. All petitions and complaints are addressed to him. In many cases the agent has money capital of his own, and advances the rent to the owner, having then to reimburse himself, principle and interest, from the tenants. In these cases the agent becomes practically irresponsible to and irremovable by his employer. It is not only the great absentee proprietors who keep agents. In their case a vice-master is perhaps unavoidable, and may, if he be well selected, be even a benefit. But the resident gentry have equally their agents; gentry residing in other parts of Ireland, and gentry *even residing on their own property*, and perhaps rarely leaving it.

They take no part in the management of their estate, but hand it all over to *the agent*, to manage and deal with as he sees fit, referring every one who may attempt to transact business with them, to him. Neither is it the large estates only that are managed by agency, but all estates almost without exception. There would seem something in the very ownership of land in Ireland that renders a man incapable of managing it, although it sometimes happens that the very man who is agent for other persons has land of his own, which another agent manages for him.

In order to form some idea of the extent to which this system is carried, I may mention the following fact: A clergyman—an old friend of mine, an Englishman, but educated and long settled in Ireland—not long ago had a living given him there, to which there was

attached about 150 acres of glebe. As soon as he came
into possession, he was immediately canvassed for the
agency, the friends of the existing agent hoping that he
would be retained, while one or two other candidates
sent their friends to him to try and get it. 'Agent!'
said my friend, laughing at these applications, 'what do
you suppose I want with an agent? Don't you think I
could myself undertake the management of a few acres
of land, none of which is more than half a mile from
my own door?' On which he was assured that he
would find it quite impossible; that he would find it
much too troublesome; that he would be pestered by
applications from the tenants for this thing and that
thing; that he would never be able to get the rent; and
so on.

Paying very little regard to these assurances, my
friend cheerfully assumed the terrific responsibilities of
the management of this awful extent of property, and
never had the least reason to repent of it. When I
visited him some little time ago, he said that, although
his three tenants were Catholics, and although he had
rather altered their holdings, taking a little more land
into his own occupation, and so on, he had never had
the least difficulty in the matter.

Frank personal intercourse and arrangements between principals, where the human electricity of tone
and manner and look assures each party of the honest

intentions and kind upright feeling of the other, worked in this case the same effect as in others, and made arrangements easy, which, if hampered and muffled and stiffened by intermediate agency, might have led to diplomatic discussions and suspicions, and perhaps to angry dissensions.

I know of other instances. A Catholic gentleman, a friend of mine, has an estate where he never resides nor ever thought of residing, nor do I know that there is a gentleman's residence on it. He employs no agent, but just runs down twice a year to the neighbourhood to receive his rents, and he has often told me that he never had the least difficulty; that he sends word when he is coming, and the tenants come to the hotel and pay him as cheerfully, apparently, as he receives it.

So confident do I feel in the effects of personal intercourse between the owners and tenants of land, that if I were to come into possession of an estate in any part of Ireland—even where Ribbonism were at his worst—I would undertake to go down; and if I escaped for the first week or two, or till I had time to make acquaintance with the people, would promise that in six months not a Ribbonman would dare to put his foot on the property. But I would have no agent—no, not the best and kindest man that ever stepped.

Men with the best intentions will do things, or will not do them, when acting as agent for another, that they would be ashamed to do, or not to do, if acting for themselves. Men, too, will allow things to be done for them by others, or will not actively interfere to prevent their being done, that they never have thought of doing for themselves.

Even the very feeling of duty may be warped under these circumstances, since the agent may think his duty towards his employer constrains him to be over strict on the one hand, and the employer may think it his duty to support his agent, and not to interfere with him in the execution of a difficult task.

Even then, when owner and agent are above the average of humanity, the system of universal agency, where it goes to the extent of divesting the owner of all sense of duty and responsibility towards his tenantry and dependents, is a most pernicious one, calculated only to deteriorate the character and sap the morality of all concerned, and utterly to destroy the ties which should bind the inhabitants of one country by one common feeling of mutual allegiance and fidelity to each other. But we know very well that men above the average of humanity must be rare, that there are many not above, and that there are some—alas! too many—below it. On such men the pernicious influence just now mentioned must act with more and more force in proportion to their want of high feeling, and render almost certain to occur the very acts which one hears of so frequently.

Acts of rascality and fraud towards employers on the
one hand, and of oppression, partiality, injustice, and
cruelty towards the tenantry on the other, are more
often talked about than mentioned in the papers, and
are doubtless more often practised than they are talked
about. These glaring and obvious evils, however, of
the ' agent' system are by no means the limit and extent of the injury done by it. The great injury to society is the very existence of the system itself—the
existence of a great barrier wall separating class from
class, rendering all sense of common feeling between
them impossible, depriving the one class of all chance
of performing those ordinary duties of protection, advice, encouragement, and guidance which the superior
in wealth, intelligence, and power owes to the inferior
class, and repressing or poisoning at their source all that
flow of gratitude, of reverence, of trust and confidence,
of attachment strengthened by early association and old
recollection, of which Nature has implanted in no race
a more abundant spring than in the breast of an Irishman. No one has less faith than I have in a 'nostrum.'
Long-established evils must necessarily have had many
causes and require many remedies and long treatment ;
but I believe that no one thing would be more effica-
cious in the regeneration of Ireland than the abolition
of the agency system. If my voice could reach the
owners of land in Ireland, I would say to them all,

'Dismiss your agents; act for yourselves. If your property be so large and your residence so distant as to make an agent necessary, let him be an agent only; but never suffer any men to take your place, or to do your duty as master. Come and see and judge and act for yourselves, if it be but once in five years, or even in ten. Let your tenants see you, know you, hear your voice, and feel your hand; let them have access to you freely, either in person or by letter. Act thus, and you need fear nothing from them, either in person or in purse.'

I have thus attempted, sir, in these letters to give you in as brief a form as possible the result of my own observations during the last eight years in Ireland, and of my reflections upon them. For the present I will not farther intrude upon you, but remain your obliged and obedient servant,

COSMOPOLITE.

Dec. 31, 1858.

GEOLOGICAL CONTROVERSIES, AND CONCLUSION.

1859–1869.

GEOLOGICAL CONTROVERSIES, AND CONCLUSION. 1859–1869.

Controversy on changes of climate—Address to British Association
at Cambridge—Discussion on mode of formation of mountain
chains and valleys—Lecture on coal-measures at Birmingham
—Royal commission on coal—Address with account of geological survey—Letters to relatives—Original views on Devonian
rocks, and letters to Professor Ramsay on the subject—Accumulation and pressure of work—Its inevitable result accel-
erated by a fall—Gradual failure of health—The shadow of
the end—Journey to Cauterets—Attempt to resume work in
the north of Ireland—Release by death.

IN 1860 a friendly controversy was carried on, in the
pages of the *Athenæum*, on the question of the 'changes
of climate in the different regions of the earth.' It was
commenced by Sir Henry James, who considered that
variations in climate were due to 'a constant change in
the position of the axis of the earth's rotation.' His
views were opposed by Professor Airy on astronomical,
and by Mr. Jukes on geological, grounds.

In 1862 the British Association met in Cambridge,
where Mr. Jukes presided over the geological section,
and delivered an address which has been already alluded
to. The subject of it was 'The external Features of

the Earth's Surface, and the method of formation of
those variations which we call mountains, hills, table-
lands, cliffs, precipices, ravines, glens, valleys, and
plains.' After showing that these features are mainly
due to the erosive action of the ocean, the rain, and the
atmosphere, and fully adopting Professor Ramsay's then
recent explanation of the formation of mountain lakes,
he thus concludes: 'Allow me to remark how curiously
the threefold physical agencies that are in simultaneous
operation on the crust of the globe, were typified in the
old heathen mythology. The atmosphere which envelopes the land and rests upon the sea, the ocean which
fills up the deeper hollows of the earth's surface, and
the nether-seated source of heat and force that lies beneath the crust of the earth, are each personified in it
as a great divinity. If one of the old Greek poets were
to revisit and clothe these ideas in his own imagery, he
would tell us in sonorous verse of Zeus (or Jupiter), the
god of the air, ruling all things upon the land with his
own absolute and pre-eminent power; of Poseidon (or
Neptune) governing the depths of the ocean, but shaking
the shores which encircle it; and of Hades (or Pluto)
confined to his own dark regions below, tyrannising
with all the sternness of a force irresistible by anything
which can there oppose it, but rarely manifesting itself
by any open action within the realms of the other di-
vinities.'

Similar views were enunciated by Mr. Jukes in a paper on ' River Valleys in the South of Ireland,' published in the *Quarterly Journal of the Geological Society* of the year 1862.

In 1864 a discussion arose between Dr. Hugh Falconer and Mr. Jukes on the ' Mode of Formation of Mountain Chains and Valleys,' and several letters passed between them in the *Reader*—Dr. Falconer taking up the old hypothesis of ' great upheavements of the ground,' and ' vast perpendicular fissures' caused thereby, Mr. Jukes arguing for their formation by the gradual rising of large tracts of land and subsequent erosion by external action. On his opponent's bringing many well-known geological names to bear against him, Mr. Jukes replies : ' I am quite conscious of the weight of authority against the views which I advocate, and more especially that I am in this matter, and almost in this matter only, now compelled to depart somewhat from the Cambridge teaching of my dear old master, Professor Sedgwick. If I can venture to dissent from him, my audacity is not likely to meet with any greater trial. But I am also aware that my present views have never been without good authority on their side, and the number and weight of those who, like myself, are now adopting them seem to be greatly on the increase.'

Such discussions as these were always carried on by Mr. Jukes in the most fair and friendly spirit, with the sole purpose of arriving at the right solution of any scientific problem; and he took great pleasure in the thought that 'the contentions and discussions of geologists have almost ever been kept down to mere means of eliciting the sparks of truth, by the collision of various intellects.'

At the meeting of the British Association in Birmingham in 1865, Mr. Jukes was requested to give the general lecture on a subject most appropriate to the time and place, viz. the coal-measures of South Staffordshire. Great interest was then excited as to the exhaustion of that coal-field, and he urged at the conclusion of his address ' the great importance of an accurate knowledge of the extent and position of the coal-fields which we know lie somewhere beneath the red rocks of the Midland counties, and that such an exploration ought to be undertaken at the national expense,' he having previously shown that the coal must certainly lie at an immense depth, and that it was only by a combination of large proprietors that such a work could otherwise be undertaken.

Soon afterwards a royal commission was appointed to inquire into the probable duration of our coal-fields. Mr. Jukes was a member of this commission, and attended several meetings; but the results of the inquiry had not been reported when his health gave way.

In 1866 Mr. Jukes delivered an address on the

occasion of the distribution of prizes by the Marquis of
Abercorn to the successful students at the Museum of
Irish Industry. In this address an account is given of
the Geological Survey, and its connection with the
Museum. It was afterwards published, and it is to
the notes then added that it is here desired to direct
special attention, since they relate to matters in themselves important, and afford an excellent specimen of
the fearless honesty of speech which distinguished their
writer.

The address was thus commenced :

' I have been requested by Sir R. Kane, the director
of this institution, to lay before you some account of
the Geological Survey: its history, its mode of operation, and its connection with the Museum of Irish
Industry. I gladly avail myself of this opportunity to
bring under your notice a great national work, which
has now been in operation for more than twenty years;
labouring under many difficulties and much discour-
agement, but still continuing to labour on, in the hope
that its utility would ultimately be recognised and its
value proved by the best of all tests—that of results
obtained.

' Let me at once guard myself by saying that I do
not mean solely or chiefly money value, or what is
called practical utility. The cost of the survey will
ultimately be repaid to the nation either in actual gain,

N N

or in the saving of fruitless expenditure, and perhaps repaid ten times over. But I should be false to my own convictions if I did not rate intellectual gain at a far higher value than any material profit, and did not put the scientific results of the survey of the heavens and the earth, carried on by the Royal Observatories and her Majesty's Geological Survey, far before their practical application to navigation and mining.'

The following is the note to these observations :

' Men of science have of late years pandered too much to the utilitarian quackery of the age, and it is time that some one should stand up to protest against it. Government and the House of Commons should be told that Science must be supported and encouraged for her own purely abstract purposes, independently of all utilitarian applications. The necessary preliminary, indeed, to these utilitarian applications is the discovery and establishment of abstract scientific truth, by men who look to that alone, and whose whole faculties and lives are devoted to it. The men who afterwards make the practical applications of it often attain, indeed, far wider reputations than the real men of science, and become to the popular gaze the representatives of science itself. The higher class are rarely much known to the public during their lives, and are not usually men who would experience any satisfaction if they were named knights, or labelled with C.B., or would feel inclined to accept any other crumbs that might fall from the table of the politically great and powerful. Nor would they commonly care much for pecuniary rewards, unless as a means for enabling them to do their work without drudging for the support of themselves or their families. They are the men, however, who in the end rule the world; and doubtless they are often sustained in their labours by a consciousness of this fact. It would manifestly conduce to the public good and the national honour if such men, when they do arise among us, should be sought out, recognised as public benefactors, and allowed means to do that work which their faculties, and theirs only, enable them to perform.

'In the mean while the public may insist that all latitude and facilities be given, in the public scientific institutions, for the carrying on of scientific research, however abstract and remote from practice it may appear to the unscientific intelligence. At present the various authorities under whom scientific public servants have to act too often take as their model the cautious and conscientious parent who required his boy to know how to swim before he let him go into the water. It is want of knowledge of the real nature and object of scientific museums that causes the House of Commons and the Government to test their value by the number of people who visit them, and the worth of scientific lectures by the number of students who attend

them. In each case numbers may be a hindrance to
their proper use, and actually tend to prevent even that
utilitarian application of them which is so much talked
about and so little understood. The test of the number
of visitors must indeed be applied to such a museum as
that of South Kensington, where the *ad captandum*
principle is avowed, since it shows whether or no the
avowed principle has been successful. But from a scientific museum all appeals to the mere taste or the
sentiments should be carefully excluded, and the abstract
intelligence alone addressed. The one museum admits
an object because it is pretty, or at all events because
it is curious ; all mere prettiness and all mere curiosi-
ties should be as far as possible kept out of the other.
The South Kensington Museum exhibits artificial products, 'the work of men's hands,' the most attractive of
them being objects of mere ornamentation. The scien-
fic collections of the British Museum contain the works
of Nature—a term I use as a reverential periphrasis for
a higher name. When scientifically arranged, they
exhibit to the trained intelligence that other revelation
of which men of science are the ministers and inter-
preters. . . .

   ' As to the lecture system attached to our institutions, large numbers of students were not contemplated ; neither, if they came, is there a demand for
their services in the world. The value of the few,

however, whom we hope to train will certainly be felt
in this and the succeeding generations to have amply
repaid the cost of that part of our institutions, though
it can never appear in a " return," or be put into a " report."

' The sole question for the statesman to ask is, as it
appears to me, " Is the instruction given sound and
real ? Is it adapted to answer a worthy and legitimate
end ?" If the answer be in the affirmative, he may, I
think, safely and wisely leave it to time to attain that
end. I should in such a case appeal to the " wisdom of
our ancestors," and ask whether they were actuated by
the narrow, short-sighted, utilitarian spirit which the
present generation boasts of, when they founded our
public schools, and the colleges in our universities,
and all the grand educational establishments on which
so much of our national greatness is based. In my
own college, St. John's, Cambridge, a sum of more than
6000*l.* is every year shared, from one source or another,
among the *undergraduate* students, in addition to the
revenues of the Fellows and Master of the College. The
other sixteen colleges are similarly provided for, according to their size. Oxford has still larger resources ;
the University of Dublin is richly endowed ; and when
we add Eton, Westminster, Harrow, Rugby, Shrewsbury, and all the other wealthy educational establishments, we shall begin to have some idea of the vast

means given by our forefathers for the advancement
and dissemination of the learning and science of their
day, and the niggardliness of the present generation in
comparison.

'"But," some person may say, "these establishments
were not paid for out of the taxation of the country."
What does that matter? They were established by so
much of the national wealth being set aside for the
purpose. If private persons do not now come forward
with their contributions, and if our sovereigns and
princes have been deprived of the means of doing so,
and their resources absorbed into the public purse, is
it not the duty of the guardians of that purse to take
care that the national greatness suffer no diminution
at their hands? They may jealously guard against all
wasteful expenditure, but they should not imitate the
miserly farmer who impoverishes his land, because he
shrinks from the expense of draining and manuring it.'

In another note Mr. Jukes says:

' So little seems to be known of the origin and
nature of our institutions, even by the authorities under
whom they have subsequently been placed, that in the
appendix to the tenth Report of the Department of
Science and Art (for 1863), in a special report made by
my Lord Granville, Sir C. Trevelyan, and the Right
Hon. R. Lowe, the Geological Survey, the School of

Mines, and the Mining Record Office are said to have been *gradually connected together as one institution.*

'Now, as they all originated simultaneously in one man's mind, and have been developed together as one system, with a tendency perhaps towards separation, but have never yet been separated, the phrase *gradually connected together as one institution* always struck me as a very curious version of the truth. But there are what I think more serious errors in that report than mere careless expressions ; one is the very common mistake, that there is an inferior kind of science good enough for practical purposes ; and the other, that our institutions can be made self-supporting.

'The report speaks of "touching on chemistry, general mechanics, physics, and natural history only so far as is required for mining purposes." Such a half-and-half scheme would be exactly the one adapted for the production of half-educated quacks, with just a sufficient smattering of knowledge to lead themselves and their employers into all sorts of expensive blunders, which it is the very object of our institutions to put an end to.

'Even in chemistry, if I am rightly informed, it is no longer possible to make the distinction between organic and inorganic chemistry, and any one who wishes to know the true nature of minerals will have to make himself master of organic chemistry, and the principles and nomenclature derived from it. It may

be, from the researches of organic chemistry that we
may at length be enabled to understand the phenomena
even of mineral veins.   As to the " self-supporting
dodge" (for I can call it by no more respectful name),
when the Government and Parliament are prepared to
pass an Act providing that no one shall practise as a
mining captain, or engineer, or manager, overlooker,
overseer, or ground bailiff of any mine of any description, without first passing through a recognised School
of Mines, and getting its certificate of competency, they
may begin to think of making such schools self-supporting.

  'The School of Mines in Jermyn-street, and the
School of Science applied to Mining and the Arts in
Dublin, are really only missionary efforts to try and
induce our miners to gain a little more scientific knowledge, and thus enable them, among other things, to
impart to us geologists a little more of the knowledge
which is only to be gained in mining, in terms which
shall be mutually intelligible.   When this interchange
of knowledge and experience really takes place, we shall
begin to create some of that farther knowledge also, of
which we are both at present destitute ; and the only
way to acquire that is, either to turn some geologists
into miners, or induce some miners to become geologists.

  ' Just at present there is this difference between us,

that geologists know and avow their own ignorance, while miners are too often quite unaware of theirs. This was the reason of the failure of the enlightened plans of the late Sir C. Lemon, who offered to provide 10,000*l.* out of his own pocket, if the mining interest of Cornwall and Devon would give an equal sum, for the formation of a mining school there. Had the plan been adopted thirty years ago, the mining districts of the west of England might perhaps have been spared the failure of their resources and the general distress which is said to be now spreading over them.'

The following are extracts from the few and hasty letters written to the editor or her husband during the years 1864-1867:

Dublin, *April* 15, 1864.·

DEAR CARA,—I send you a bundle of *Readers*, containing the beginning of the war and Falconer's letters. My last will appear in the *Reader* to-morrow.

Falconer is a very good fellow; very clever, very genial and kind-hearted, and everybody likes him. We are perfectly good friends all the time, as the enclosed note from him will show, if you can read it. The secret of his pepperiness in public is, that for many years he was the acknowledged head of the scientific world in Calcutta, and accustomed to give law without contradiction.

I knew perfectly well what a fume my first letter
would put him into, and was prepared accordingly. In
his first private letter to me about it, he said I had
' come down on him like a squashed pumpkin,' and as
he really thought (with Murchison, Sedgwick, and most
of the old geologists) that our new ideas were not only
heresies, but mere absurdities, he was indignant accordingly. I know that almost all the older fellows
think us crazy—as they considered my address to the
Geological Section at Cambridge—but in the long run
they will give in. . . .

Dublin, Feb. 4, 1866.

March 7 is the night fixed for reading my paper on
Devon. . . . We are getting still tremendous gales and
storms of rain; but I get a ride almost every day, which
keeps me going in spite of the work I have in hand. . . .
The 'Fenians' seem to be blowing over, though sympathy with them is very widely spread; ninety-nine out
of every hundred of the entire population of Ireland go
wholly with them, so far as detestation of English
government is concerned. All ranks and classes, sects
and parties, Orangemen, Protestants, Catholics, priests,
clergymen of the established Church, fellows of Trinity
College, as well as the peasantry, look on England as

a foreign country; on the English government as a
foreign government, ruling Ireland by no right but that
of the strongest force; and thoroughly convinced that,
if it were possible, Ireland ought to be independent as
a nation, with its own king or president, taking rank
as an independent country like Belgium, for instance.
People may laugh at the notion, but it is in the heart
of every Irishman, high or low. Fenianism is merely
an exaggerated expression of this universal feeling. . . .
The reason why more men of that stamp were not mixed
up with it was, that they saw the absurdity of it and its
utter helplessness—not that they disapproved of it if
there were any chance of its being carried out.

Every one would have respected Irishmen more if
the rebellion of '98 had been successful. It will take
generations, and a total alteration of the mode of government, to make the Irish nation feel one with the
English nation. . . .

Dublin, Dec. 6, 1866.

. . . You will have seen the talk of the Fenian row
expected. The authorities are really alarmed, and the
military fully expect disturbance; but it is impossible
to let the loyal subjects come forward and arm themselves on account of the Orangemen. Their ferocious
hatred of anything called Catholic is such, that, if once

the loyal subjects were allowed to make a demonstration, they who consider themselves the only loyal subjects would come forward and prove it by shooting all the ' Romans' they could catch sight of. —— was telling a story the other day of an Orangeman on his death-bed, who was visited by a friend anxious about his soul, who begged him, if he still preserved the right faith, to make some sign or utter some words to assure him of it; on which the dying man faintly whispered, ' D-d-damn the Pope !' to the great satisfaction of his friend.

March 14, 1867.

I have been thinking of writing to you often lately, but there has been such a succession of botherations pressing on me, I have hardly known which way to turn. The Department of Science and Art is a great deal worse than any Fenian insurrection. They coolly suppress the office of director of our museum, dismiss Sir R. Kane, and propose to govern us through a secretary, one of their science inspectors. . . . As to what is to take place next Sunday, no one knows; but the military authorities firmly believe there is much work in store for them. There is a strong rumour that we are all to get up murdered next Monday morning, nevertheless Augusta, C., and Miss B. are making preparations to dance at St. Patrick's ball on Monday night. . . .

The 'Battle of Tallaght' took place within four miles of
us, at which fourteen policemen routed several hundred
Fenian shop-boys, who left the country strewed with
arms and ammunition. I rode out the next evening over
the ground and up on to the hills, but could not see a
Fenian ghost. An old farmer spoke of them as ' Fools,
fools, fools!' About ten o'clock that night, Miss B.
not having been on an outside car, I took her and C.
on one all through Dublin to see the fun ; but we could
not find any, and Miss B. declared Dublin to be dis-
gustingly tranquil. . . . I give my last lecture to-day,
and contemplate a dash down a coal-pit in Staffordshire
next week.

Castlebar, Co. Mayo, April 29, 1867.

Your letter followed me here, where I am inspect-
ing work in a more dreary miserable country than I
think you could form any idea of without seeing it—
miles of brown bog with low gravel mounds at intervals, occupied by wretched fields, separated by low stone
walls, with large pools that are only approachable at
intervals for boggy marshes, but are called lakes. This
morning there is a howling wind and pelting rain, and
I have to do thirty miles of that country on an outside
car, with only a dirty ill-kept public-house, that calls
itself a hotel, at the end of it. Such is my country life
in Ireland, and I have now been a month at that work.

I shall be in Staffordshire some time in May, but everything is at sixes and sevens on the survey, so that I can't tell. . . . I shall be heartily glad when I can escape with a pension. . . .

---

72 *Upper Leeson-street, Dublin,*
July 14, 1867.

We were greatly shocked by the death of Mr. Scholefield. I thought him looking rather better than he had been when I was in town last. Our plans still depend for their execution on South Kensington, but some time in August I hope to get away into Devonshire with O'Kelly, to go on with my Devonian investigations. . . . I see no chance of any real holiday for the next four years, except that my field-work is a change from that indoors. I am at work now every day from six A.M. to eleven at the third edition of my *Student's Manual*, and then till five o'clock P.M. at the office on maps, &c. . . .

Dartmouth, Aug. 28, 1867.

I have been for a fortnight in South Devon with Mr. O'Kelly, one of our Irish staff, hard at work on the Devonian rocks, about which I am at war with all the geological world, and shall have to astonish them still more before I have done. We are bothered with regattas, which interfere with the hotels and upset our arrangements. We go to-morrow to Torcross and Start Point, and expect to be in Plymouth on Sunday; then on to East Looe in Cornwall, and so back to Dublin about the middle of September. We have here rocky cliffs and crags three times as high as yours, and grand woods down to the water's edge, and the whole country a mass of hills and wooded dells, with trout-streams running into tidal rivers, and the sea clear enough to see the rocks under a fathom of water. However, I don't want to make you envious.

Dublin, Sept. 8, 1867.

I had to get home in a great hurry from Falmouth, and got a great tumblification round the Land's End. Our lords and masters at South Kensington have cap-sized everything, and introduced so many alterations and botherations, that I feel like a cork under a cata-ract. I landed at six this morning, and find letters from Augusta that say she is well and enjoying her-self with E. Y. in a château in Brittany, where, how-ever, they see no strangers, except a troop of poor people, sick of fever and ague, who come to be doctored with quinine by Alice Forbes. I fancy that is some-thing new in French country life. Dublin looks dread-fully dull; but I have work in the field, and shall only

be here a day or two. . . . W—— is not at all one of
those whom I spoke of as not being gentlemen. He is
one of Nature's gentlemen, and a most trustworthy man
in every respect. _____

The few remaining letters to Professor Ramsay
dwell chiefly upon the topic to which Mr. Jukes devoted much thought during the last few years of his
life, viz. the classification or grouping of the rocks in
North Devon and West Somerset. As he was firmly
convinced that he had found in the south-west of Ireland the key to the difficulties of that problem, he endeavoured to bring his views before the scientific world
by papers read to the geological societies of Dublin and
London, and to secure the attention of his scientific
brethren by having these papers printed separately
in extenso, and transmitted to every member of the
two societies. With the untiring pursuit of truth by
which he was always characterised, he had worked at
this question in every interval he could snatch from
his duties in Ireland ; and there is little doubt that his
anxiety and labour to solve it, and to place his views
on record, greatly contributed to overtask those powers,
which had been already too severely tried. Not that
the papers in which he discusses the subject evince any
failure of mental energy — on the contrary, they are
among the clearest and most powerful of his writings—

but each strain against the tide of dissent, together with the pressure of work, caused a greater reaction and subsequent exhaustion.

He left his solution of the Devonian question as a legacy to the geological world, in full confidence that time and farther research would prove it correct; a confidence shared by those who, though perhaps ignorant of its exact bearing, knew well with what industry and foresight he was wont to test any new geological theory before giving it to the world, how thoroughly and exhaustively he sifted the previously-known facts, and for what a length of time he had prosecuted his researches into this dim page of the earth's history.

In the introductory statement to his *Additional Notes* in 1867, he says :

'I have gone into the foregoing personal detail in order to show that my present attempt at putting straight the Devonian puzzle is not a hasty or ill-considered one, but is the fulfilment of a long-cherished design, dating from the year 1852 at the latest, and that, in spite of the many necessary distractions imposed by the conduct of the survey and the labours of my lectureship in Ireland, I have kept this purpose steadily in view ever since, and left unused no means of acquiring information respecting it that happened to present themselves. I expect to be able to show that the true Devonian rocks are everywhere synchronous with the Carboniferous Limestone, their palæontological differences depending upon *habitat* and *province,* not upon *time.*

'Doubtless it may appear to some of very little consequence to the science of geology what rocks we include under this designation of "Devonian." The whole of our present nomenclature is so obviously provisional and temporary, that it may hardly seem worth the trouble to give it an exact signification. The terms Cambrian, Silurian, Devonian, Carboniferous, &c. are merely part of the scaffolding, useful in building the science of geology, but no part of the structure itself, and will doubtless be thrown aside by our successors as old lumber, just like the terms Greywacke, transition, &c. have been by ourselves.

'It is, however, not the mere nomenclature I wish to correct, but to show that two distinct things have been confounded under one name, and that, by whatever name it be called, the Old Red Sandstone is a group of rocks distinct from those which contain the marine Devonian fossils, and altogether below them.

'I would also wish to urge upon the attention of geologists, that there may have been two distinct contemporaneous assemblages of animals in very close neighbourhood to each other, the one either sending "colonies," as Barrande calls them, that is, temporary occupants, into the province inhabited by the other, or

else one set finally spreading over the other, and thus
producing the appearance of absolute *succession*, and
obscuring the fact of their long *contemporaneity.*'

Mr. Jukes' arguments gained the attention, if they
did not bring conviction to the minds, of many eminent
geologists, as was shown at the meeting of the British
Association at Exeter, in 1869, when Professor Hark-
ness presided over the Geological Section, and brought
the subject prominently forward, taking occasion also
to pay a graceful tribute to Mr. Jukes' memory, in
speaking of 'the earnest love of science and kindness
of heart which so distinguished him and caused him
to be beloved by all who had the pleasure of knowing
him.'

Taunton, Aug. 1, 1866.

MY DEAR RAMSAY,—From a letter received from Sir
R. here, I find that they have not yet sanctioned the
employment of the survey on the coal commission.

It upsets my plans, because, if I am to do South
Staffordshire, the end of this month and the rest of
the autumn would be by far the most convenient time
for me to do it in. One day more over the south end
of the Quantocks will finish my work here, and I hope
to reach Dublin on Friday morning. . . . There is one
consideration I would put before you, which will perhaps make you look a little more cordially on my opera-

tions here than you have hitherto appeared to do, somewhat to my disappointment. The consideration is this : that it shows the necessity for the survey of the United Kingdom being kept together as one body of men. No one could have made out the Irish and Scotch Silurians till Wales had been done. In like manner no one could make out Devon and Cornwall except by the aid of the south-west of Ireland.

My work here is of course most sketchy. It would take a man three years to insert all the dips, determine which was the real boundary-line between the marine fossiliferous beds above, and the genuine Old Red below, and trace it through all its convolutions. I have, however, by the light of my Irish experience—and trusting to that alone, in spite of all changes of lithological character and other difficulties—made out three great areas of genuine Old Red Sandstone, covered in each case by corresponding masses of beds (chiefly grey slates) containing marine fossils. These slates (which are perhaps 5000 feet thick) you may call Devonian if you like, or carboniferous slate, or whatever you please ; but they are clearly above and wholly above the Old Red Sandstone, of which 3000 feet are shown at least, and Heaven knows how much more, for it undulates over considerable areas. Now, whoever re-surveys the west of England, will have to take my present work as the key to the puzzle and the basis of his operations, or else he will have to spend a year or two in the south-west of Ireland, to learn his lesson where the rocks are clearly shown, and where they are free from the bedevilment of igneous rocks and mineral veins, which obscure them and alter them.

The whole operation, then, ought to be under the general direction of one head, who could sanction this transference of a man from England to Ireland, and *vice versâ.*

Ireland, in fact, shows the old red plants better than either England or Scotland, though Scotland has some of these mingled with the fish remains (Coccosteus, &c.). Ireland has a few of the fish remains mingled with the plants, and also has in one connected area, the change of type from the conglomeritic and soft Old Red Sandstone of South Wales to the slaty Old Red of Devon.

So that the whole story will only come out plain by the union of knowledge gained in the three kingdoms, and therefore the whole ought to be *kept together under one director-general.*—Yours, in haste,

J. BEETE JUKES.

Dublin, Dec. 8, 1866.

My DEAR RAMSAY,—As I read their lordships' letter, I am immediately to get about fourteen new men. They

say each district surveyor and senior geologist is to
have three assistants under him. Sir Roderick has
already named Dunoyer a district surveyor. We have
only two assistant geologists now, and therefore, as
their number is expressly unlimited—but each senior
must have three at least, both in Scotland and Ireland,
under him—it seems to be an indefinite extension of
the survey. I rejoice at it, and shall expressly thank
them for it; but I shall also point out that the ' unsatisfactory progress' they speak of is not our fault, but
theirs. When we remonstrated ten or twelve years
ago against all our best men going to Australia, India,
&c., were we not told that it was good for the colonies,
and that the mother country could afford to wait ?

I have just heard that the council of the Geological
Society decline to publish my last paper except in abstract without illustrations.* . . . .

Globe Hotel, Newton Abbott, Devon,
Aug. 18, 1867.

O'Kelly and I have had two days round here now,
taking it easy, and have been kept in a mixture of
amusement and bewilderment. . . . . The great limestone band near Chudleigh is full of Stringocephalus. I

* The paper here referred to was subsequently printed in full
by Mr. Jukes under the title of ' Additional Notes on the Grouping of the Rocks of North Devon and West Somerset.'

did not exactly understand what Sedgwick meant when I stopped with him there in 1841, and he went into the quarry, and said, ' Ah, I thought De la Beche had made a mistake ;' but I recollect his knocking out a shell, which, I have no doubt, was a stringocephalous. I believe there's a big north-west and south-east fault—downthrow to north-west—cuts off granite, and passes through Chudleigh, cutting the limestone; but this is a mere guess, and would take weeks of work to prove. If you have a man you could spare to do this district over again, and put it a little straighter, I think it would be wise. All this is totally independent of my views of classification.

We have got Bronteus flabellifer, Stringocephalus, &c. in the upper limestone ; I believe, if anything, it is rather above the ordinary Carboniferous Limestone than below. Barrande, I see, says that the place of the siphon in Cephalapods is not worth a halfpenny, and that Clymenia is merely a Goniatite whose siphon has happened to get to his belly, instead of his back. The ' unconformability of Coal-measures on these limestones' is merely one hypothesis out of many which may be used to explain queer dodges in the beds. The whole country is full of queer wrenches—capital grounds for a good youngster, if you have one to spare. I want one or two more days at it, then I go to Dartmouth, and work on towards Start Point to look for the real Old Red.

After making a tolerable examination of the neighbourhood of Dartmouth, we came on here yesterday, and
examined all the shore round the Start Point before
dinner. I expected of course to find mica schist, gneiss,
or something of that sort. Dr. Harvey Holl (who is
not going to Dundee) talked of its being Laurentian.
Why, my dear fellow, it is all blue and black carboniferous slate—just what you saw when with me in Bantry Bay—and ' divil a taste' of mica schist or anything
of the sort. When we got to Hall sands, within a
mile of Start Point, the slates were much crumpled in
the way mica schists are sometimes, and they were full
of small quartz veins in places, like those which occur
in mica schist, and sometimes the surfaces of those
quartz veins were coated with a micaceous glaze, so that
pieces of them, lying about in the fields, might be supposed to have come out of mica schist, and I suppose
that Sedgwick and Murchison took some of these for
mica schist. But the blue-black slate between the
quartz veins is no more micacised than it is among the
limestones at Torquay. As to the general geology,
there is a belt of Glengariff Old Red Sandstone five miles
wide, running east and west from Dartmouth. It dips
south at 60° or so. If it is all regular, it must be
25,000 feet thick. I believe it is an inverted anticlinal.
Then there is a space three miles wide where you see

nothing as you go south ; then there is three miles of carboniferous slate, dipping north at 60°, or *at* the old red, which dips south *at it.* What is the explanation I don't yet know. We start directly for Kingsbridge, where perhaps we shall find out; but that metamorphic area must be kicked over the walls of creation into infinite space.—Yours truly,

J. BEETE JUKES.

Dublin, *Sept.* 13, 1867.

MY DEAR RAMSAY,—I suppose you have by this time escaped from the British Association, and that this will find you recovered from the effects.

I write to ask if you know anything of the reason why the ground round Veryan Bay in Cornwall is considered to be Silurian. I suppose it must be on account of some fossils found there. It was done, I fancy, since Sir Henry's time. As the carboniferous slate runs up to and apparently into that country without any obvious change, either in the nature of the rocks, the aspect of the country, or any other obvious character, I should like to know the evidence for it, and I don't know where to look. I had only the few hours between a morning and afternoon train to drive from St. Austell down to Garran Haven and back, so of course can't speak with authority on the subject, beyond the fact that I could

see no appearance of change, and that the vertical black
slates of Mevogissey and Portmellin are certainly our
Irish carboniferous slates.   I am not at all surprised at
nobody having made out South Devon, and everybody
having gone wrong about it.   There must be hosts of
faults, but besides that, there is contortion and inver-
sion to an unknown amount.   I actually saw this at
Bovisand Bay [section] : a sandstone in grey slate, bent
back on itself, and forming apparently two horizontal
beds, about ten feet apart, lying in grey slate, and running regularly, apparently, as two beds for twenty or
thirty yards in the base of the cliff, and then faulted so
that the top piece was brought down to the lower piece.
To build any conclusion on the apparent stratigraphical
position of the rocks there is hopeless.   They dip
steadily south at 40° to 60° for miles together, both on
the strike and the dip.   If you are to believe the apparent dips, and build a stratigraphical section on them,
they must be 50,000 or 60,000 feet thick altogether.
With the Irish clue, I believe a year or two's hard work
of one or two good men would put it straight; but
without a good extraneous clue to the true stratigraphical position of the rocks, no geologist that ever walked
could do more than get himself into a hopeless muddle
of confusion.   I am now in a position to prepare the
rough notes for a good critical paper on all the old
work I can lay my hands on, and as I believe my Irish

work has given me the true clue, I shall think it my duty to the Science, and to this branch of the survey, to put that view forward. I will take care not to do so offensively, but certainly I cannot consent to suppress truth out of consideration for the feelings of any one whatever.—Yours very truly,

J. BEETE JUKES.

Carlow, Sept. 19, 1867.

MY DEAR RAMSAY,—Yours of the 17th catches me here at work for the coal commission. I found out all about that Silurian. It is based on a paper by Sedgwick and M'Coy in the eighth volume of the *Q. J. G. S.*, which said paper, as well as M'Coy's big book on Paleozoic fossils, is, I find, full of matter, not only confirming my views on Devonian, but anticipatory of them. This explains what I heard from Selwyn, that M'Coy was very sore at my not having alluded to him. In fact, I find passages in De la Beche, Sedgwick and Murchison, Sedgwick and M'Coy, Phillips, and others, together with Barrande's recent work, all explainable and made coherent by our Irish work, which gives in reality the clue to the mystery. Whoever re-surveys Cornwall must come and learn his lesson in Munster. It shows the necessity for a general survey of the United Kingdom by one set of men. . . . I must say I do not fully understand and appreciate men *having feelings* on such points. It is the oddest and most inexplicable thing to me that any one should be annoyed or feel hurt, as if it were a personal injury to himself. Surely any man who is above a spoilt child in such a matter would be more glad that the truth should be arrived at, than sorry that he himself should be shown to have fallen involuntarily into error. I must say that I have an involuntary contempt for any man who could have *feelings* in such a matter.

I shall be back in Dublin in a day or two.—Yours very truly,   J. BEETE JUKES.

A summary of Mr. Jukes' views on the Devonian formation, extracted from his notes on the subject, printed in 1868, will appropriately conclude this series of letters.

'I now beg leave—not in any spirit of dictation to my brethren of the hammer, but simply as a statement of the conclusion to which some years of thought and work have gradually conducted me—deliberately to affirm that the old red sandstone cannot be classed with the Devonian slates and limestones without a violation of all the rules and logic which govern our geological nomenclature. They are two formations, as distinct from each other as are the old red sandstone and the

carboniferous limestone. I believe that the old red
sandstone lies wholly below the Devonian slates and
limestones, which contain marine fossils both in Munster and Devon. As to the exact relations between the
carboniferous limestone and those Devonian slates and
limestones, I do not wish to dogmatise upon a question which I believe to be open to farther investigation. Where they both come together in the county
of Cork, the carboniferous limestone is uppermost; but
neither of them have any such thick development there
as in neighbouring districts, where it occurs alone.

'Both groups of rock have coal-measures resting
upon them, and both rest upon the old red sandstone.
These are stratigraphical facts which admit of clear
ocular demonstration even in Devonshire, and are obvious in Munster. The exact interpretation to be put on
the palæontological facts, when those are also established beyond all reasonable doubt, is, I think, a question for the future.

'In the mean time I must hold to the opinion that,
after the period during which the old red sandstone of
the British Islands was formed, a sea spread over these
districts, in parts of which mechanically-formed rocks
—mud and sand—were accumulated, so as to form a
thickness of several thousand feet; and that during
either a part, or the whole, of the time, other parts of
that sea to which the mud and sands did not extend

became the receptacle of the carboniferous limestone, formed chiefly from the growth of marine animals; and that in some places these accumulations produced a mass 2000 or 3000 feet in thickness.

' Subsequently to the formation of these groups of rock came the period in which the coal-measures were spread over them both, having a very variable structure in different places, but chiefly consisting of mechanically-formed muds and sands.

' I can only briefly allude to the various igneous rocks which accompanied the formation of these aqueous rocks in Devon, or to those that have been subsequently intruded among them. The granites of Devon and Cornwall are obviously included in the latter category, as well as the Elvans and other granitic, and some trappean, varieties of rock. Numerous and large fissures and dislocations have subsequently traversed both igneous and aqueous rocks. Some of these have more recently become the receptacles of spars and ores, and are now mineral veins. It is difficult to say how often these actions have been repeated; neither is it easy even to guess at the greater or less amount of disturbance that accompanied the intrusion of the granite; or, if they accompanied that intrusion, they may have been greater in the districts intermediate between the present surface bosses of granite than they were in their immediate neighbourhood.

'These and many similar problems—some, perhaps, yet even undreamed of—lie awaiting the future labours of geologists in the west of England, and if the results of the labours of my colleagues and myself—in addition to those of the other Irish geologists, among whom Sir R. J. Griffith takes the first place—in working out the geological structure of Munster, contribute to clear the ground for the commencement of future researches in the west of England, and in making more perfect the classification of these rocks in the rest of the world, it will be felt as the highest reward to which we could aspire.'

In addition to the Directorship of the Irish survey (of itself a weight quite sufficient for one man to bear), Mr. Jukes had, during the time he held that office, written many papers and published several works on geology. The mental labour of these years was immense, and the rest he took wholly inadequate for the restoration of his energies. He was an habitual early riser, usually in his study before six o'clock, thus securing some uninterrupted time, but making the working hours of the day far too long. And though the indoor employment was occasionally relieved by field work, this relief was more apparent than real, the physical toil being frequently too severe to serve as a relaxation, and a considerable amount of thought being involved in the duties of a director, while overlooking the work of the field geologists.

From certain passages in his later letters it would appear that he had some consciousness of the 'shadow of the dark night coming down,' though to his imagination it seems to have taken the form of an undefined incapacity for his work, rather than the more gracious but more solemn shape of death. His connection with the survey lasted for nearly twenty-three years, during which he had with his colleagues been engaged in unceasing labour for the public benefit—labour of a nature little appreciated during the lifetime of the men on whom it devolved. But in future years the foundation they have laid for the advancement of knowledge and for the prosperity of their country will doubtless be recognised; and those who build on this foundation will not fail to admire the patient toil of their predecessors, the geological surveyors of Great Britian and Ireland. Nor must it be supposed that Mr. Jukes ever regretted his devotion to Science, though fully aware of the sacrifices she demands of her faithful votaries. Respecting the work of the survey he thus speaks in 1867 :

'No amount of ability, either mental or bodily, will fit a man for our work, unless he have that natural instinct for it which shall induce him, like Professor Ramsay and myself, and many other officers of the

survey, knowingly to abandon all hope of wealth, all
love of ease, and all ambition of advancement in life,
in order to devote his whole energies, thoughts, and
aspirations to the pursuit of geology. . . . It is, in
fact, what occurs in all the natural sciences. Had the
Astronomers Royal, or had Professors Owen, Huxley,
and Tyndal, and the many other distinguished professors of science, gone to the bar, or the church, or the
practice of physic, would not the same ability, labour,
and perseverance, which have raised them to a barren
eminence in their several pursuits, have elevated them
to honour and affluence long before ? The *over-master-
ing bent of their inclinations* alone determined them to
science as the pursuit of their lives. Nor do I suppose
that they ever regretted it; but without that over-
mastering bent of inclination, they would never have
submitted to the toils and privations they have had to
encounter.'

Among the publications above referred to were the
following: in 1853 *Popular Physical Geology,* published
in Lovell Reeve's series, and the completion of the
Memoir of the South Staffordshire Coal-field. In 1856
the article on geology in the *Encyclopædia Britannica.*
In 1857 the first edition of the *Student's Manual of
Geology;* and in 1858 a new and greatly enlarged edition of the *Memoir of the South Staffordshire Coal-field.*
But the work which tried him most was doubtless the

second edition of his *Student's Manual*, published in
1863. The comprehensive nature of this work, his
great desire to bring it up to the requirements of the
time, and to add to it an index which should be partly
a glossary, involved an amount of labour which only
those who have been engaged in similar undertakings
can appreciate. The pecuniary benefit from the work
was very insignificant, owing to the expense of the
illustrations and the compliance of the publishers with
the author's desire that the price should be as moderate
as possible. The value to students of a manual, which
combined an immense amount of information with the
most lucid arrangement, was speedily felt both at home
and on the Continent, a third edition being soon called
for, and a proposal made for its translation into French.
In 1868 he made another effort to render a third edition
more perfect than the preceding; but found himself
quite unequal to the task, and was glad to resign the
completion of it to his colleague Mr. Geikie, director of
the Scotch survey, under whose able editorship it is now
(1871) passing through the press. It was, moreover,
during his directorship that (as Professor Huxley states
in his obituary notice at the Geological Society) Mr.
Jukes edited and largely contributed to no fewer than
forty-two memoirs explanatory of the geological maps
of the southern, eastern, and western parts of Ireland,
executed by the survey.

It will have been seen from some of even the earlier
letters to Professor Ramsay, written in Ireland, that
mental anxiety and bodily toil had begun to undermine
Mr. Jukes' health. These symptoms might perhaps
have been counteracted, had they led him to take the only
remedy of any use in such cases—absolute rest for
months or even years; but in the autumn of 1864 they
were greatly aggravated by an accident which occurred
while he was overlooking survey work in the south-west
of Ireland. The only available quarters in these wild
districts were the small inns, with their dingy apart-
ments and rude accommodation, and it was whilst staying at one of these that he missed his footing near the
bottom of an awkward staircase, and fell on the limestone flags of the hall, striking the back of his head
with such force as completely stunned him for a time,
and caused much alarm to the people of the house.
When he recovered, however, he thought so little of
the accident, that he jumped into the car waiting at
the door to convey him to his work of inspection, which
he continued for some days, being anxious to finish it
before commencing his autumn holiday. There is little
doubt that this fall had a serious effect on his health;
for he found an incapacity for continuous thought gradually creeping over him, and soon became so unwell
that he gave up the attempt, and hastened to join Mrs.
Jukes at Coblentz. Still, he did not connect his in-

disposition with the fall, and did not even mention it
to her. During his stay at Coblentz he became seriously ill, and the physician who was called in, questioned
him closely as to his having undergone some shock or
fall, and then the accident was recollected. From this
time his health gradually failed ; and the different excursions and periods of rest, recommended so earnestly
by his friends and medical advisers, though refreshing
him for a time, did not arrest the disease, which made
slow but fatal progress. In 1868 a journey to Cauterets
was advised, and, as far as practicable, a sea voyage in
preference to railway travelling. He and Mrs. Jukes
therefore left Dublin by steamer for Portsmouth. The
following letters, the last of any length received from
him, relate to this excursion and to their tedious and
dangerous voyage on returning from Bordeaux during
the great storm of August in that year.

Haslar Hospital, Gosport, July 19, 1868.

DEAR CARA,—I am not here as a patient, but as the
guest of my old messmate Shadwell, who is captain
superintendent. Our steamer put into Portsmouth
yesterday, after a delightful voyage of three days from
Dublin, in which we visited the environs of Falmouth,
Plymouth, and Southampton before we came here, and
as we found there was no steamer to Havre till Monday, and Shadwell wished to see us, here we are. The

houses of the officers are behind the hospital, with very pleasant gardens and a kind of park round all. It is the quietest sort of retreat, half surrounded by the sea, and the rest by batteries and dockyard arrangements, with the Isle of Wight appearing in the openings among the trees at the back, in beautiful vistas. We go on board the Havre steamer to-morrow night (where I am afraid we shall not get a cabin to ourselves, as we did in the Dublin one), and mean to get to the top of the Pyrenees before we come back.

Hôtel de Paris, Cauterets, Hautes Pyrenees,
Aug. 13, 1868.

We had a delightful voyage to Havre, and a very pleasant trip by Rouen (where we spent a night) to Paris, where we spent two or three quiet days (for Paris), and I saw Marçon, Barrande, and some other celebrities. Then we travelled all night to Bordeaux in the most tremendous storm of thunder and lightning I ever witnessed, and stayed there three or four days with our friends. We got very comfortably to Landes at the foot of the Pyrenees, and luckily had secured places in the coupé of the diligence up to Cauterets, but I was so ill with diarrhœa that I really began to think I should never get there. We had not written for a room, and the place was crowded, so we were lucky in getting a bedroom in a *succursale*, or house used by the hotel, at which the diligence stopped, and here we have been ever since.

Cauterets is in a crevice of the Pyrenees, itself at about the height of Snowdon, but with steep pine-covered crevassed hills rising up on each side of the crevice to a height of five or six thousand feet above the town; so that there are patches of snow visible on their green tops from our bedroom window, notwithstanding the intense heat of the weather. The baths are about a mile from us (at least the one we were directed to use) and about 400 feet above the town. There is an omnibus every ten minutes, which, however, we did not use after the first two days. I am certainly better, and sleep much more like a top than I did, and they say that the diarrhœa is inevitable; for every spring of cold water—and they are numerous—contains some abominable mineral solution or other, and they are so cold and pleasant it is difficult to avoid drinking when the sun is broiling you above, with a granite cliff for a heater on each side. We went up one day to Lac Gaube, which is about 5700 feet above the sea, surrounded by peaks, one of which is the third loftiest in the Pyrenees. A little dirty wretch of a glacier was melting away on the side of it, the summit being 10,800 feet high. To-day the mountains are covered with clouds, and it feels quite chilly. We propose to leave the day after to-morrow and go into the next valley, which is said to be the most beautiful in the Pyrenees, and then in a day or two back to Bordeaux. I hope the clouds will have the goodness to clear away again, and let us have one peep at the valley of Luz and St. Sauveur. Hitherto it has been too clear and too blazing; and the Pyrenees are not like the Alps, with parallel longitudinal ranges and intermediate valleys, into which the rivers descend from each side, till the rivers at length cut through one or other; but merely one ridge cut by a multitude of ravines down each slope, these ravines leading into one another as they descend the slopes. . . .

Gudig, Pembray, South Wales,
Aug. 25, 1868.

You will be surprised at the above address; but the fact is, we are glad to have any address at all other than the sea bottom. The steamer by which we left Bordeaux was a small screw, and after getting into the Bay of Biscay we found she had very little cargo and very little ballast, and rolled as if she had been built for that purpose. . . . When the fair wind died away she could only make three knots an hour with her screw, and we crawled out of the Bay of Biscay and across the mouth of the English Channel into the mouth of the

Bristol Channel during Wednesday, Thursday, and Friday, till about twelve o'clock on Friday night we were some thirty-five miles to the westward of Lundy Island. The only other female on board was a middle-aged Scotchwoman, who was going home from a family in whose service she had been in the south of France. There was also a gentleman of the name of Fox coming home from Spain, who had been recommended to try that route, to whóm the captain had given up his berth in his cabin.

We learnt, just time enough, that the ship was a teetotaller, and had neither wine, beer, nor spirits on board; so I sent off half a dozen of claret, which the steward put loose in a locker, where they were all broken the second day. Luckily Mr. Fox brought some in his carpet-bag, which he shared with us. The ship was as dirty as it well could be, and they were always cleaning the machinery with naphtha, causing such a smell as alone to make Augusta feel ill. . . The captain was very much of an old foozle; said very long graces, both before breakfast and dinner; and though it was a teetotal vessel, did not hesitate to take his share of Mr. Fox's claret. We were promised to be landed in Dublin about midday on Saturday; but towards three o'clock that morning there came up a furious gale from the west, with as heavy a rolling sea as I ever saw in my life. We were shortly obliged to heave to, and lay all that morning pitching and rolling, but luckily not taking much water on board. This was the more fortunate, as it was then found that the pumping gear was out of order, and would not work properly, so that if the captain put the helm up and attempted to run before the gale towards Bristol, we should probably be pooped by some heavy sea, and gradually swamped from inability to pump it out. Luckily about noon the gale moderated a little, so as to enable us to set the mainsail, and gradually to crawl up to the northward. Mr. Fox and I were constantly at the captain to make for Milford Haven if he could possibly fetch it; and though he would not yield to us, yet in the middle of Saturday night, finding he had only one day's more coal on board, he did put his helm up, and about six A.M. on Sunday morning Augusta called to me, for I had fallen asleep, that we were in Milford Haven. I need not say that I dressed as quickly as possible, and that as soon as we came to an anchor, and found a boat alongside, Mr. Fox and ourselves tumbled into it, went to the railway station, and came here by the first train. I shall cross to Waterford to-morrow in the mail-boat, as I have work to do in Galway in September. We can see from the window of this room between the sand-hills the bow of a brig high and dry. She came ashore in the gale without a soul in her, and we hear of two other vessels in the same case in this neighbourhood, and of vessels driving past Ilfracombe and other places in a perfectly helpless condition. They say, indeed, they have not known such a gale at this time of year for many years.

Augusta behaved in a first-rate manner through all the storm, when everybody felt that though there was no imminent downright danger, it all depended on many circumstances whether we got through it or not. I could only regret that her anxiety to accompany and take care of me brought her into such a mess. I am certainly better now, and hope to be able to take care of myself. —With our united kind love to all, ever your affectionate brother,

BEETE.

It was not until the spring of 1869 that his wife and immediate relatives gave up the hope that his excellent constitution might eventually enable him to rally from the attacks which had alarmed them. In the early part of the year he was much grieved by the loss of one of his colleagues, Mr. Du Noyer, who died of scarlet fever. The following is part of the last letter received by his friends in England:

Downpatrick, Feb. 27, 1869.

. . . Here I am in this little country town in the north of Ireland, endeavouring to supply the place of

poor Du Noyer, who died of scarlet fever at the beginning of the year. I have done some days of inspection
duty, which I think has done me good ; but it *snew* last
night, and the country is all white this morning. Still,
I think the country air is better for me than that of
Dublin.

But neither pure air nor quiet rest, neither the skill
of a faithful friend and physician nor the love and care
of a devoted wife, could now avail. The work of life
was done, the burden and heat of the day had been
bravely borne ; and instead of the ' peaceful hermitage' to
which we look forward in ' weary age,' he found that
deeper rest, where neither toil nor care can evermore
disturb ' the heaven of our repose.'

Almost the last words of conscious utterance addressed to his wife a few days before his death were that
God was going to take him home, recalling Wordsworth's
beautiful line,

' True to the kindred points of heaven and home.'

He died on July 29, 1869, and was buried in the
churchyard of Selly Oak, near his native town of Birmingham. His colleague and successor, Mr. Hull, writes
thus feelingly of him in the *Geological Magazine* for
September of that year :

' We can recall the massive form, the penetrating

glance, and the sturdy step of our late friend, as, many
years ago, we accompanied him amongst the mountains
of North Wales, and had the advantage of his instruction when attempting to unravel the intricate structure
of their rocks. How hard it is to realise that that head
is now laid at rest in a quiet churchyard of central
England !'

The portrait accompanying this volume will, it is
hoped, convey something of the noble contour and expression of that head—the massive brow, where reigned
the majesty of thought, the serene grey-blue eyes, the
finely-chiselled nose, the full mouth and large white
teeth, in part concealed by abundant growth of brown
beard. A friend who knew him in the time of fullest
manhood—ere toil and care had brought that look of
weariness into the eyes and thinned the dark-brown
locks—used to say, ' Jukes looks as if he might have
been a Roman emperor !' And he certainly possessed
several of the qualities which we associate with the con-
querors of the ancient world—calm courage in danger,
energy of purpose and action, power of organisation, and,
above all, straightforward honesty and truthfulness.
Truth was, in fact, his idol. A want of it in others, or
anything mean and underhand, was the one fault for
which he had no toleration. This absolute worship of
truth in speech and action is apt to be termed *imprudence*. Would that it prevailed more universally ! For

surely it is not from the imprudence of speaking the truth that the world has so long suffered, but from the far greater imprudence of concealing it. Must we still exclaim with Iago,

> ' Take note, take note, O world,
> To be direct and honest is not safe' !

Let us rather say with Pope, though in a more prosaic strain, ' An honest man's the noblest work of God,' and believe that safety is found only in the Truth ' which liveth and conquereth evermore.'

The Editor now closes her attempt to set in an appropriate framework the autographic portrait of her brother, which is here presented to his friends in lifelong remembrance of his many noble gifts of mind and heart.

It has been to her most truly a labour of love, and has served during many months to bring the companion of her early days so vividly before her, that she has appeared to be living two lives—one in the present, with its daily duties and pleasures, the other in the past, sharing his voyages and adventures in Newfoundland and Australia, or breathing with him the free air of the Welsh hills. Should this volume serve to recall her brother to the remembrance of his friends, and to convey to those who were unacquainted with him, a faint picture of his arduous labours in the cause of science, the Editor's object will be attained.

CHRONOLOGICAL LIST OF BOOKS, PAPERS, &c.
BY J. BEETE JUKES.

1838.

A Popular Sketch of the Geology of the County of Leicester. Analyst, vol. viii. p. 1.

A Popular Sketch of the Geology of Derbyshire. Ibid. p. 214.

1839.

On the Position of the Rocks along the South Boundary of the Pennine Chain. Rep. Brit. Assoc. for 1838, Trans. of Sections, p. 79.

Sketch of the Geology of Derbyshire, second part. Analyst, vol. ix. p. 1.

On the Geology of the Northern Part of the County of Stafford. Ibid. p. 233.

1840.

Report on the Geology of Newfoundland. Folio.

Ditto. Edin. New Phil. Journ. vol. xxix. p. 103.

1842.

Address to the Lit. and Phil. Soc. Wolverhampton.

Inaugural Address to the Wolverhampton Branch of the Dudley and Midland Geol. Soc.

Excursions in and about Newfoundland during the years 1839 and 1840. 2 vols. 8vo, London.

The Geology of Charnwood Forest. An Appendix to T. R. Potter's 'History and Antiquities of Charnwood Forest.' 4to, Nottingham.

1843.

General Report of the Geological Survey of Newfoundland. 8vo, London.

A few Remarks on the Nomenclature and Classification of Rock

Formations in new Countries. Tasmanian Journ. Nat. Science, vol. ii. p. 1.

1847.

Narrative of the Surveying Voyage of H.M.S. Fly . . . in Torres Strait, New Guinea, and other Islands of the Eastern Archipelago, . . . together with an Excursion into the Interior of the Eastern Part of Java. 2 vols. 8vo, London.

Notice of some Tertiary Rocks in the Islands stretching from Java to Timor. Rep. Brit. Assoc. for 1846, Trans. of Sections, p. 67.

Sketch of the Geological Structure of Australia. Ibid. p. 68.

Notice of the Aborigines of Newfoundland. Ibid. p. 114.

Notes on the three Races of Men inhabiting the Islands of the Indian and Pacific Oceans. Ibid. p. 114.

Notes on the Palæozoic Formations of New South Wales and Van Dieman's Land. Quart. Journ. Geol. Soc., vol. iii. p. 241.

1848.

Notes on the Geology of the Coasts of Australia. Quart. Journ. Geol. Soc., vol. iv. p. 142.

(With Mr. A. H. Selwyn) Sketch of the Structure of the Country extending from Cader Idris to Moel Siabod, North Wales. Ibid. p. 300.

1850.

A Sketch of the Physical Structure of Australia, so far as it is at present known. 8vo, London.

On the Relations between the New Red Sandstone, the Coal-measures, and the Silurian Rocks of the South Staffordshire Coal-field. Rep. Brit. Assoc. for 1849, Trans. of Sections, p. 55.

1851.

On the Geology of the South Staffordshire Coal-field. Journ. Geol. Soc., Dublin, vol. v. p. 114.

Sketch of the Geology of the County of Waterford. Ibid. p. 147.

1852.

Australia and its Gold Regions. Chambers' Repository of Instructive and Amusing Tracts.

1853.

On Devonian Rocks in the South of Ireland. Rep. Brit. Assoc. for 1852, Trans. of Sections, p. 51.

Popular Physical Geology. 12mo, London.
On the Geology of the South Staffordshire Coal-field. Records
of the School of Mines, vol. i. part ii. p. 149 (Geological Sur-
vey).
On the Geology of Australia, with especial reference to the Gold
Regions. Lectures on Gold delivered at the Museum of Prac-
tical Geology. 8vo, London.
On the Occurrence of Caradoc Sandstone at Great Barr, South
Staffordshire. Quart. Journ. Geol. Soc., vol. ix. p. 179.

1854.
(With Mr. A. Wyley.) On the Structure of the North-eastern
Part of the County Wicklow. Journ. Geol. Soc. Dublin, vol.
vi. p. 28.
Annual Address delivered before the Geological Society of Dublin,
February 1854. Journ. Geol. Soc. Dublin, vol. vi. p. 61.
On the Barometrical Measurement of the Peak of Teneriffe. Proc.
Roy. Irish Acad., vol. vi. p. 89.

1855.
Annual Address delivered before the Geological Society of Dublin,
February 1855. Journ. Geol. Soc. Dublin, vol. vi. p. 252.
(With Mr. J. W. Salter.) Notes on the Classification of the
Devonian and Carboniferous Rocks of the South of Ireland.
Journ. Geol. Soc. Dublin, vol. vii. p. 63.

1856.
Note to Mr. F. Foot's paper on the Trappean Rocks in the neighbourhood of Killarney. Ibid. vol. vii. p. 172.
The Article 'Geology, or Mineralogical Science,' in the eighth
edition of the Encyclopædia Britannica.

1857.
The Student's Manual of Geology. 8vo, Edinburgh.
On the Alteration of Clay-slate and Grit-stone into Mica-schist
and Gneiss by the Granite of Wicklow, &c. Rep. Brit. Assoc.
for 1856, Trans. of Sections, p. 68.
Notes on the Calp of Kilkenny and Limerick. Journ. Geol. Soc.
Dublin, vol. vii. p. 277.

1858.
Notes on the Old Red Sandstone of South Wales. Rep. Brit.
Assoc for 1857, Trans. of Sections, p. 73.

QQ

On the Igneous Rocks of Arklow Head. Journ. Geol. Soc. Dublin, vol. viii. p. 17.
Iron Ores of the South Staffordshire Coal-field. Geological Survey.

1859.

On the Progress of Geology. Quarterly Review, July.
The South Staffordshire Coal-field, second edition. Memoirs of the Geological Survey of Great Britain.
(With Prof. Haughton.) On the Lower Palæozoic Rocks of the South-east of England, and their associated Igneous Rocks. Trans. Roy. Irish Acad., vol. xxiii. p. 563.
(With Mr. G. V. du Noyer.) On the Geological Structure of Cahercourll Mountain, ten miles west of Tralee. Journ. Geol. Soc. Dublin, vol. viii. p. 106.
On the Occurrence of some isolated Patches of the Lower Coal-measure Shales in the Northern Part of the County of Dublin. Ibid. vol. viii. p. 162.

1861.

On the Igneous Rocks interstratified with the Carboniferous Limestones of the Basin of Limerick. Rep. Brit. Assoc. for 1860, Trans. of Sections, p. 84.

1862.

On the Mode of Formation of some of the River Valleys in the South of Ireland. Quart. Journ. Geol. Soc., vol. xviii. p. 378.
Note on the Way in which the Calamine occurs at Silver-Mines, County of Tipperary. Journ. Geol. Soc. Dublin, vol. x. p. 11.
Inventory Catalogue of the Specimens illustrating the Composition, Structure, and other Characters of the Irish, British, and Foreign Rocks, in the Collection of the Museum of Irish Industry, Dublin. 8vo, Dublin.
The Student's Manual of Geology, second edition. 8vo, Edinburgh.

1863.

Address to the Geol. Section (Cambridge Meeting). Rep. Brit. Assoc. for 1862, Trans. of Sections, p. 54.
The School Manual of Geology. 8vo, Edinburgh.

1864.

On some Indentations in Bones of a Cervus Megaceros, found
in June 1863 underneath a Bog near Legan, County Long-
ford. Journ. Geol. Soc. Dublin, vol. x. p. 127.
On the Flint Implements found in the Gravel of St. Acheul, near
Amiens, and their Mode of Occurrence. Proc. Roy. Irish
Acad., vol. viii. p. 220.

1865.

Notes for a Comparison between the Rocks of the South-west of
Ireland and those of North Devon and Rhenish Prussia.
Journ. Roy. Geol. Soc. Ireland.
Sketch of the Geological Structure of the South Staffordshire
Coal-field. 8vo, Birmingham.
Former Extension of the Coal-measures, (Letter on the). Geol.
Mag. vol. ii. p. 135.
The Outlier of Carboniferous Limestone near Corwen. Ibid.
p. 326.
Glaciation in N. Devon and its Rocks, (Letter on). Ibid. p. 473.
Further Notes on the Classification of the Rocks of N. Devon.
Read before the Geol. Soc. of Ireland.

1866.

Some Descriptive Notes in Professor Ramsay's Memoir ' On the
Geology of North Wales.' Memoirs of the Geol. Survey, vol. iii.
On the Carboniferous Slate (or Devonian Rocks), and the Old
Red Sandstone of South Ireland and North Devon. Quart.
Journ. of Geol. Soc., vol. xxii. p. 320.
Atmospheric *v.* Marine Denudation, (Letter on). Geol. Mag.,
vol. iii. p. 232.
Reply to Mr. G. P. Scrope's Article on the Origin of Valleys, &c.
Ibid. p. 331.

1867.

Additional Notes on the Grouping of the Rocks of North Devon
and West Somerset. Read to Geol. Soc. London. 8vo, Dub-
lin (privately printed).
H.M. Geological Survey of the United Kingdom, and its connec-
tion with the Museum of Irish Industry and that of Prac.
Geology in London. An Address delivered before the Lord
Lieutenant of Ireland, Dec. 31st, 1866, and published, with
Notes.

Letter on Glaciation in Devon and its Borders. Geol. Mag., vol. iv. p. 41.

Letter on the Devonian Rocks of Devonshire. Ibid. p. 87 (Corrections, p. 133).

On the Gorge of the Avon at Clifton. Ibid. p. 444.

1868.

Notes on Parts of South Devon and Cornwall, with Remarks on the true Relations of the Old Red Sandstone to the Devonian Formation. Journ. Roy. Geol. Soc. Ireland, vol. ii. p. 67.

The Chalk of Antrim. Geol. Mag., vol. v. p. 345.

Geological Map of Ireland: scale eight miles to an inch. London, Stanford.

Forty-two Memoirs explanatory of the Maps of the South, East, and West of Ireland were edited and in a great measure written by Mr. Jukes during the progress of the Irish Survey from 1850 to 1869.

Part Author of various sheets of the Map of the Geological Survey of England (Wales, Staffordshire, &c.).

THE END.

LONDON:

ROBSON AND SONS, PRINTERS, PANCRAS ROAD, N.W.

Printed in the United States
By Bookmasters